Molecular and Cellular Mechanisms of Antibody Activity

Falk Nimmerjahn
Editor

Molecular and Cellular Mechanisms of Antibody Activity

Springer

Editor
Falk Nimmerjahn
Institute of Genetics
Department of Biology
University of Erlangen-Nuremberg
Erlangen, Germany

ISBN 978-1-4614-7106-6 ISBN 978-1-4614-7107-3 (eBook)
DOI 10.1007/978-1-4614-7107-3
Springer New York Heidelberg Dordrecht London

Library of Congress Control Number: 2013938377

Printed on acid-free paper

Springer is part of Springer Science+Business Media (www.springer.com)

Preface

The humoral immune system, that is, antibodies produced by B cells, is an essential component of our defense against microbial infections. An impaired antibody response against bacterial, viral, or fungal microorganisms results in a heightened level of infections. This book brings together leading experts in the field of immunoglobulin activity and function to provide the reader with up-to-date information on the mechanisms of action of this class of molecules. During the last years, tremendous progress has been made in understanding how the different immunoglobulin isotypes mediate their activity. This book concentrates on IgM, IgA, and the four different IgG subclasses. We will not discuss the activity and function of IgE, which is covered in detail by other textbooks and reviews. The book is meant for readers with a basic understanding of how the immune system works. We will not introduce the molecular mechanism of how B cells develop and become activated and how immunoglobulin class switching is accomplished. Instead, we will focus on how immunoglobulin molecules recruit the humoral and cellular effector functions of the innate immune system. In Chap. 1, Mikael Karlsson from the Karolinska Institute in Stockholm in Sweden, will discuss the role of IgM and IgD in infection and inflammatory diseases. Jenny Woof from the University College of Dundee in the UK will give an in-depth overview in Chap. 2 of how IgA, the immunoglobulin subclass that dominates the inner body surfaces such as the gut, lung, and saliva, is critical to prevent microorganisms from invading the body at these sites. Chapters 3–8 will then switch to the family of immunoglobulin G subclasses, which are indispensable for protective antimicrobial immune responses. Peter Sondermann, head of research at SuppreMol GmbH in Munich, will start with a detailed insight into the IgG structure (Chap. 3), followed by a chapter (Chap. 4) of Jeanette Leussen from the University of Utrecht in the Netherlands and myself on the molecular and cellular mechanisms involved in the activity of IgG which, on the one hand, prevents infections but, on the other hand, is responsible for the destruction of healthy tissues during autoimmune diseases. Besides this pro-inflammatory activity, IgG is used since many years as an anti-inflammatory treatment to suppress autoimmune pathology. The tremendous advances we have made in understanding this "other side" of IgG activity are summarized in Chap. 5 by myself. Shozo Izui from the

University of Geneva in Switzerland will then provide in Chap. 6 a detailed example of how red blood cell or immunoglobulin-specific autoantibodies (so-called rheuma factors) mediate their activity. Michael Karsten and Jörg Köhl from the University of Lübeck in Germany will introduce the intricate interplay between cellular and humoral effector functions triggered by IgG (Chap. 7), followed by a detailed overview of how IgG half-life is controlled by the neonatal FcRn by Kristi Baker, Timo Rath, and Richard Blumberg from Harvard University in Boston (Chap. 8). Besides connecting the adaptive with the innate immune system, antibodies also feedback on B cells to regulate their own production, which will be introduced in Chap. 9 by Birgitta Heyman from Lund University in Sweden. Finally, Christian Kellner and Matthias Peipp from the University of Kiel in Germany will introduce in Chap. 10 how the function of IgG can be enhanced and which novel formats of therapeutic antibodies are currently used and may be used in the near future.

Last but not least, I would like to sincerely thank all the authors for their contributions and efforts for making this book possible.

Erlangen, Germany Falk Nimmerjahn

Contents

Contributors

Kristi Baker Division of Gastroenterology, Hepatology and Endoscopy, Brigham and Women's Hospital and Harvard Medical School, Boston, MA, USA

Richard S. Blumberg Division of Gastroenterology, Hepatology and Endoscopy, Harvard Medical School and Brigham and Women's Hospital, Boston, MA, USA

Birgitta Heyman Department of Medical Biochemistry and Microbiology, Uppsala University, Uppsala, Sweden

Shozo Izui Department of Pathology and Immunology, University of Geneva, Geneva, Switzerland

Mikael Karlsson Department of Medicine, Karolinska University Hospital/ Karolinska Institutet, Stockholm, Sweden

Christian M. Karsten Institute for Systemic Inflammation Research, University of Lübeck, Lübeck, Germany

Christian Kellner Division of Stem Cell Transplantation and Immunotherapy, 2nd Medical Department, Christian-Albrechts-University Kiel, Kiel, Germany

Jörg Köhl Institute for Systemic Inflammation Research, University of Lübeck, Lübeck, Germany

Division of Cellular and Molecular Immunology, Cincinnati Children's Hospital and University of Cincinnati, College of Medicine, Cincinnati, OH, USA

Jeanette H.W. Leusen Department of Immunology, Laboratory for Immunotherapy, University Medical Centre Utrecht, Utrecht, EA, The Netherlands

Falk Nimmerjahn Institute of Genetics, Department of Biology, University of Erlangen-Nuremberg, Erlangen, Germany

Matthias Peipp Division of Stem Cell Transplantation and Immunotherapy, 2nd Medical Department, Christian-Albrechts-University Kiel, Kiel, Germany

Timo Rath Division of Gastroenterology, Hepatology and Endoscopy, Brigham and Women's Hospital and Harvard Medical School, Boston, MA, USA

Peter Sondermann CSO, SuppreMol GmbH, Martinsried, Germany

Jenny M. Woof Division of Cancer Research, Medical Research Institute, Jacqui Wood Cancer Centre, Ninewells Hospital and Medical School, University of Dundee, Dundee, UK

Chapter 1
IgM and IgD in Infection and Inflammatory Diseases

Mikael Karlsson

Abstract The IgM and IgD antibodies were the first to evolve and are also part of the first line of immune defense against invading pathogens. They are multifunctional proteins that both function as receptors mediating activating signaling on the cell surface and when released, they take on a different role mediating many important functions in the immune system. The effectiveness and usefulness of these proteins is proved by the fact that they have been modified by evolution to give rise to all the other subclasses of antibodies in our immune system. In inflammatory diseases IgM and IgD can have both pathogenic and protective roles depending on the type of disease and tissue affected. In this chapter these dual roles of IgM and IgD is discussed as well as their role in B cell development and infectious disease.

Keywords IgM • IgD • Inflammatory diseases • Immune system • Adaptive immune system • Immunoglobulin domain • Innate immunity

1.1 Evolution of IgM and IgD

1.1.1 Evolution of the Antibody Isotypes IgM and IgD

Even though the principal building blocks of the immune system are well defined, much is still not known about when and how the adaptive immune system evolved to work in cooperation with innate immunity (Pancer and Cooper 2006). The development of antibodies was preceded by the evolution of the immunoglobulin (Ig) domain which is used not only in antibodies but also in, among others, the TCR and

M. Karlsson, Ph.D (✉)
Department of Medicine, Karolinska University Hospital/Karolinska Institutet,
KS L2:04, 17176, Stockholm, Sweden
e-mail: Mikael.Karlsson@ki.se

F. Nimmerjahn (ed.), *Molecular and Cellular Mechanisms of Antibody Activity*,
DOI 10.1007/978-1-4614-7107-3_1, © Springer Science+Business Media New York 2013

MHC proteins. The Ig domain is also found in *Drosophila,* for example, in the Dscam proteins that aid in opsonization of bacteria and immune defense. Just as in the mammal antibody coding locus, the so-called Ig locus, the *Dscam* gene is composed of a number of alternative exons that can be put together to create molecular diversity (Ghosh et al. 2011). This recent discovery shows that diversification using Ig domains is used also in invertebrates and that these rely on more than innate immunity and pattern recognition for their immune defense. Furthermore, similar molecules have evolved separately in lamprey and hagfish (agnathans) using leucine-rich repeats to make up variable lymphocyte receptors (Alder et al. 2005). Importantly, this shows a strong selective pressure to evolve this type of diversification. The IgM subclass was the earliest antibody to evolve followed later by IgD (Hadge 1985). IgM is present in jawed vertebrates (gnathostomes) that existed already during the Devon period over 400 million years ago. The IgD subclass is first found in bony fish (teleosts) which is the only organism expressing it besides mammals. However, somatic hypermutation (SHM), a way to further diversify the immune repertoire, has so far only been found in vertebrates (Stavnezer and Amemiya 2004). This mutagenic process can be found in fish and coincides with occurrence of the activation-induced deaminase (AID), essential for this reaction (Muramatsu et al. 2000). AID is also crucial for class-switch recombination (CSR), to other isotypes than IgM and IgD, a process that did not occur until much later in evolution. It has been suggested that this is more due to evolutionary changes in the Ig locus rather than alterations of the AID enzyme itself. Amphibians are the most primitive vertebrates known to use DNA recombination to switch antibody classes. This remarkable diversification process has been kept in every group of animals that developed thereafter, suggesting a strong evolutionary advantage of having CSR.

1.1.2 Evolution of Natural Antibodies and the Germline Repertoire

Natural antibodies are defined as inherited and germline-encoded antibodies produced by naïve mature B cells that have gone through selection but that have not yet encountered external antigenic stimuli (Hooijkaas et al. 1984). The existence of these antibodies stems from studies of antibodies found in germ-free mice as well as in human cord blood (Chou et al. 2009). Some of the natural antibodies including IgM (nIgM) are part of the pool of antibodies that are described as polyreactive, meaning that they individually react to a number of unrelated antigens, including lipopolysaccharide (LPS) and self-antigens, such as dsDNA and insulin (Satoh et al. 1983; Dighiero et al. 1983). This type of “promiscuity,” binding both self and non-self, exists both in the IgM and IgG compartments, despite that switching to IgG developed much later than IgM. Much of the studies of the natural IgG repertoire have been done on intravenous immunoglobulin (IVIG) preparations used for therapy, which is purified IgG antibodies pooled from thousands of individuals (Nimmerjahn and Ravetch 2008). These IgGs are composed essentially by the same repertoire of specificities as the primordial IgM (Marchalonis et al. 2006). One underlying reason for this is that evolution of the VH and VL as well as joining (J) segments is highly conserved, and as an example, the VH regions in shark show 50 % identity to the human VH germline prototype gene. Still, even though the basic gene segments are conserved over millions of years, how they are used to create antibodies shows great individual diversity within vertebrates (Marchalonis et al. 2006). The extremes are the individual IgM gene cassettes in sharks and the translocation processes and VDJ recombination that occur in mammals. Furthermore, diversification in chickens depends almost completely on gene conversion, showing yet another way to create diversity within the immune system (Ratcliffe 2006).

1.2 Structure and Expression of IgM and IgD

1.2.1 Molecular Genetics, Selection, and Survival

The production of the IgM heavy chain (μHC) is the first antibody subclass specific gene expression to occur in the developing B cell in the bone marrow (BM) (Martensson et al. 2007; Raff et al. 1976; Burrows et al. 1979). The assembly is ordered; first upregulation of RAG genes leads to D to J gene segment ligation already in lymphoid progenitor cells, with the ability to produce both B and T cells (Schatz and Swanson 2011). The subsequent V to DJ and μ-constant region rearrangement requires additional expression of the transcription factor Pax5 which is essential in the initial developmental stages for B cell linage commitment in the bone marrow (Medvedovic et al. 2011). Furthermore, Pax5 is needed throughout B cell development, to maintain cell identity, and is later also involved in regulating activation of mature B cells (Medvedovic et al. 2011). After being expressed, μHC is involved in B cell selection in the bone marrow, where it first is part of the preBCR, together with the surrogate light chain and associated with Igα– and Igβ-chains on the cell surface. The Igα and Igβ contain the signaling motifs (ITAM) of the BCR, whereas the μHC adds little to the signaling (Martensson et al. 2007; Reth 1989). In the next step of development, the μHC is paired with λ or κ light chains and participates in receptor-editing processes where up to 50 % of the B cells escape negative selection by rearranging their IgM-BCR (Gay et al. 1993). These steps and the distinct development stages in the BM assure a functional BCR and set a first threshold for autoreactivity (Luning Prak et al. 2011). The δ heavy chain (δHC) is expressed when the naïve B cell has gone through selection and leaves the bone marrow. These naïve mature B cells express both IgM and IgD on the surface with identical VDJ, making up their BCRs (Fig. 1.1). This is achieved by a process where the δHC is transcribed on the same mRNA unit as μHC (Gritzmacher 1989). The relative surface expression of IgM- and IgD-BCRs is then regulated by alternative transcriptional termination and splicing, and not by DNA recombination. When IgM and IgD are part of the BCR, the accompanying Igα and Igβ chains mediate the

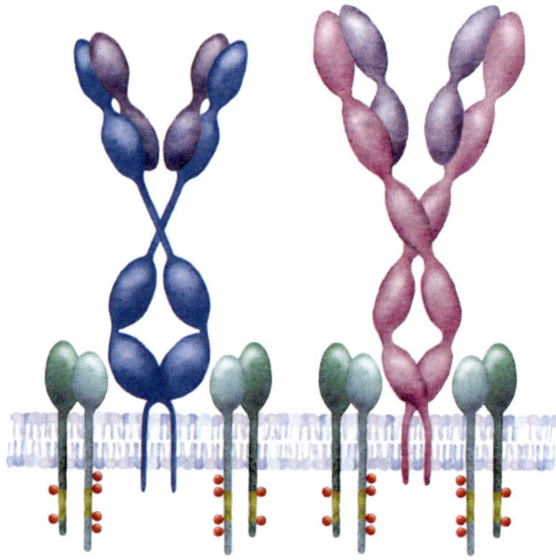

Fig. 1.1 IgD and IgM as B cell receptors. IgD (*left*) and IgM (*right*) are first expressed as B cell receptors when the B cell leave the bone marrow. They are anchored in the plasma membrane through a transmembrane region and associate with two Iga and Igb chains that provide ITAM-containing signaling motifs that mediate activation when the B cell meets its antigen

signaling from the receptor following interaction with its cognate antigen. These side chains are also responsible for internalization of the BCR for antigen uptake to present peptides on MHC II (Jang et al. 2010). After activation, the B cells will have the ability to permanently switch and become producers of soluble IgM and IgD molecules as well as other subclasses. Importantly, in the periphery constant signaling from the BCR is needed for B cell survival. If the BCR is experimentally deleted, B cells die within 5 days. Rescue from death can be achieved by inducing tonic PI3 kinase signaling through activation of the FOXO1 transcription factor (Srinivasan et al. 2009). The source for the tonic signaling in non-manipulated cells is not fully understood, but could be mediated via weak BCRs binding to self-ligands or even independently of BCR phosphorylation. In the final stage of B cell development, the fully activated cells terminally differentiate into plasma cells. This is accompanied by loss of expression of the BCR as well as Pax5 and upregulation of, among others, Pax5-suppressed genes such as *Blimp-1*, *Xbp1,* and J chain, all essential for plasma cell differentiation as well as assembly of soluble IgM (Kallies and Nutt 2007).

1.2.2 Subpopulations of Naïve B Cells Expressing IgM and IgD B Cell Receptors

Naïve B cells leaving the BM, at this time expressing BCR receptors made up with IgM and IgD, continue to develop in the periphery and go through a number of transitional (T) stages called T1, T2, and T3 to end up in the naïve mature B cell repertoire (Allman et al. 2001; Sims et al. 2005). Interestingly, at the T1 stage about 40 % of the B cells have reactivity to self-antigens. This drops to around 20 % as the cell develops further, showing that an additional peripheral selection step

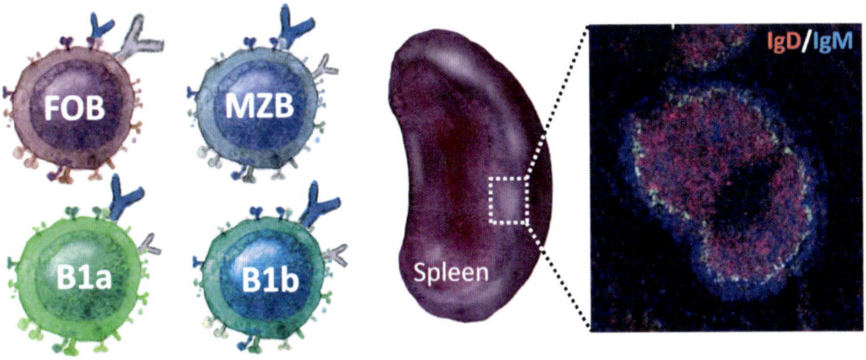

Fig. 1.2 The naïve B cell repertoire expresses different levels of IgM and IgD. *Left*: the naïve repertoire of B cells consists of follicular B cells (*FOB*), marginal zone B cells (*MZB*), and B1a and B1b B cells. Of these the B1a and B1b B cells express high levels of IgM BCRs and reside predominantly in peritoneal and pleural cavities. FOBs and MZBs on the other hand are found in the spleen where they are organized such as MZBs (IgM high) are in the marginal zone, whereas FOB(IgD high) are found within the follicle (*right*)

(Meffre and Wardemann 2008; Wardemann et al. 2003). B cells also develop into different naïve subpopulations called B1 or B2 cells (Fig. 1.2). B1 cells got their name because they develop first in ontogeny and they can be further divided into two subpopulations, B1a and B1b (Baumgarth 2011). Both of these are the main B cell population in the peritoneal and pleural cavities, whereas B1a cells can also be found in small numbers in the spleen. However, none of these are found in lymph nodes. The development of B1 cells differs from B2 cells in that they arise also during fetal hematopoiesis as opposed to adult hematopoiesis that produce both B1 and B2 cells. B cells of the B2 lineage develop into two distinct populations that are called follicular B cells (FOB) and marginal zone B cells (MZB) based on the microanatomical location where they are enriched (Lopes-Carvalho et al. 2005; Pillai and Cariappa 2009) (Fig. 1.2). FOBs are, in majority, found within follicles of lymph nodes and spleen as well as in the blood. MZBs on the other hand are specific to the spleen where they migrate back and forth between the follicle and marginal sinus that aligns the marginal zone to transport antigen for deposition on Follicular Dendritic cells (FDCs). The B cells are distinctive in their expression of a number of receptors (markers) including IgM- and IgD-BCRs, where MZBs are IgM high and FOBs are IgD high (Fig. 1.2). The B1a and B1b cells are also high expressers of IgM, and together with MZBs these are sometimes designated "innate like" in that they have been connected to production of natural and polyreactive antibodies (Kearney 2005).

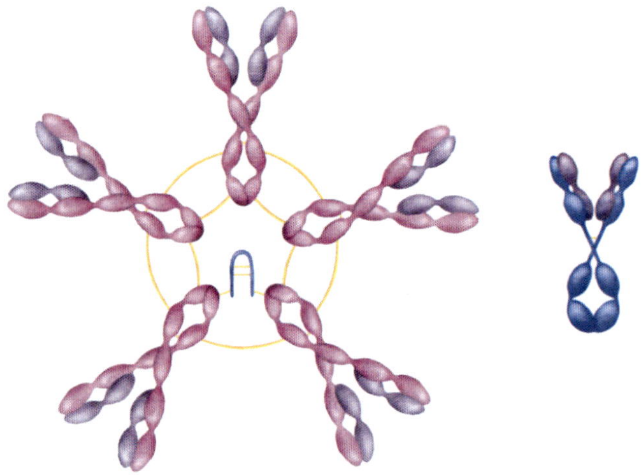

Fig. 1.3 Soluble forms of IgM and IgD. When IgM (*left*) and IgD (*right*) are secreted, IgM exists as a pentameric protein assembled with a J chain, whereas IgD are expressed as a highly glycosylated monomer

1.2.3 Protein Heterogeneity

Both in a noninfectious steady state and during an immune response, IgM and IgD antibodies are produced as soluble forms in addition to their function in the BCR (Fig. 1.3). The transition from cell-bound IgM and IgD as part of the BCR to soluble proteins is regulated by alternative splicing, removing the transmembrane domain (Rogers et al. 1980; Early et al. 1980). Soluble IgM and IgD (sIgM, sIgD) are as all antibodies glycoproteins, and their function is greatly regulated by the degree of specific glycosylation. The sIgD is being produced as a monomer, whereas the vast majority of sIgM is pentameric by additional processing that adds the J chain. The J chain is attached to two of the monomers by disulfide bounds, and monomeric sIgM can only be found in trace amounts in serum (Niles et al. 1995). Occasionally sIgM can also be found as a hexamer lacking the J chain but in principal is found as a 970 kDa pentameric protein with a concentration of 1.5 mg/ml and half-life of around 10 days (Ehrenstein and Notley 2010). Even though IgD-producing B cells have been shown to express the J chain, this protein is not attached to IgD (Chen and Cerutti 2010). Instead, the Fc part and hinge regions of sIgD have a number of specific glycosylation sites, and the sIgD is heavily glycosylated (Arnold et al. 2004), and the O-linked and N-linked glycans can compose up to 15 % of its molecular weight. The extent of glycosylation varies a lot, and thus the molecular weight of sIgD varies in serum. Although variable, the average molecular weight is about 184 kDa, and the concentration is much lower than IgM: 0.03 mg/ml with a half-life of 3 days. sIgM is also glycosylated, with 5 N-linked glycans on each heavy chain, which are essential for the pentameric assembly and addition of the J chain (de Lalla et al. 1998). In addition, sIgM can be associated with a protein called the secretory component (SC) in a J chain-dependent manner (Mostov and Blobel 1982). The SC is attached to sIgM through non-covalent interactions during transportation to mucosal surfaces (see Sect. 3.2.1). Structurally, sIgM and sIgD also have distinct characteristics compared to the Y-shaped structure of other subclasses. The long hinge regions of IgD make it take on a more T-shaped structure rather than the Y-shaped structure of other subclasses (Sun et al. 2005). The IgM pentamer, on the other hand, has been shown to take on a mushroom-type shape, and not planar, where the c-terminal Fc domains form a steem sticking out of the plane of the Fab regions (Czajkowsky and Shao 2009). Finally, neither IgM nor IgD is being transferred over placenta since the receptor involved in this process does not bind these subclasses (Roopenian and Akilesh 2007). Therefore, neonates' immune protection is composed by transferred IgG antibodies from the mother, and their IgM and IgD antibodies are encoded from their own genome.

1.3 Effector Mechanisms: IgM

1.3.1 Complement Activation

The complement system is composed of over 25 serum proteins (including C1–C9) and cell-surface receptors binding complement. It was first discovered in 1901 by Jules Bordet who later won the Nobel Prize (Bordet and Genou 1901; Lachmann 2006). The major source for many of the complement components is the liver, but during an immune response, myeloid cells can produce many of the proteins locally (Roozendaal and Carroll 2007). Over a century of research has shown that the complement system is an important part of the innate immune system and its activation is mediated through three major pathways (Roozendaal and Carroll 2007; Muller-Eberhard 1988). These are called the classical pathway, the lectin pathway, and the alternative pathway. The classical pathway involves antibodies and is activated when the C1q protein binds to its ligands, including sIgM and sIgG immune complexes (Kojouharova et al. 2010). The binding of C1q is restricted to pentameric or hexameric sIgM, whereas it has no affinity for the monomeric protein. On the other hand, a single pentameric sIgM molecule is enough to activate the complement cascade in in vitro assays, making it 1,000-fold more efficient than sIgG (Cooper et al. 1983; Borsos and Rapp 1965). After the initial trigger, complement activation proceeds as a cascade of proteolytic events that terminates in the membrane attack complex, capable of destroying, for example, invading bacteria. As an important additional effect of the response, small split products, referred to as anaphylatoxins, of the C3, C4, and C5 proteins are released and can bind to specific receptors on a number of immune and nonimmune cells and regulate their activation. Cleavage products of C3 tag microbial pathogens which facilitate their recognition and removal by phagocytes (Carroll 2008). Complement is also involved in the removal of dead cells and cellular debris and by this can take part in tissue homeostasis

(Fishelson et al. 2001). Complement-tagged antigens are also more potent activators of B cells than antigens alone. Mechanistically, co-ligation of complement receptors (CD21) on B cells lowers the threshold of activation and enhances the adaptive immune response. This was elegantly shown in a seminal paper where tagging of one or more cleavage products of C3 (C3d) enhanced the response stepwise to the protein antigen hen egg lysozyme (HEL) (Dempsey et al. 1996). This directly shows the efficiency of complement as an endogenous adjuvant. In addition to C1q, the structurally related mannose-binding lectin (MBL), belonging to the collectin family of proteins, has also been shown to bind certain glycoforms of sIgM (Arnold et al. 2005). MBL initiates the lectin pathway of complement activation and does this in absence of sIgM, as it is a pattern recognition protein in its own right, recognizing among other things bacteria and apoptotic cells (Ip et al. 2009). In humans, MBL is bound to about 20 % of serum sIgM, but it does not seem to bind antigen-bound sIgM, as the glycan structures on the Fc part needed for this binding are not exposed (Arnold et al. 2005). Therefore, it has been proposed that MBL depletes serum aggregates of sIgM as a scavenging function. Acting as a natural adjuvant, binding of IgM enhances the response to an antigen in a complement-dependent manner. This is however only true for large and especially particulate antigens as the response to small protein antigens, such as OVA, is not altered in absence of sIgM (Getahun and Heyman 2006).

1.3.2 The IgM Fc Receptor

Even though IgM was identified in the 1930s, and has since been extensively studied, the existence of a sIgM receptor, analogous to Fc receptors for IgG, IgA, and IgE, remained elusive for a long time. However, in recent years three receptors in the same gene cluster on chromosome 1 have been identified, and the list now includes the poly-Ig receptor (pIgR), the Fcα/μR, and the FcμR (Klimovich 2011) (Fig. 1.4). In addition to these, there are other receptors that have been linked to soluble IgM binding.

1.3.2.1 pIgR

Immune protection of mucosal surfaces is the first line of defense against invading microorganism, and these sites are constantly under pressure by environmental antigens. Antibodies are an important part of this defense, and both sIgM and sIgA are actively transported to these sites. The J chain and part of the IgM Fc together associates with the pIgR that, in turn, mediates the so-called epithelial transcytosis of the antibody (Kaetzel 2005). Once the pIgR has transported IgM to the mucosal side, it is cleaved and the receptor remains as the secretory component (SC) non-covalently attached to the IgM pentamer, via the J chain (Johansen et al. 2000). The pIgR belongs to the Ig protein superfamily and is composed of five extracellular domains

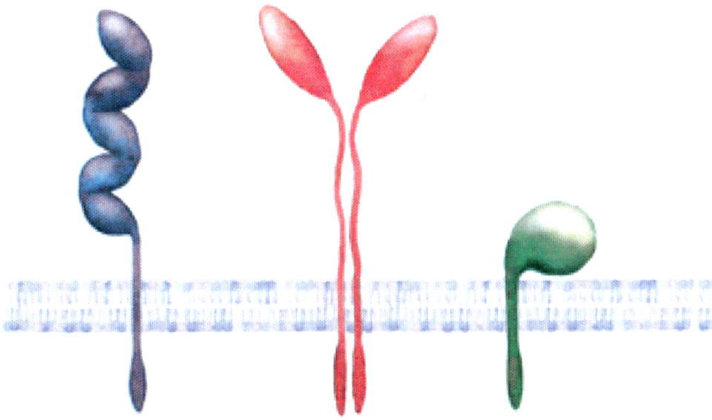

Fig. 1.4 The IgM Fc receptors. Three Fc receptors for IgM have been described. These are all encoded on chromosome 1 and are named the poly-Ig receptor (pIgR), the Fcα/μR, and the FcμR

linked by disulfide bonds (Krajci et al. 1989). The receptor is heavily glycosylated, but these residues are not involved in the ligand binding. It is expressed as a precursor on the basolateral surface of endothelial cells where it binds IgM and IgA and is then processed as it passes through endosomes. Finally, as it reaches the apical side of the cell, it is cleaved off (Mostov and Blobel 1982). In addition to this, pIgR has also been shown to be able to transport immune complexes in the other direction, at least in the lamina propria of the intestine. Thus, the receptor enables sampling of antigens from the environment. The pIgR is upregulated at mucosal sites by inflammation and inflammatory cytokines such as IFN-γ, TNF, and IL-1 (Blanch et al. 1999; Phillips et al. 1990). The regulation of pIgR is also connected to innate immunity and pattern recognition where LPS and dsRNA binding to TLRs induces expression of the receptor through NF-κB signaling pathways (Schneeman et al. 2005). pIgR-deficient mice have no mucosal Igs and have reduced protection against influenza A and B virus after intranasal inoculation (Asahi et al. 2002). On the other hand data from these mice show no difference in clearance of pathogenic bacteria or alterations of commensal bacteria in the gut, suggesting that secreted IgM and IgA had no major role in these types of responses. The pIgR was however shown to protect from pathology induced by bacterial toxins when cholera toxin was feed to deficient mice and their controls (Uren et al. 2005; Sait et al. 2003). It should be noted that in most of these studies, the relative contribution of IgM versus IgA has not been addressed.

1.3.2.2 Fcα/μR

The Fcα/μR belongs, just as pIgR, to the Ig superfamily of receptors by binding both sIgM and IgA even though it has one order of magnitude less affinity to IgA compared to sIgM (Ohno et al. 1990; Shibuya et al. 2000). The receptor is composed of one Ig domain linked to a mucin stalk and a transmembrane region, expressed as a homodimer on the cell surface. Cross-linking of the receptor in vitro has been shown to mediate phagocytosis of both IgM-opsonized beads and *S. aureus* (Shibuya et al. 2000). In mice it is expressed both lymphoid (primarily spleen) and nonlymphoid tissues. On a cellular level its expression can be detected on B cells and macrophages but not on T cells. The wide expression found in mice is not seen in humans, where Fcα/μR is primarily expressed on FDCs (Kikuno et al. 2007; Kubagawa et al. 2009). Fcα/μR is also connected to mucosal immunology where it is found to be expressed in the lamina propria, on intestinal macrophages, and in germinal centers (GC). This implies that it takes part both in local and systemic mucosal immunity. Mice deficient in the Fcα/μR have normal immune responses to T cell-dependent (TD) and T cell-independent (TI) antigens but, interestingly, show increased GC formation after injections with the TI-II antigen Ficoll. These GCs generated large amounts of IgG3 that had gone through affinity maturation resulting in increased affinity (Honda et al. 2009). These data suggest that Fcα/μR expression on either MZBs or FDCs alters the balance in the response to TI-II antigens from an extra follicular response towards a GC response. In line with this, the authors found increased retention of the antigen on these cells.

1.3.2.3 FcμR/Toso

The FcμR was first cloned and named Toso as a receptor that blocks FAS-mediated apoptosis through regulation of caspase activity (Hitoshi et al. 1998). It has further been shown to set the threshold for apoptosis by regulating NF-κB levels and that it is essential for TNF-α-mediated liver damage (Nguyen et al. 2011). Unlike pIgR and Fcα/μR the FcμR is reported to only recognize IgM and not IgA (Kubagawa et al. 2009). This is mirrored in that even though the ligand-binding domain of the three receptors resembles each other, FcμR is less similar to the other two. Also, unlike the other FcRs for IgM, the FcμR is expressed on lymphocytes including B cells, NK cells, and T cells in humans, whereas in mice its expression is confined to mature B cells only (Shima et al. 2010). It should be noted that FcμR is not expressed on phagocytes, much different from the other Fc receptors (Nimmerjahn and Ravetch 2006). Still, in a study by Shima and co-workers, FcμR transcripts were detected in macrophages and dendritic cells from the spleen even though no surface receptor was expressed (Shima et al. 2010). Thus, it is possible that a particular activation signal will induce surface expression of the receptor. FcμR has a long cytoplasmic tail with several tyrosine and serine residues that can be targets of phosphorylation upon engagement of IgM immune complexes. In response to cross-linking, it is downregulated and surface expression is therefore dependent on sIgM in the environment. It shuttles to the lysosome upon engagement, and it can also be downregulated on mRNA level after TLR7/9 stimulation (Bérengère et al. 2012). Mice deficient in FcμR show increased basal levels of serum IgM and IgG3, but not other IgG subclasses (Ouchida et al. 2012; Honjo et al. 2012). The mice also display increased numbers of B1a and reduce MZB cells in the spleen, and when immunized, the mice display enhanced responses to selective bacterial antigen determinants such as phosphorylcholine, whereas the response to haptenated antigens was lower. In addition the mice accumulated autoantibodies suggesting a selective role for the FcμR in tolerance and responses to modified self-determinants.

The ability to mediate uptake of IgM immune complexes suggests that this receptor could be involved in antigen presentation by B cells and affect adaptive immunity including T cell activation/regulation. It should be noted that one group has failed to repeat the binding of IgM to Toso and the exact function of this receptor in IgM-mediated immunity remains to be fully defined (Nguyen et al. 2011). Future studies using the FcμR/Toso-knockout mice will most likely resolve this issue.

1.3.2.4 Other Receptors

Other putative receptors for sIgM include the glycoprotein CD22. It is expressed on B cells and mediates negative regulation of BCR signaling via recruitment of SHP-1 to an ITIM motif in its intracellular domain (Doody et al. 1995). Through seven immunoglobulin domains, CD22 binds α2,6-linked sialic acid-containing glycoconjugates and is part of the sialic acid-binding Ig-like lectin family (Siglecs) (Crocker et al. 2007). This sugar modification of proteins is quite abundant and is present on the BCR itself, as well as on sIgM. CD22 could thereby function as a receptor for sIgM, and recently this was demonstrated that this may be the case (Adachi et al. 2012). This suggests that CD22 could have a similar negative regulatory function on B cells for IgM immune complexes as FcγRIIB has for IgG immune complexes (Ono et al. 1996). In support for IgM immune complexes directly regulating BCR signaling, C3 deposition was shown to occur locally on the B cell surface after antigen binding (Manderson et al. 2006).

1.3.3 IgM, Tissue Homeostasis, and Regulation of Commensal Mucosal Microbiota

The relative polyreactivity and self-reactivity in the natural sIgM repertoire provides the basis for its participation in homeostatic processes. Natural sIgM has the ability to bind altered or modified self-antigens occurring on cells undergoing apoptosis or oxidized LDL (oxLDL). There is also an overlap between self- and non-self-reactivity where, although the antibodies recognize different antigens, they still recognize the same oxidation-specific epitope (Chou et al. 2009). As an example, the E06/T15 sIgM antibody that recognizes oxLDL specifically binds to the phosphorylcholine (PC) headgroup of oxLDL. This antigenic determinant is also found on the surface of some bacteria as well as on apoptotic cells. The generation of these epitopes is linked to vital parts of metabolism and signal transduction where oxygenation is an important process. The natural oxidative processes also create reactive oxygen species (ROS) which damage DNA, lipids, and proteins. sIgM recognizes these altered self-antigens and thus has the ability to counteract oxidative damage together with other molecules recognizing these. Thus, IgM aids in clearance by macrophages and other scavenging cells (Miller et al. 2011). Apart from this function of sIgM, the secreted antibody can together with IgA shape and regulate the maintenance and establishment of symbiotic microbiota (Shulzhenko et al. 2011). For IgA this has been shown to be an important part of the regulation of metabolism, and many studies on how the microbiome affects disease and even obesity are now emerging (Turnbaugh et al. 2006).

1.4 IgM in Immune Defense

Many studies have defined an important role for both natural (nIgM) and immune sIgM in defense against various pathogens, including bacteria, parasites, fungi, and viruses (Table 1.1).

A key tool in these experiments has been transfer of specific sIgM, as well as the usage of mice that are deficient in sIgM (Boes et al. 1998a, b). These studies have also shown that the effective source of sIgM from either the natural or immune pool as well as cellular origin (B1 vs. B2 B cells) varies depending on the pathogen and model used (Baumgarth et al. 2000a; Rapaka et al. 2010; Choi and Baumgarth 2008; Gil-Cruz et al. 2009; Alugupalli et al. 2003; Baumgarth et al. 2000b). One of the key features of IgM's ability to limit infection is its direct binding to the pathogen through the inherited polyreactivity (Miller et al. 2011). This affinity and ability to aggregate antigens is also greatly enhanced by the pentameric structure by providing ten antigenic binding sites per molecule. The polyreactivity and high valency also allow for binding of different structures on the same antigen. In humans selective IgM deficiency is a very rare immunodeficiency (less than 300 cases reported) but nevertheless is associated with severe disease with recurrent infections (Moise et al. 2010).

1.4.1 IgM in Viral Defense

The early studies on the role of nIgM versus pathogen-induced sIgM in viral defense gave somewhat conflicting results. Passive transfer of virus-specific sIgM was protective when given prior to infection but not after (Palladino et al. 1995). On the other hand, the early virus-induced sIgM was shown to be important for the protection against vesicular stomatitis virus (VSV) (Coutinho et al. 1995). Thus, these data suggested that IgM is important in memory responses to influenza, whereas for VSV the early pre-GC production is important. It was later shown that both natural

Table 1.1 The role of IgM in pathogen defense

	Pathogen	The role of IgM	References
Virus	Influenza virus	Both nIgM and immune IgM are protective and regulate IgG resp	Baumgarth et al. 2000a; Palladino et al. 1995
	West Nile virus	sIgM is important early in the infection to stop viremia in CNS	Diamond et al. 2003
	Choriomeningitis virus (LCMV)	Virus-specific IgM transgenic mice are protected from infection	Seiler et al. 1998
	Vesicular stomatitis virus	Transfer of nIgM confers protection	Ochsenbein et al. 1999
	Mouse hepatitis virus	T cell-dependent regulation of nIgM stop virus-induced MS like disease	Gil-Cruz et al. 2012
	HIV	Transfer of broadly neutralizing IgM anti-HIV blocked mucosal transmission	Wolbank et al. 2003
Bacteria	*Borrelia hermsii*	Immune IgM from MZBs give protection	Belperron et al. 2005; Malkiel et al. 2009
	Nocardia brasiliensis	Transfer of induced polyreactive IgM is protective	Salinas-Carmona and Perez-Rivera 2004
	Ehrlichia muris	T cell-independent IgM production stop infection	Bitsaktsis et al. 2007
	Streptococcus pneumoniae	nIgM activation of complement is necessary for protection	Brown et al. 2002
	Pseudomonas aeruginosa	Innate activation of B1b cells to produce IgM is protective	Kinoshita et al. 2006
Parasite	*Toxoplasma gondii/evansi*	nIgM and immune IgM induce complement activation mediate protection	Kaneko et al. 2004; Couper et al. 2005a
	Trypanosoma evansi/congo	nIgM needed for defense and IgM mediate complement-dependent phagocytosis	Pan et al. 2006; Baral et al. 2007
	Brugia pahangi	IgM mediates phagocytosis in absence of functional complement activation	Rajan et al. 2005
	Plasmodium falcip./chab.	Transfer of immune IgM to sIgM-deficient mice reduces parasitemia	Couper et al. 2005b
	Strongyloides stercoralis	IgM mediated complement activation and direct killing of the parasite	Brigandi et al. 1996

(continued)

Table 1.1 (continued)

	Pathogen	The role of IgM	References
Fungi	*Cryptococcus neoformans*	sIgM enhance TH1 polarization and macrophage clearance	Subramaniam et al. 2010
	Pseudomonas murina	nIgM influence DC recognition and T cell activation	Rapaka et al. 2010
	Candida albicans	Vaccine-induced immune IgM induces protection	Xin and Cutler 2011

IgM and sIgM are needed for proper response against influenza where natural IgM gave an initial protection; virus-induced sIgM is needed for proper viral-specific IgG responses (Baumgarth et al. 2000a). These data suggest overlapping mechanisms where sIgM take part in both direct clearance of virus particles but also participate in directing the adaptive antiviral immune response (Ehrenstein and Notley 2010). Direct neutralization/clearance of virus in vivo by sIgM has been shown for a number of viruses including VSV, influenza, and lymphocytic choriomeningitis virus (Baumgarth et al. 2000a; Ochsenbein et al. 1999). sIgM can block virus from entering cells by blocking but also induce neutralization via complement activation or increased phagocytosis (Jayasekera et al. 2007). One common feature of a number of antiviral sIgM and subsequent IgG responses is that they retain their polyreactivity with cross-reactivity to autoantigens (Mouquet and Nussenzweig 2011). This holds true for a number of viruses including HIV, hepatitis viruses, Epstein-Barr virus, and vaccinia virus (Kataaha et al. 1985; Misra et al. 1989; Toh et al. 1979; Dales et al. 1983). The autoreactive feature of viral antibody responses has been known for a long time and has been the base for studies of autoimmunity in relation to viral infection (Munz et al. 2009). Still little is known why viruses can give rise to this type of polyreactive antibodies and how this is beneficial in fighting viral infection for the host. One of the most well-defined and documented case for polyreactivity comes from studies of antibodies elicited against HIV, but the underlying reason remained elusive for a long time (Mouquet and Nussenzweig 2011). Recently, however, it was shown that many of the germline precursors of anti-HIV antibodies were polyreactive (Mouquet et al. 2010). This suggests that polyreactive naïve IgM-expressing B cells were preferentially recruited in response to HIV. One potential mechanistic explanation for this is that viral antigens, such as HIV gp160 which is scarce on the viral surface, provide relatively few epitopes (Mouquet et al. 2010). Thus, the polyreactivity of the antibody increases the affinity for the virus by so-called heteroligation, where one arm binds gp160 and the other, or others in case of IgM, other epitopes.

1.4.2 IgM in Bacterial Defense

The ability of IgM to kill bacteria is closely connected to activation of the complement system that got its name because it "complements" IgM in its function (Lachmann 2006). The activation of the classical pathway, initiated by C1q binding IgM, ends in formation of the membrane attack complex that leaves holes in the membrane of bacteria and lead to lysis. This "lytic path," via IgM, is efficient and the high concentration of sIgM and complement (>1 mg/ml) in human serum makes this a potent pathway that extends also to other microorganisms such as virus. In the first study using sIgM-knockout mice, IgM was needed to clear bacteria. The mice displayed decreased TNF production and neutrophil infiltration (Boes et al. 1998b). Still, even though the IgM-mediated pathway of complement activation has been shown to be important to clear a number of infections, including *S. pneumoniae,* there are also important roles for IgM-independent complement activation (Brown et al. 2002; Connolly and Benach 2001). In addition, Fc receptor-mediated functions could be important in vivo, such as the binding of IgM-coated *S. Aureus* by the Fcα/μR on B cells (Shibuya et al. 2000). For bacterial defense, the B cells recruited to produce IgM differ depending on the pathogen. For *P. aeruginosa,* IgM production of B1b cells in the liver was shown to be essential, whereas for *Borrelia*, IgM production from MZBs is important (Malkiel et al. 2009; Kinoshita et al. 2006). Besides being involved in the response to extracellular bacteria, IgM has been shown to be important during certain intracellular bacterial infections. Something that traditionally was thought to be taken care of by T cell activation and production of IFN-γ, TNF-α, and IL-12. In the case of *E. muris,* T cells were dispensable, but not B cells. Furthermore, infected B cell-deficient mice could be rescued by transfer of IgM from both wild-type and MHC II-deficient mice, suggesting a role for nIgM (Bitsaktsis et al. 2007). Just as polyreactive IgM is found to neutralize virus, polyreactive IgM was shown to be protective against *N. brasiliensis* in transfer experiments (Salinas-Carmona and Perez-Rivera 2004). This is true for protective antibodies to a number of bacterial pathogens, such as reactivity to capsular polysaccharides including LPS and 1–3 dextran. Responses to these antigens are a large part of the innate T cell-independent response (Wardemann et al. 2003; Mahmoud et al. 2011). T cell independency usually suggests that the response is fast and short lived, but it has been shown that IgM-secreting cells can be long lived and that IgM memory cells can be elicited from the initial non-GC immune response (Foote et al. 2012; Taylor et al. 2012).

1.4.3 IgM in Defense Against Parasites

To evade parasite infection, a potent Th2-dominated response is needed, traditionally including the production of antibodies of the IgE isotype. However, an important function of IgM in this response is to direct phagocytes against the invading pathogen. This function is mediated by several pathways where sIgM but not complement activation was needed for macrophage-dependent phagocytosis of *B. pahangi* (Rajan et al. 2005). In the clearance of *T. congolense,* on the other hand, IgM-mediated complement activation and subsequent phagocytosis through the CR3 receptor were needed (Pan et al. 2006). Parasites, similarly to bacteria, can also be directly killed by IgM-mediated complement activation (Brigandi et al. 1996). In the response to trypanosomes, IgM has been reported to be a key factor in controlling parasitemia. In these experiments in infected mice, low IgM and high IgG coincided with uncontrolled parasite expansion (Baral et al. 2007). Also, in the case of an infection with parasites such as *Toxoplasma* the source of IgM, either nIgM or immune IgM has been reported to be important (Kaneko et al. 2004; Couper et al. 2005a). sIgM-deficient mice are highly susceptible to *P. chabaudi* infection. Serum transfer from naïve wild-type mice to the sIgM-deficient mice did not lead to protection, whereas transfer of serum from immunized mice, containing mainly *P. chabaudi*-specific IgM, gave protection (Couper et al. 2005b). This suggests that the natural IgM repertoire is not recognizing this parasite, in contrast to reports for other parasites. Interestingly, in these experiments the early IgM response in wt mice were followed by a specific IgG1 and IgG2a response, whereas the same subclasses in sIgM ko mice were preceded by an IgD response. This suggests different origin and selection of specific B cell precursors in the response to this parasite. Parasites, being large multicellular organisms, can also carry endosymbionts in the form of specific bacterial species. One of these bacterial families, *Wolbachia,* is symbiotic with its worm host, like *O. volvulus,* to the extent that the parasite depends on the bacteria for its development of embryos and larvae. Thus, when the immune system fights these infections, an effective response can be directed both at the parasite itself and also at the bacteria they depend upon. That this is the case has been shown where *Wolbachia*-specific antibodies are a large part of the response that clears *O. volvulus* (Brattig et al. 2004). These findings have also resulted in the use of antibiotics in parasite infections to clear the endosymbionts essential for parasite survival.

1.4.4 IgM in Fungal Defense

Opportunistic fungal infections can cause severe and often fatal disease in immunosuppressed individuals, and currently there are no vaccines available. Fungal organisms are numerous but are relatively conserved, and the majority of these have cell walls that consist of the polysaccharides: β-glucan, mannan, and chitin. These are all targets for multiple innate host defense pathways including nIgM, and they can also induce direct nonantigen-specific or mitogenic activation of B cells (Diamantstein et al. 1982; Chai et al. 2011). That antibodies can confer protection to fungal infections has been shown using a mannan-based vaccine that stopped candidiasis. In the same study a single, monoclonal, IgM antibody was enough to induce protection (Xin and Cutler 2011). Natural IgM also suppressed fugal burden and spread as well as function to modulate DC recognition of fungi to support and modulate subsequent T cell activation (Subramaniam et al. 2010; Rapaka et al. 2010).

1.5 IgM in Autoimmunity and Disease

The term autoantibodies suggests that the antibodies could be part of the pathology of autoimmune diseases. Even though pathogenic antibodies may have affinity for the same epitopes as their natural antibody counterparts, they have usually undergone CSR and affinity maturation in GCs. As will be discussed below, very few studies have indicated a role for sIgM in driving autoimmunity. Instead the vast majority of data suggests a protective role, with the most important function being sIgM's ability to function as a scavenger. Both B1a/b cells and MZBs have been implicated in autoimmunity as early producers of IgM, and the antibodies have been suggested to be part of the pathology. As these cells are fast responders to innate stimuli and, in the case of MZBs, very good antigen-presenting cells, it is to be expected that they are part of inflammatory responses. This is something to bear in mind, and as described below, innate self-reactive B cell activation does not imply that the IgM produced is necessarily pathological.

1.5.1 IgM and Maintenance of Tolerance

There is evidence that the ability of IgM to aid in clearance mechanisms is universal for the majority of inflammatory diseases to maintain tolerance. Supporting this, sIgM-deficient mice have a higher incidence of SLE-like autoimmunity (Ehrenstein and Notley 2010). Natural IgM is thought to assist in clearance of dying apoptotic cells, thus preventing secondary necrosis, which is pro-inflammatory and could induce autoreactive immune responses (Peng et al. 2005; Silverman et al. 2008). In SLE patients, nIgM with different specificities for self-antigens, such as PC or MDA-LDL, has been shown to correlate with lower disease activity (Gronwall et al. 2012). Antibodies to PC are also significantly higher in SLE patients without cardiovascular complications, and low IgM anti-PC levels have been suggested to be used as basis for specific therapy in cardiovascular disease (Caidahl et al. 2012). IgM anti-DNA and anticardiolipin has further been shown to be higher in SLE patients without renal disease (Gronwall et al. 2012). This suggests that different sIgM specificities can protect against distinct disease mechanisms. The protective scavenging function also extends to atherosclerosis, where atherosclerosis-prone, LDL receptor-deficient mice show enhanced disease when crossed to sIgM- and C1q-deficient mice (Lewis et al. 2012). The disease in mice lacking sIgM was more severe than in mice lacking C1q, suggesting that protection is only in part regulated by the classical pathway of complement activation. IgM also has a protective role in rheumatoid arthritis. Injection of apoptotic cells in a model of inflammatory arthritis reduced disease with activation of IL-10-producing lymphocytes (Notley et al. 2011). This effect was absent in sIgM-deficient mice but could be reconstituted by infusion of IgM. Also, injection of PC-specific IgM alone inhibits arthritis (Chen et al. 2009a). Interestingly, the heritability of IgM anti-PC levels is only moderate (40 %), showing that environmental factors play a predominant role (Rahman et al. 2011). The specific environmental factors enhancing this specific IgM are unknown, but finding ways to enhance IgM anti-PC is an attractive strategy for therapy. An alternative way through which IgM might prevent autoimmunity is its ability to prevent infections and keep the balance with microbes in the environment. In absence of this protection by IgM, chronic inflammation has the possibility to promote autoimmune disease development (Munz et al. 2009; Bach 2005).

1.5.2 IgM and Innate B Cells in Loss of Tolerance

The scavenging and protective function of IgM is inherited in its pentameric structure. As mentioned earlier, healthy individuals show undetectable serum levels of monomeric IgM. However, some patients with SLE or RA can show increased levels of the monomeric form correlating with more severe disease (Harisdangkul et al. 1984; Theofilopoulos et al. 1974). This is in line with data from IgM monomer-expressing mice that show accelerated disease when crossed to lupus-prone mice (Youd et al. 2004). In contrast to the described protective role of IgM, pentameric sIgM can instead promote pathology together with autoantibodies of other subclasses, such as rheumatoid factor (RF) in RA (Hardy et al. 1987). IgM-RF is used for diagnostics and can give information on disease activity as well as erosiveness of the RA (Vencovsky et al. 2003). Recently, the use of IgM-RF as a diagnostic marker has been challenged by the discovery of autoantibodies against citrullinated proteins (CCP). Diagnostics using anti-CCP antibodies have high sensitivity and specificity and are predictive of disease. However, their relation to pathology remains to be determined (Conrad et al. 2010). Smoking has been shown to be a strong environmental factor in development of anti-CCP antibodies, and the smoke-induced inflammation in the lung leads to direct formation of citrulline modifications and promotion of anti-CCP antibodies, including IgM (Klareskog et al. 2011). Also, IgM anti-CCP antibodies have been reported to have a different B cell origin as they recognize a different set of modified epitopes compared to IgG anti-CCP (Suwannalai et al. 2011). IgM anti-CCP antibodies are not pathogenic when injected into mice alone, but do enhance the response to citrullinated collagen type II when injected together with other arthritis-inducing antibodies (Kuhn et al. 2006; Uysal et al. 2009).

Polyreactive nIgM produced by B1 cells has been shown to induce tissue damage in ischemia-reperfusion injury (Zhang and Carroll 2007). This inflammatory response follows an ischemic event (hypoxic stress) where reperfusion causes severe pathology. In this response, nIgM antibodies cause tissue damage by binding to apoptotic and altered tissue antigens, probably in an attempt to restore tissue homeostasis. Local IgM production by B1 cells was also shown to be important in the early phase of a delayed-type hypersensitivity (DTH) response in the skin. The local IgM production activates complement and leads to mast cell- and platelet-dependent release of vasoactive substances that promote recruitment of T cells. Similarly, B1 cells have also been found to migrate to the pancreas and promote T cell infiltration in a mouse model for diabetes (Ryan et al. 2010). In diabetic NOD mice, MZBs have been found to be expanded during disease development and migrate to the pancreas to present insulin to T cells, possibly adding to autoimmunity. In line with these results, depletion of B1 cells reduces autoimmunity in lupus-prone mice (Murakami et al. 1995).

1.5.3 Primary Immunodeficiencies Involving IgM

A number of primary human immunodeficiencies exist, where genes affecting B cell development and antibody production are mutated (van der Burg et al. 2012). Much of what we know about B cell biology, especially in humans, stems from studies on these patients and subsequent animal models made to carry these mutations. Hyper IgM syndrome is a primary immunodeficiency caused by a number of different genetic defects leading to increased IgM and absence, or very low concentration, of other immunoglobulin subclasses (Davies and Thrasher 2010). These mutations affect functions such as class-switch recombination and somatic hypermutation, for example, defects in signaling through CD40 or expression of the AID enzyme. Complications include, besides increased risk for infection, development of various inflammatory diseases and autoimmunity. A variety of autoantibodies including antinuclear and cardiolipin antibodies can be detected and have been suggested to add to the complications in these patients (Gulino and Notarangelo 2003).

1.6 Effector Functions: IgD

Whereas IgM has been studied quite extensively and its effector functions have been well defined, IgD has remained a mystery (Chen and Cerutti 2010). As described earlier, IgD is first expressed by alternative splicing together with IgM at the transitional and mature B cell stages and takes part in B cell development. However, there are IgD+/IgM- (IgD^{alone}) B cells that are a result of AID-dependent class switching, suggested to be the result of a B cell that has encountered its cognate antigen (Chen et al. 2009b). In humans this is a relatively rare population in circulation (about 3 %), whereas they are found to be enriched in the upper respiratory tract mucosa (about 20–25 % of the B cells). A distinct feature of secreted IgD from these B cells is also that up to 90 % use λ light chains, whereas the norm for IgD+/IgM+B cells is the use of κ light chain (Arpin et al. 1998).

1.6.1 A Putative FcδR

Secreted IgD can bind a yet unknown calcium-mobilizing receptor (FcδR) on a number of lymphocytes including mast cells and basophils (Chen et al. 2009b). sIgD also, albeit to a lesser degree, has affinity for neutrophils, monocytes, macrophages, and dendritic cells. The receptor is distinct from the IgG, IgA, and IgE receptors and has relatively high affinity as IgD was only able to be replaced after treatment with trypsin. IgD immune complexes were also found to be taken up by basophils and cross-linking using an anti-IgD antibody-induced calcium flux suggesting signaling properties of the receptor. Interestingly, cross-linking of surface-bound IgD also induced production of B cell-activating cytokines including BAFF, APRIL, IL-3, and IL-4. In addition, activation of IgD-armed basophils induced secretion of antimicrobial factors such as β-defensin and cathelicidin.

1.6.2 IgD as a Pattern Recognition Molecule

Unlike IgM, the IgD genes have undergone a large number of alterations through evolution, and this has given rise to a multitude of variants in different species. The heavy chain can vary a lot in size and composition, and together with the extensive differences in glycosylation, it is evident that IgD has had a distinct evolutionary path from IgM (Edholm et al. 2011). Also, the fact that IgW of sharks is an ancient orthologue of IgD shows the early origin of this isotype that has been altered through evolution. In 2009 Chen et al. showed that IgD bound pathogens regardless of its antigen specificity but by using its Cδ1 domain and not the antigen binding site of the antibody. This together with findings that many IgD positive plasmablasts display hypermutated VH and VL regions predicted to be incapable of Ag binding suggests that IgD has an inherited function as a pattern recognition molecule (Liu et al. 1996). Also, indicative of this function is that catfish IgD even lacks the VH region (Edholm et al. 2010).

1.7 IgD in Immune Defense

An important finding in the quest for a role of IgD in immune defense was that IgD^{alone} B cells are enriched in the upper respiratory mucosa. This together with the fact that IgD has the ability to bind microbial virulence factors and pathogenic respiratory viruses and bacteria (such as *H. influenza*) suggests a specific function for IgD in mucosal immunity. In addition, the ability of IgD to bind to immune cells inducing so-called arming of basophils suggests that this subclass is important for innate and adaptive immune defense. Also, supporting a role in mucosal immunity is the fact that patients with selective IgA deficiency compensate by having increased numbers of IgD^{alone} B cells in their mucosa, possibly replacing the role for

IgA-mediated protection (Chen and Cerutti 2011). Certain pathogens also promote specific IgD production and activation of IgD^{alone} B cells such as the parasite *P. chabaudi* and bacteria *N. lactamica* (Couper et al. 2005b; Vaughan et al. 2010).

1.8 IgD in Autoimmunity and Disease

Whereas IgM is expressed in the immature stages of B cell development in the bone marrow, IgD is first upregulated on transitional B cells in the spleen (Loder et al. 1999). It is not known why naïve B cells have this dual expression and why it differs between subtypes of B cells. One current hypothesis is that since IgD is more flexible due to a large hinge region, it binds antigen more easily as opposed to the IgM-BCR, which would only bind antigen with high affinity (Chen and Cerutti 2010). The IgD-BCR associates with a distinct set of molecules on the surface of the B cell, and it has been suggested that it has a function in delivering tolerogenic or even apoptotic signals even though the data available are somewhat contradictive (Peckham et al. 2001). To support a suppressive role, self-reactive anergic B cells from both humans and mice have been shown to express more IgD than IgM. Also in humans, B cells isolated from 10 donors were shown to use primarily VH4-34 heavy chain, enriched for charged amino acids, which has been associated with autoimmunity before (Zheng et al. 2004). Likewise, in double-transgenic mice expressing genes for the model antigen hen egg lysozyme (HEL) as well as a high-affinity anti-HEL antibody, it was found that surface expression of IgM was greatly reduced whereas IgD expression remained high (Goodnow et al. 1988). These data suggest a difference between IgM and IgD-BCRs with respect to B cell tolerance. Cell intrinsically, the IgD-BCR can even block IgM-BCR-mediated activation and stop antibody secretion (Phillips and Klaus 1993). However, IgD-deficient mice show no major difference in B cell development or responses to T cell-dependent and independent antigens (Nitschke et al. 1993; Roes and Rajewsky 1993). Still, when IgD ko mice were crossed to lupus-prone lpr mice, the mice developed a more severe disease including elevated autoantibody levels, increased immune complex deposition, and nephritis in the kidneys (Guo et al. 2011).

The role for secreted IgD has likewise been a mystery, and its role is poorly defined in most cases. In the clinic it has only been used as an indicator for multiple myeloma and predicts aggressive forms of the cancer (Vladutiu and Netto 1982). Still, sIgD is found at elevated levels in inflammatory diseases such as allergy (including asthma and rhinitis) and autoimmune diseases (such as rheumatoid arthritis and SLE) (Chen and Cerutti 2010). The sIgD has also been shown to be self-antigen specific where antinuclear specificities are found in SLE patients (Ritchie 1968). Also in RA, sIgD can contribute to disease pathology by being a target for autoantibodies, mainly of the IgA and IgG classes giving rise to RF-induced pathology (Kantor et al. 1970; Pope et al. 1982). A role for secreted IgD in pro-inflammatory pathways is supported by the fact that patients with hyper-IgD syndrome (HIDS) have high IgD levels in serum accompanied with periodic fevers (Chen and Cerutti 2011).

1.9 Concluding Remarks

IgM and IgD are antibodies at the front line of immunity. They evolved first and they are the first antibodies to recognize novel pathogens to fight infection. They are also important in selection and maturation of the repertoire of lymphocytes that protects us from infection. Inherited in the function of IgM and IgD is their ability to balance the immune response where their self-reactivity is as important as their ability to sense foreign antigens. A task for the future is to harness this ability and to be able to steer immune responses and regulation via IgM and IgD towards homeostasis and away from imbalance and autoimmune disease.

Acknowledgements I would like to thank Malin Winerdahl for artwork and Linda Kvastad for providing the histology picture. I would also like to thank Emilie Grasset and Fredrik Wermeling for critically reading the manuscript. The B cell biology group at Karolinska Institutet is supported by grants from the Swedish Research Council, the Swedish Cancer Foundation, the Torsten Söderberg foundation, the Cardiovascular Research Program at Karolinska Institutet, Gustav V's 80-Year Foundation, and the Swedish Rheumatism Association.

References

Adachi T, Harumiya S, Takematsu H, Kozutsumi Y, Wabl M, Fujimoto M et al (2012) CD22 serves as a receptor for soluble IgM. Eur J Immunol 42(1):241–247

Alder MN, Rogozin IB, Iyer LM, Glazko GV, Cooper MD, Pancer Z (2005) Diversity and function of adaptive immune receptors in a jawless vertebrate. Science 310(5756):1970–1973

Allman D, Lindsley RC, DeMuth W, Rudd K, Shinton SA, Hardy RR (2001) Resolution of three nonproliferative immature splenic B cell subsets reveals multiple selection points during peripheral B cell maturation. J Immunol 167(12):6834–6840

Alugupalli KR, Gerstein RM, Chen J, Szomolanyi-Tsuda E, Woodland RT, Leong JM (2003) The resolution of relapsing fever borreliosis requires IgM and is concurrent with expansion of B1b lymphocytes. J Immunol 170(7):3819–3827

Arnold JN, Radcliffe CM, Wormald MR, Royle L, Harvey DJ, Crispin M et al (2004) The glycosylation of human serum IgD and IgE and the accessibility of identified oligomannose structures for interaction with mannan-binding lectin. J Immunol 173(11):6831–6840

Arnold JN, Wormald MR, Suter DM, Radcliffe CM, Harvey DJ, Dwek RA et al (2005) Human serum IgM glycosylation: identification of glycoforms that can bind to mannan-binding lectin. J Biol Chem 280(32):29080–29087

Arpin C, de Bouteiller O, Razanajaona D, Fugier-Vivier I, Briere F, Banchereau J et al (1998) The normal counterpart of IgD myeloma cells in germinal center displays extensively mutated IgVH gene, Cmu-Cdelta switch, and lambda light chain expression. J Exp Med 187(8):1169–1178

Asahi Y, Yoshikawa T, Watanabe I, Iwasaki T, Hasegawa H, Sato Y et al (2002) Protection against influenza virus infection in polymeric Ig receptor knockout mice immunized intranasally with adjuvant-combined vaccines. J Immunol 168(6):2930–2938

Bach JF (2005) Infections and autoimmune diseases. J Autoimmun 25(Suppl):74–80

Baral TN, De Baetselier P, Brombacher F, Magez S (2007) Control of *Trypanosoma evansi* infection is IgM mediated and does not require a type I inflammatory response. J Infect Dis 195(10):1513–1520

Baumgarth N (2011) The double life of a B-1 cell: self-reactivity selects for protective effector functions. Nat Rev Immunol 11(1):34–46
Baumgarth N, Herman OC, Jager GC, Brown LE, Herzenberg LA, Chen J (2000a) B-1 and B-2 cell-derived immunoglobulin M antibodies are nonredundant components of the protective response to influenza virus infection. J Exp Med 192(2):271–280
Baumgarth N, Chen J, Herman OC, Jager GC, Herzenberg LA (2000b) The role of B-1 and B-2 cells in immune protection from influenza virus infection. Curr Top Microbiol Immunol 252:163–169
Belperron AA, Dailey CM, Bockenstedt LK (2005) Infection-induced marginal zone B cell production of Borrelia hermsii-specific antibody is impaired in the absence of CD1d. J Immunol 174(9):5681–5686
Bérengère V, David A, Wiestner A (2012) TOSO, the Fcμ receptor, is highly expressed on chronic lymphocytic leukemia B cells, internalizes upon IgM binding, shuttles to the lysosome, and is downregulated in response to TLR activation. J Immunol 187:4040–4050
Bitsaktsis C, Nandi B, Racine R, MacNamara KC, Winslow G (2007) T-Cell-independent humoral immunity is sufficient for protection against fatal intracellular ehrlichia infection. Infect Immun 75(10):4933–4941
Blanch VJ, Piskurich JF, Kaetzel CS (1999) Cutting edge: coordinate regulation of IFN regulatory factor-1 and the polymeric Ig receptor by proinflammatory cytokines. J Immunol 162(3):1232–1235
Boes M, Esau C, Fischer MB, Schmidt T, Carroll M, Chen J (1998a) Enhanced B-1 cell development, but impaired IgG antibody responses in mice deficient in secreted IgM. J Immunol 160(10):4776–4787
Boes M, Prodeus AP, Schmidt T, Carroll MC, Chen J (1998b) A critical role of natural immunoglobulin M in immediate defense against systemic bacterial infection. J Exp Med 188(12):2381–2386
Bordet J, Genou O (1901) Sur L'existence de substances sensibilisatrices dans la plupart des serum antimicrobiens. Ann Inst Pasteur 15:289–302
Borsos T, Rapp HJ (1965) Complement fixation on cell surfaces by 19S and 7S antibodies. Science 150(3695):505–506
Brattig NW, Bazzocchi C, Kirschning CJ, Reiling N, Buttner DW, Ceciliani F et al (2004) The major surface protein of Wolbachia endosymbionts in filarial nematodes elicits immune responses through TLR2 and TLR4. J Immunol 173(1):437–445
Brigandi RA, Rotman HL, Yutanawiboonchai W, Leon O, Nolan TJ, Schad GA et al (1996) Strongyloides stercoralis: role of antibody and complement in immunity to the third stage of larvae in BALB/cByJ mice. Exp Parasitol 82(3):279–289
Brown JS, Hussell T, Gilliland SM, Holden DW, Paton JC, Ehrenstein MR et al (2002) The classical pathway is the dominant complement pathway required for innate immunity to *Streptococcus pneumoniae* infection in mice. Proc Natl Acad Sci USA 99(26):16969–16974
Burrows P, LeJeune M, Kearney JF (1979) Evidence that murine pre-B cells synthesise mu heavy chains but no light chains. Nature 280(5725):838–840
Caidahl K, Hartford M, Karlsson T, Herlitz J, Pettersson K, de Faire U et al (2012) IgM-phosphorylcholine autoantibodies and outcome in acute coronary syndromes, Int J Cardiol. doi:10.1016/j.ijcard.2012.01.018
Carroll MC (2008) Complement and humoral immunity. Vaccine 26(Suppl 8):I28–I33
Chai LY, Vonk AG, Kullberg BJ, Verweij PE, Verschueren I, van der Meer JW et al (2011) Aspergillus fumigatus cell wall components differentially modulate host TLR2 and TLR4 responses. Microbes Infect 13(2):151–159
Chen K, Cerutti A (2010) New insights into the enigma of immunoglobulin D. Immunol Rev 237(1):160–179
Chen K, Cerutti A (2011) The function and regulation of immunoglobulin D. Curr Opin Immunol 23(3):345–352
Chen Y, Khanna S, Goodyear CS, Park YB, Raz E, Thiel S et al (2009a) Regulation of dendritic cells and macrophages by an anti-apoptotic cell natural antibody that suppresses TLR responses and inhibits inflammatory arthritis. J Immunol 183(2):1346–1359

Chen K, Xu W, Wilson M, He B, Miller NW, Bengten E et al (2009b) Immunoglobulin D enhances immune surveillance by activating antimicrobial, proinflammatory and B cell-stimulating programs in basophils. Nat Immunol 10(8):889–898
Choi YS, Baumgarth N (2008) Dual role for B-1a cells in immunity to influenza virus infection. J Exp Med 205(13):3053–3064
Chou MY, Fogelstrand L, Hartvigsen K, Hansen LF, Woelkers D, Shaw PX et al (2009) Oxidation-specific epitopes are dominant targets of innate natural antibodies in mice and humans. J Clin Invest 119(5):1335–1349
Connolly SE, Benach JL (2001) Cutting edge: the spirochetemia of murine relapsing fever is cleared by complement-independent bactericidal antibodies. J Immunol 167(6):3029–3032
Conrad K, Roggenbuck D, Reinhold D, Dorner T (2010) Profiling of rheumatoid arthritis associated autoantibodies. Autoimmun Rev 9(6):431–435
Cooper NR, Nemerow GR, Mayes JT (1983) Methods to detect and quantitate complement activation. Springer Semin Immunopathol 6(2–3):195–212
Couper KN, Roberts CW, Brombacher F, Alexander J, Johnson LL (2005a) *Toxoplasma gondii-specific* immunoglobulin M limits parasite dissemination by preventing host cell invasion. Infect Immunol 73(12):8060–8068
Couper KN, Phillips RS, Brombacher F, Alexander J (2005b) Parasite-specific IgM plays a significant role in the protective immune response to asexual erythrocytic stage *Plasmodium chabaudi* AS infection. Parasite Immunol 27(5):171–180
Coutinho A, Kazatchkine MD, Avrameas S (1995) Natural autoantibodies. Curr Opin Immunol 7(6):812–818
Crocker PR, Paulson JC, Varki A (2007) Siglecs and their roles in the immune system. Nat Rev Immunol 7(4):255–266
Czajkowsky DM, Shao Z (2009) The human IgM pentamer is a mushroom-shaped molecule with a flexural bias. Proc Natl Acad Sci USA 106(35):14960–14965
Dales S, Fujinami RS, Oldstone MB (1983) Infection with vaccinia favors the selection of hybridomas synthesizing autoantibodies against intermediate filaments, one of them cross-reacting with the virus hemagglutinin. J Immunol 131(3):1546–1553
Davies EG, Thrasher AJ (2010) Update on the hyper immunoglobulin M syndromes. Br J Haematol 149(2):167–180
de Lalla C, Fagioli C, Cessi FS, Smilovich D, Sitia R (1998) Biogenesis and function of IgM: the role of the conserved mu-chain tailpiece glycans. Mol Immunol 35(13):837–845
Dempsey PW, Allison ME, Akkaraju S, Goodnow CC, Fearon DT (1996) C3d of complement as a molecular adjuvant: bridging innate and acquired immunity. Science 271(5247):348–350
Diamantstein T, Klos M, Osawa H, Chen ZC (1982) Chitin: an immunological adjuvant and a polyclonal B-lymphocyte activator. Int Arch Allergy Appl Immunol 68(4):377–381
Diamond MS, Sitati EM, Friend LD, Higgs S, Shrestha B, Engle M (2003) A critical role for induced IgM in the protection against West Nile virus infection. J Exp Med 198(12):1853–1862
Dighiero G, Lymberi P, Mazie JC, Rouyre S, Butler-Browne GS, Whalen RG et al (1983) Murine hybridomas secreting natural monoclonal antibodies reacting with self antigens. J Immunol 131(5):2267–2272
Doody GM, Justement LB, Delibrias CC, Matthews RJ, Lin J, Thomas ML et al (1995) A role in B cell activation for CD22 and the protein tyrosine phosphatase SHP. Science 269(5221):242–244
Early P, Rogers J, Davis M, Calame K, Bond M, Wall R et al (1980) Two mRNAs can be produced from a single immunoglobulin mu gene by alternative RNA processing pathways. Cell 20(2):313–319
Edholm ES, Bengten E, Stafford JL, Sahoo M, Taylor EB, Miller NW et al (2010) Identification of two IgD+ B cell populations in channel catfish, *Ictalurus punctatus*. J Immunol 185(7):4082–4094
Edholm ES, Bengten E, Wilson M (2011) Insights into the function of IgD. Dev Comp Immunol 35(12):1309–1316

Ehrenstein MR, Notley CA (2010) The importance of natural IgM: scavenger, protector and regulator. Nat Rev Immunol 10(11):778–786
Fishelson Z, Attali G, Mevorach D (2001) Complement and apoptosis. Mol Immunol 38(2–3):207–219
Foote JB, Mahmoud TI, Vale AM, Kearney JF (2012) Long-term maintenance of polysaccharide-specific antibodies by IgM-secreting cells. J Immunol 188(1):57–67
Gay D, Saunders T, Camper S, Weigert M (1993) Receptor editing: an approach by autoreactive B cells to escape tolerance. J Exp Med 177(4):999–1008
Getahun A, Heyman B (2006) How antibodies act as natural adjuvants. Immunol Lett 104(1–2):38–45
Ghosh J, Lun CM, Majeske AJ, Sacchi S, Schrankel CS, Smith LC (2011) Invertebrate immune diversity. Dev Comp Immunol 35(9):959–974
Gil-Cruz C, Bobat S, Marshall JL, Kingsley RA, Ross EA, Henderson IR et al (2009) The porin OmpD from nontyphoidal Salmonella is a key target for a protective B1b cell antibody response. Proc Natl Acad Sci USA 106(24):9803–9808
Gil-Cruz C, Perez-Shibayama C, Firner S, Waisman A, Bechmann I, Thiel V et al (2012) T helper cell- and CD40-dependent germline IgM prevents chronic virus-induced demyelinating disease. Proc Natl Acad Sci USA 109(4):1233–1238
Goodnow CC, Crosbie J, Adelstein S, Lavoie TB, Smith-Gill SJ, Brink RA et al (1988) Altered immunoglobulin expression and functional silencing of self-reactive B lymphocytes in transgenic mice. Nature 334(6184):676–682
Gritzmacher CA (1989) Molecular aspects of heavy-chain class switching. Crit Rev Immunol 9(3):173–200
Gronwall C, Akhter E, Oh C, Burlingame RW, Petri M, Silverman GJ (2012) IgM autoantibodies to distinct apoptosis-associated antigens correlate with protection from cardiovascular events and renal disease in patients with SLE. Clin Immunol 142(3):390–398
Gulino AV, Notarangelo LD (2003) Hyper IgM syndromes. Curr Opin Rheumatol 15(4):422–429
Guo L, Tian J, Guo Z, Zheng B, Han S (2011) The absence of immunoglobulin D B cell receptor-mediated signals promotes the production of autoantibodies and exacerbates glomerulonephritis in murine lupus. Clin Exp Immunol 164(2):227–235
Hadge D (1985) Evolution of the immunoglobulins. Allerg Immunol (Leipz) 31(4):231–243
Hardy RR, Hayakawa K, Shimizu M, Yamasaki K, Kishimoto T (1987) Rheumatoid factor secretion from human Leu-1+ B cells. Science 236(4797):81–83
Harisdangkul V, Barnes TY, Songcharoen S, Pennebaker JB (1984) Clinical significance of low molecular weight IgM in patients with systemic lupus erythematosus. J Rheumatol 11(5):638–643
Hitoshi Y, Lorens J, Kitada SI, Fisher J, LaBarge M, Ring HZ et al (1998) Toso, a cell surface, specific regulator of Fas-induced apoptosis in T cells. Immunity 8(4):461–471
Honda S, Kurita N, Miyamoto A, Cho Y, Usui K, Takeshita K et al (2009) Enhanced humoral immune responses against T-independent antigens in Fc alpha/muR-deficient mice. Proc Natl Acad Sci USA 106(27):11230–11235
Honjo K, Kubagawa Y, Jones DM, Dizon B, Zhu Z, Ohno H et al (2012) Altered Ig levels and antibody responses in mice deficient for the Fc receptor for IgM (FcmuR). Proc Natl Acad Sci USA 109(39):15882–15887, Epub 2012/09/18
Hooijkaas H, Benner R, Pleasants JR, Wostmann BS (1984) Isotypes and specificities of immunoglobulins produced by germ-free mice fed chemically defined ultrafiltered antigen-free diet. Eur J Immunol 14(12):1127–1130
Ip WK, Takahashi K, Ezekowitz RA, Stuart LM (2009) Mannose-binding lectin and innate immunity. Immunol Rev 230(1):9–21
Jang C, Machtaler S, Matsuuchi L (2010) The role of Ig-alpha/beta in B cell antigen receptor internalization. Immunol Lett 134(1):75–82
Jayasekera JP, Moseman EA, Carroll MC (2007) Natural antibody and complement mediate neutralization of influenza virus in the absence of prior immunity. J Virol 81(7):3487–3494

Johansen FE, Braathen R, Brandtzaeg P (2000) Role of J chain in secretory immunoglobulin formation. Scand J Immunol 52(3):240–248
Kaetzel CS (2005) The polymeric immunoglobulin receptor: bridging innate and adaptive immune responses at mucosal surfaces. Immunol Rev 206:83–99
Kallies A, Nutt SL (2007) Terminal differentiation of lymphocytes depends on Blimp-1. Curr Opin Immunol 19(2):156–162
Kaneko Y, Takashima Y, Xuaun X, Igarashi I, Nagasawa H, Mikami T et al (2004) Natural IgM antibodies in sera from various animals but not the cat kill *Toxoplasma gondii* by activating the classical complement pathway. Parasitology 128(Pt 2):123–129
Kantor GL, Van Herle AJ, Barnett EV (1970) Auto-antibodies of the IgD class. Clin Exp Immunol 6(6):951–962
Kataaha PK, Mortazavi-Milani SM, Russell G, Holborow EJ (1985) Anti-intermediate filament antibodies, antikeratin antibody, and antiperinuclear factor in rheumatoid arthritis and infectious mononucleosis. Ann Rheum Dis 44(7):446–449
Kearney JF (2005) Innate-like B, cells. Springer Semin Immunopathol 26(4):377–383
Kikuno K, Kang DW, Tahara K, Torii I, Kubagawa HM, Ho KJ et al (2007) Unusual biochemical features and follicular dendritic cell expression of human Fcalpha/mu receptor. Eur J Immunol 37(12):3540–3550
Kinoshita M, Shinomiya N, Ono S, Tsujimoto H, Kawabata T, Matsumoto A et al (2006) Restoration of natural IgM production from liver B cells by exogenous IL-18 improves the survival of burn-injured mice infected with *Pseudomonas aeruginosa*. J Immunol 177(7):4627–4635
Klareskog L, Malmstrom V, Lundberg K, Padyukov L, Alfredsson L (2011) Smoking, citrullination and genetic variability in the immunopathogenesis of rheumatoid arthritis. Semin Immunol 23(2):92–98
Klimovich VB (2011) IgM and its receptors: structural and functional aspects. Biochemistry (Mosc) 76(5):534–549
Kojouharova M, Reid K, Gadjeva M (2010) New insights into the molecular mechanisms of classical complement activation. Mol Immunol 47(13):2154–2160
Krajci P, Solberg R, Sandberg M, Oyen O, Jahnsen T, Brandtzaeg P (1989) Molecular cloning of the human transmembrane secretory component (poly-Ig receptor) and its mRNA expression in human tissues. Biochem Biophys Res Commun 158(3):783–789
Kubagawa H, Oka S, Kubagawa Y, Torii I, Takayama E, Kang DW et al (2009) Identity of the elusive IgM Fc receptor (FcmuR) in humans. J Exp Med 206(12):2779–2793
Kuhn KA, Kulik L, Tomooka B, Braschler KJ, Arend WP, Robinson WH et al (2006) Antibodies against citrullinated proteins enhance tissue injury in experimental autoimmune arthritis. J Clin Invest 116(4):961–973
Lachmann P (2006) Complement before molecular biology. Mol Immunol 43(6):496–508
Lewis MJ, Malik TH, Ehrenstein MR, Boyle JJ, Botto M, Haskard DO (2012) Immunoglobulin M is required for protection against atherosclerosis in low-density lipoprotein receptor-deficient mice. Circulation 120:417–426
Liu YJ, de Bouteiller O, Arpin C, Briere F, Galibert L, Ho S et al (1996) Normal human IgD+IgM- germinal center B cells can express up to 80 mutations in the variable region of their IgD transcripts. Immunity 4(6):603–613
Loder F, Mutschler B, Ray RJ, Paige CJ, Sideras P, Torres R et al (1999) B cell development in the spleen takes place in discrete steps and is determined by the quality of B cell receptor-derived signals. J Exp Med 190(1):75–89
Lopes-Carvalho T, Foote J, Kearney JF (2005) Marginal zone B cells in lymphocyte activation and regulation. Curr Opin Immunol 17(3):244–250
Luning Prak ET, Monestier M, Eisenberg RA (2011) B cell receptor editing in tolerance and autoimmunity. Ann N Y Acad Sci 1217:96–121
Mahmoud TI, Schroeder HW Jr, Kearney JF (2011) Limiting CDR-H3 diversity abrogates the antibody response to the bacterial polysaccharide alpha 1→3 dextran. J Immunol 187(2):879–886

Malkiel S, Kuhlow CJ, Mena P, Benach JL (2009) The loss and gain of marginal zone and peritoneal B cells is different in response to relapsing fever and Lyme disease Borrelia. J Immunol 182(1):498–506
Manderson AP, Quah B, Botto M, Goodnow CC, Walport MJ, Parish CR (2006) A novel mechanism for complement activation at the surface of B cells following antigen binding. J Immunol 177(8):5155–5162
Marchalonis JJ, Adelman MK, Schluter SF, Ramsland PA (2006) The antibody repertoire in evolution: chance, selection, and continuity. Dev Comp Immunol 30(1–2):223–247
Martensson IL, Keenan RA, Licence S (2007) The pre-B-cell receptor. Curr Opin Immunol 19(2):137–142
Medvedovic J, Ebert A, Tagoh H, Busslinger M (2011) Pax5: a master regulator of B cell development and leukemogenesis. Adv Immunol 111:179–206
Meffre E, Wardemann H (2008) B-cell tolerance checkpoints in health and autoimmunity. Curr Opin Immunol 20(6):632–638
Miller YI, Choi SH, Wiesner P, Fang L, Harkewicz R, Hartvigsen K et al (2011) Oxidation-specific epitopes are danger-associated molecular patterns recognized by pattern recognition receptors of innate immunity. Circ Res 108(2):235–248
Misra R, Venables PJ, Plater-Zyberk C, Watkins PF, Maini RN (1989) Anti-cardiolipin antibodies in infectious mononucleosis react with the membrane of activated lymphocytes. Clin Exp Immunol 75(1):35–40
Moise A, Nedelcu FD, Toader MA, Sora SM, Tica A, Ferastraoaru DE et al (2010) Primary immunodeficiencies of the B lymphocyte. J Med Life 3(1):60–63
Mostov KE, Blobel G (1982) A transmembrane precursor of secretory component. The receptor for transcellular transport of polymeric immunoglobulins. J Biol Chem 257(19):11816–11821
Mouquet H, Nussenzweig MC (2011) Polyreactive antibodies in adaptive immune responses to viruses. Cell Mol Life Sci 69:1435–1445
Mouquet H, Scheid JF, Zoller MJ, Krogsgaard M, Ott RG, Shukair S et al (2010) Polyreactivity increases the apparent affinity of anti-HIV antibodies by heteroligation. Nature 467(7315):591–595
Muller-Eberhard HJ (1988) Molecular organization and function of the complement system. Annu Rev Biochem 57:321–347
Munz C, Lunemann JD, Getts MT, Miller SD (2009) Antiviral immune responses: triggers of or triggered by autoimmunity? Nat Rev Immunol 9(4):246–258
Murakami M, Yoshioka H, Shirai T, Tsubata T, Honjo T (1995) Prevention of autoimmune symptoms in autoimmune-prone mice by elimination of B-1 cells. Int Immunol 7(5):877–882
Muramatsu M, Kinoshita K, Fagarasan S, Yamada S, Shinkai Y, Honjo T (2000) Class switch recombination and hypermutation require activation-induced cytidine deaminase (AID), a potential RNA editing enzyme. Cell 102(5):553–563
Nguyen XH, Lang PA, Lang KS, Adam D, Fattakhova G, Foger N et al (2011) Toso regulates the balance between apoptotic and nonapoptotic death receptor signaling by facilitating RIP1 ubiquitination. Blood 118(3):598–608
Niles MJ, Matsuuchi L, Koshland ME (1995) Polymer IgM assembly and secretion in lymphoid and nonlymphoid cell lines: evidence that J chain is required for pentamer IgM synthesis. Proc Natl Acad Sci USA 92(7):2884–2888
Nimmerjahn F, Ravetch JV (2006) Fcgamma receptors: old friends and new family members. Immunity 24(1):19–28
Nimmerjahn F, Ravetch JV (2008) Anti-inflammatory actions of intravenous immunoglobulin. Annu Rev Immunol 26:513–533
Nitschke L, Kosco MH, Kohler G, Lamers MC (1993) Immunoglobulin D-deficient mice can mount normal immune responses to thymus-independent and -dependent antigens. Proc Natl Acad Sci USA 90(5):1887–1891
Notley CA, Brown MA, Wright GP, Ehrenstein MR (2011) Natural IgM is required for suppression of inflammatory arthritis by apoptotic cells. J Immunol 186(8):4967–4972

Ochsenbein AF, Fehr T, Lutz C, Suter M, Brombacher F, Hengartner H et al (1999) Control of early viral and bacterial distribution and disease by natural antibodies. Science 286(5447):2156–2159
Ohno T, Kubagawa H, Sanders SK, Cooper MD (1990) Biochemical nature of an Fc mu receptor on human B-lineage cells. J Exp Med 172(4):1165–1175
Ono M, Bolland S, Tempst P, Ravetch JV (1996) Role of the inositol phosphatase SHIP in negative regulation of the immune system by the receptor Fc(gamma)RIIB. Nature 383(6597):263–266
Ouchida R, Mori H, Hase K, Takatsu H, Kurosaki T, Tokuhisa T et al (2012) Critical role of the IgM Fc receptor in IgM homeostasis, B-cell survival, and humoral immune responses. Proc Natl Acad Sci USA 109(40):E2699–E2706, Epub 2012/09/19
Palladino G, Mozdzanowska K, Washko G, Gerhard W (1995) Virus-neutralizing antibodies of immunoglobulin G (IgG) but not of IgM or IgA isotypes can cure influenza virus pneumonia in SCID mice. J Virol 69(4):2075–2081
Pan W, Ogunremi O, Wei G, Shi M, Tabel H (2006) CR3 (CD11b/CD18) is the major macrophage receptor for IgM antibody-mediated phagocytosis of African trypanosomes: diverse effect on subsequent synthesis of tumor necrosis factor alpha and nitric oxide. Microbes Infect 8(5):1209–1218
Pancer Z, Cooper MD (2006) The evolution of adaptive immunity. Annu Rev Immunol 24:497–518
Peckham D, Andersen-Nissen E, Finkelman FD, Stunz LL, Ashman RF (2001) Difference in apoptosis induction between surface IgD and IgM. Int Immunol 13(3):285–295
Peng Y, Kowalewski R, Kim S, Elkon KB (2005) The role of IgM antibodies in the recognition and clearance of apoptotic cells. Mol Immunol 42(7):781–787
Phillips C, Klaus GG (1993) Cross-linking of surface IgM, but not surface IgD receptors, by soluble monoclonal antibodies primes murine B cells to secrete immunoglobulin in response to lymphokines. Eur J Immunol 23(2):574–577
Phillips JO, Everson MP, Moldoveanu Z, Lue C, Mestecky J (1990) Synergistic effect of IL-4 and IFN-gamma on the expression of polymeric Ig receptor (secretory component) and IgA binding by human epithelial cells. J Immunol 145(6):1740–1744
Pillai S, Cariappa A (2009) The follicular versus marginal zone B lymphocyte cell fate decision. Nat Rev Immunol 9(11):767–777
Pope RM, Keightley R, McDuffy S (1982) Circulating autoantibodies to IgD in rheumatic diseases. J Immunol 128(4):1860–1863
Raff MC, Megson M, Owen JJ, Cooper MD (1976) Early production of intracellular IgM by B-lymphocyte precursors in mouse. Nature 259(5540):224–226
Rahman I, Atout R, Pedersen NL, de Faire U, Frostegard J, Ninio E et al (2011) Genetic and environmental regulation of inflammatory CVD biomarkers Lp-PLA2 and IgM anti-PC. Atherosclerosis 218(1):117–122
Rajan B, Ramalingam T, Rajan TV (2005) Critical role for IgM in host protection in experimental filarial infection. J Immunol 175(3):1827–1833
Rapaka RR, Ricks DM, Alcorn JF, Chen K, Khader SA, Zheng M et al (2010) Conserved natural IgM antibodies mediate innate and adaptive immunity against the opportunistic fungus Pneumocystis murina. J Exp Med 207(13):2907–2919
Ratcliffe MJ (2006) Antibodies, immunoglobulin genes and the bursa of Fabricius in chicken B cell development. Dev Comp Immunol 30(1–2):101–118
Reth M (1989) Antigen receptor tail clue. Nature 338(6214):383–384
Ritchie RF (1968) Two new antinuclear antibodies: their relationship to the homogeneous immunofluorescent pattern. Arthritis Rheum 11(1):37–43
Roes J, Rajewsky K (1993) Immunoglobulin D (IgD)-deficient mice reveal an auxiliary receptor function for IgD in antigen-mediated recruitment of B cells. J Exp Med 177(1):45–55
Rogers J, Early P, Carter C, Calame K, Bond M, Hood L et al (1980) Two mRNAs with different 3' ends encode membrane-bound and secreted forms of immunoglobulin mu chain. Cell 20(2):303–312

Roopenian DC, Akilesh S (2007) FcRn: the neonatal Fc receptor comes of age. Nat Rev Immunol 7(9):715–725
Roozendaal R, Carroll MC (2007) Complement receptors CD21 and CD35 in humoral immunity. Immunol Rev 219:157–166
Ryan GA, Wang CJ, Chamberlain JL, Attridge K, Schmidt EM, Kenefeck R et al (2010) B1 cells promote pancreas infiltration by autoreactive T cells. J Immunol 185(5):2800–2807
Sait L, Galic M, Strugnell RA, Janssen PH (2003) Secretory antibodies do not affect the composition of the bacterial microbiota in the terminal ileum of 10-week-old mice. Appl Environ Microbiol 69(4):2100–2109
Salinas-Carmona MC, Perez-Rivera I (2004) Humoral immunity through immunoglobulin M protects mice from an experimental actinomycetoma infection by Nocardia brasiliensis. Infect Immun 72(10):5597–5604
Satoh J, Prabhakar BS, Haspel MV, Ginsberg-Fellner F, Notkins AL (1983) Human monoclonal autoantibodies that react with multiple endocrine organs. N Engl J Med 309(4):217–220
Schatz DG, Swanson PC (2011) V(D)J recombination: mechanisms of initiation. Annu Rev Genet 45:167–202
Schneeman TA, Bruno ME, Schjerven H, Johansen FE, Chady L, Kaetzel CS (2005) Regulation of the polymeric Ig receptor by signaling through TLRs 3 and 4: linking innate and adaptive immune responses. J Immunol 175(1):376–384
Seiler P, Kalinke U, Rulicke T, Bucher EM, Bose C, Zinkernagel RM et al (1998) Enhanced virus clearance by early inducible lymphocytic choriomeningitis virus-neutralizing antibodies in immunoglobulin-transgenic mice. J Virol 72(3):2253–2258
Shibuya A, Sakamoto N, Shimizu Y, Shibuya K, Osawa M, Hiroyama T et al (2000) Fc alpha/mu receptor mediates endocytosis of IgM-coated microbes. Nat Immunol 1(5):441–446
Shima H, Takatsu H, Fukuda S, Ohmae M, Hase K, Kubagawa H et al (2010) Identification of TOSO/FAIM3 as an Fc receptor for IgM. Int Immunol 22(3):149–156
Shulzhenko N, Morgun A, Hsiao W, Battle M, Yao M, Gavrilova O et al (2011) Crosstalk between B lymphocytes, microbiota and the intestinal epithelium governs immunity versus metabolism in the gut. Nat Med 17(12):1585–1593
Silverman GJ, Srikrishnan R, Germar K, Goodyear CS, Andrews KA, Ginzler EM et al (2008) Genetic imprinting of autoantibody repertoires in systemic lupus erythematosus patients. Clin Exp Immunol 153(1):102–116
Sims GP, Ettinger R, Shirota Y, Yarboro CH, Illei GG, Lipsky PE (2005) Identification and characterization of circulating human transitional B cells. Blood 105(11):4390–4398
Srinivasan L, Sasaki Y, Calado DP, Zhang B, Paik JH, DePinho RA et al (2009) PI3 kinase signals BCR-dependent mature B cell survival. Cell 139(3):573–586
Stavnezer J, Amemiya CT (2004) Evolution of isotype switching. Semin Immunol 16(4):257–275
Subramaniam KS, Datta K, Quintero E, Manix C, Marks MS, Pirofski LA (2010) The absence of serum IgM enhances the susceptibility of mice to pulmonary challenge with *Cryptococcus neoformans*. J Immunol 184(10):5755–5767
Sun Z, Almogren A, Furtado PB, Chowdhury B, Kerr MA, Perkins SJ (2005) Semi-extended solution structure of human myeloma immunoglobulin D determined by constrained X-ray scattering. J Mol Biol 353(1):155–173
Suwannalai P, Willemze A, van Toorn L, Stoeken-Rijsbergen G, Levarht N, Drijfhout JW et al (2011) The fine specificity of IgM anti-citrullinated protein antibodies (ACPA) is different from that of IgG ACPA. Arthritis Res Ther 13(6):R195
Taylor JJ, Pape KA, Jenkins MK (2012) A germinal center-independent pathway generates unswitched memory B cells early in the primary response. J Exp Med 209(3):597–606
Theofilopoulos AN, Burtonboy G, LoSpalluto JJ, Ziff M (1974) IgM rheumatoid factor and low molecular weight IgM. An association with vasculitis. Arthritis Rheum 17(3):272–284
Toh BH, Yildiz A, Sotelo J, Osung O, Holborow EJ, Kanakoudi F et al (1979) Viral infections and IgM autoantibodies to cytoplasmic intermediate filaments. Clin Exp Immunol 37(1):76–82

Turnbaugh PJ, Ley RE, Mahowald MA, Magrini V, Mardis ER, Gordon JI (2006) An obesity-associated gut microbiome with increased capacity for energy harvest. Nature 444(7122):1027–1031
Uren TK, Wijburg OL, Simmons C, Johansen FE, Brandtzaeg P, Strugnell RA (2005) Vaccine-induced protection against gastrointestinal bacterial infections in the absence of secretory antibodies. Eur J Immunol 35(1):180–188
Uysal H, Bockermann R, Nandakumar KS, Sehnert B, Bajtner E, Engstrom A et al (2009) Structure and pathogenicity of antibodies specific for citrullinated collagen type II in experimental arthritis. J Exp Med 206(2):449–462
van der Burg M, van Zelm MC, Driessen GJ, van Dongen JJ (2012) New frontiers of primary antibody deficiencies. Cell Mol Life Sci 69(1):59–73
Vaughan AT, Brackenbury LS, Massari P, Davenport V, Gorringe A, Heyderman RS et al (2010) Neisseria lactamica selectively induces mitogenic proliferation of the naive B cell pool via cell surface Ig. J Immunol 185(6):3652–3660
Vencovsky J, Machacek S, Sedova L, Kafkova J, Gatterova J, Pesakova V et al (2003) Autoantibodies can be prognostic markers of an erosive disease in early rheumatoid arthritis. Ann Rheum Dis 62(5):427–430
Vladutiu AO, Netto D (1982) Is quantitation of serum IgD clinically useful? Clin Chem 28(6):1409–1410
Wardemann H, Yurasov S, Schaefer A, Young JW, Meffre E, Nussenzweig MC (2003) Predominant autoantibody production by early human B cell precursors. Science 301(5638):1374–1377
Wolbank S, Kunert R, Stiegler G, Katinger H (2003) Characterization of human class-switched polymeric (immunoglobulin M [IgM] and IgA) anti-human immunodeficiency virus type 1 antibodies 2F5 and 2G12. J Virol 77(7):4095–4103
Xin H, Cutler JE (2011) Vaccine and monoclonal antibody that enhance mouse resistance to candidiasis. Clin Vaccine Immunol 18(10):1656–1667
Youd ME, Luus L, Corley RB (2004) IgM monomers accelerate disease manifestations in autoimmune-prone Fas-deficient mice. J Autoimmun 23(4):333–343
Zhang M, Carroll MC (2007) Natural IgM-mediated innate autoimmunity: a new target for early intervention of ischemia-reperfusion injury. Expert Opin Biol Ther 7(10):1575–1582
Zheng NY, Wilson K, Wang X, Boston A, Kolar G, Jackson SM et al (2004) Human immunoglobulin selection associated with class switch and possible tolerogenic origins for C delta class-switched B cells. J Clin Invest 113(8):1188–1201

Chapter 2
Immunoglobulin A: Molecular Mechanisms of Function and Role in Immune Defence

Jenny M. Woof

Abstract Our immune system produces more immunoglobulin A (IgA) than all of the other antibody classes combined. Much of this synthesis is directed towards protection of the mucosal surfaces which form a vast and vulnerable interface with the environment. As the predominant immunoglobulin class at these surfaces, IgA is an important first line of defence. In addition, IgA is also a major serum immunoglobulin. Both monomeric IgA in serum and polymeric secretory forms of IgA mediate a wide range of protective functions through interaction with numerous receptors and other mediators. It is telling that in order to gain better opportunities for invasion, certain pathogens have evolved mechanisms to thwart IgA function. An improved understanding of this multifaceted immunoglobulin is likely to inform strategies for improved treatments for infections and other diseases.

Keywords Immunoglobulin A • Immune defence • IgA • IgA1 • IgA2 • Glycosylation • Dimeric IgA • IgA receptors

2.1 Introduction

IgA was first recognised as an immunoglobulin isotype at the end of the 1950s by Heremans and colleagues, when it became apparent that the carbohydrate-rich antibody species present in the β-globulin fraction of serum was distinct from the isotypes that had already been defined, namely, IgG and IgM (reviewed by Mestecky et al. 2005). Tomasi and co-workers shortly afterwards showed that IgA was present in a polymeric form in many external secretions (reviewed by Mestecky et al. 2005).

J.M. Woof, B.Sc. Ph.D (✉)
Division of Cancer Research, Medical Research Institute, Jacqui Wood Cancer Centre, Ninewells Hospital and Medical School, University of Dundee, James Arrott Drive, DD1 9SY Dundee, UK
e-mail: j.m.woof@dundee.ac.uk

F. Nimmerjahn (ed.), *Molecular and Cellular Mechanisms of Antibody Activity*, DOI 10.1007/978-1-4614-7107-3_2, © Springer Science+Business Media New York 2013

Although in some secretions, e.g. those of the lower respiratory and female genital tracts, the level of IgG may at times be quite high and even exceed that of IgA, it is clear that IgA is the predominant antibody class in most external secretions of humans. Indeed, it is now known that the majority of the body's immunoglobulin production is directed towards the IgA class. In fact, the average human produces more IgA per day than all of the other antibody classes combined. Typically, over 60 mg IgA is produced per kg of body weight each day (Mestecky et al. 1986; Conley and Delacroix 1987). Most of this IgA is localised to the mucosal tissues, especially those of the gastrointestinal, respiratory and genitourinary tracts, where it is generated by the abundant plasma cells in the mucosal subepithelium (Conley and Delacroix 1987; Mestecky et al. 1991).

Considerable energy expenditure is presumably necessary to maintain such prolific IgA production. The maintenance of such high energy cost during human evolution suggests that IgA serves as a keystone of a mucosal immune defence system that is critical for survival. Mucosal surfaces have an immense collective surface area, typically 400 m^2 in the human adult (Childers et al. 1989). These surfaces are thus the main area of exposure to the environment and represent a key point of vulnerability to attack by pathogens. A highly effective system for their protection is clearly essential. IgA makes a critical contribution to this process, helping to prevent breaching of the mucosal surfaces by pathogens, while also having a role to play in the circulation (Woof and Russell 2011). This chapter will consider structure and functional aspects of both serum and secretory forms of IgA, concentrating chiefly on the human system.

2.2 Multiple Molecular Forms of IgA

In human serum, IgA is the second most prevalent immunoglobulin class after IgG. IgA is normally present at concentrations of 2–3 mg/ml, i.e. about one fifth that of IgG. Because it does not interact with FcRn, the receptor responsible for rescuing IgG from degradation, IgA has a much faster turnover than IgG, being metabolised about five times faster. Therefore, the relative concentrations and rates of metabolism suggest that the production rates of serum IgG and serum IgA are similar.

Serum IgA exists principally as a monomer with a molecular mass of approximately 170 kDa (Fig. 2.1). In marked contrast, for IgA in the mucosal secretions (secretory IgA, SIgA), the majority is in dimeric form (Fig. 2.1). Some higher molecular weight species, mainly trimers and tetramers, are also present. As mentioned above, SIgA is present in the secretions that bathe the respiratory, gastrointestinal and genitourinary tracts. It is also the predominant immunoglobulin in milk, colostrum, tears and saliva. As will be discussed in more detail later, SIgA is formed following the interaction of dimeric IgA, produced locally at mucosal sites, with the polymeric immunoglobulin receptor (pIgR) (Mostov et al. 1980). During transfer of the dimeric IgA-pIgR complex across the epithelium, the receptor is cleaved. The major cleavage product of the receptor, termed secretory component (SC), remains

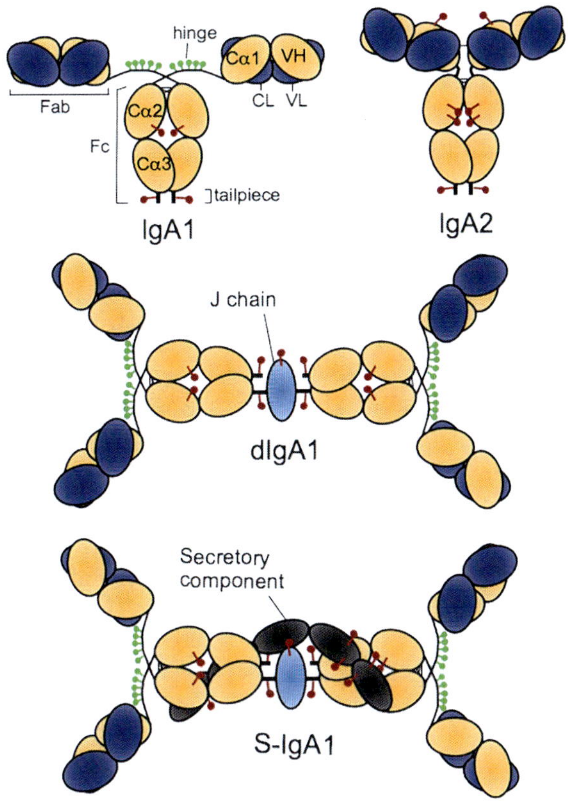

Fig. 2.1 *Schematic representation of the monomeric forms of human IgA1 and IgA2 and the dimeric (dIgA1) and secretory (SIgA1) forms of IgA1.* Heavy chains are shown in gold and light chains in *purple*. J chain is shown in *blue* and secretory component in *dark grey*. O-linked sugars on the IgA1 hinge are shown in *green*, while N-linked oligosaccharides are shown in *red*

bonded to the IgA dimer on release into the mucosal secretions, serving as an integral component of SIgA (Fig. 2.1).

2.3 IgA Structure

2.3.1 Monomeric Unit

Similar to the other immunoglobulin classes, IgA molecules are based on a monomer unit comprising two heavy chains (in this case, α-chains) and two light chains. The latter are common to all immunoglobulin classes, while the α-heavy chains are unique to IgA and essentially define the antibody class. Each chain is folded up into a number of globular domains, four for the heavy chain (named from the N-terminus VH, Cα1, Cα2 and Cα3) and two for the light chain (known as VL and CL) (Fig. 2.1). A hinge region separates the Cα1 and Cα2 domains of the heavy chain.

All the domains fold up into essentially the same secondary architecture typical of all immunoglobulins, known as the immunoglobulin fold. The domains each comprise around 110 amino acids and are arranged into two β-sheets, made up of antiparallel β-strands, which form a sandwich stabilised by a characteristic disulphide bond. In common with the other immunoglobulin classes, each domain is encoded by a separate exon. In humans but not all species, IgA is unusual in that the hinge is encoded by a region located at the 5′ end of the Cα2 exon rather than having its own distinct exon or exons as is seen with IgG, for example.

At the C-terminus of the α-heavy chain of IgA, there is an 18 amino acid extension known as the tailpiece. This element is central to IgA's ability to polymerise. The IgA tailpiece shares close similarity to that present at the C-terminus of the IgM heavy chain. IgM is similarly capable of polymerisation, in its case primarily into pentamers. Tailpiece equivalents are lacking in IgG and IgE with the result that these antibodies exist solely as monomers.

Like other immunoglobulins, the heavy and light chains of IgA are arranged into two antigen-binding Fab regions that are linked through the hinge region to the Fc region (Fig. 2.1). This arrangement is stabilised through an extensive array of non-covalent interactions, primarily hydrogen bonds and van der Waals contacts, between paired domains from different polypeptide chains. Thus, in the Fab region, the VH domains pair closely with the VL domains and the Cα1 domains with the CL domains, while in the Fc region, the Cα3 domains of the two heavy chains are closely paired. Such close association between domains contributes to stability by removing substantial surface areas of the domains from contact with solvent. Some further stabilisation is contributed by interaction between neighbouring domains in the same polypeptide chain. An exception to the domain pairing is seen with the Cα2 domains, which do not form a close pair. Instead, the "interior" surfaces of these domains lie apart from each other but are partially removed from solvent exposure by N-linked oligosaccharides that are attached to these surfaces (see Sect. 2.3.3).

As in all antibody molecules, the paired VH and VL domains at the tips of the two Fab regions of IgA form the sites responsible for binding to antigen. In each case, the antigen-binding site itself is formed by the association of the three hypervariable loops or complementarity determining regions (CDRs) of the VH domain with the three CDRs of the VL domain. The precise three-dimensional conformation of the antigen-binding site formed through the coming together of these six loops at the tips of the VH and VL domains, unique for each different antibody, determines the specificity for antigen. Thus, binding of high affinity will only occur when the molecular structure of the antigen (or subcomponent of the antigen) is the exact three-dimensional complement of the antigen-binding site.

While the Fab arms dictate antigen binding, the Fc region is responsible for the effector function of the IgA molecule. IgA fulfils its role as an adaptor molecule by linking these two critical functions together, thereby enabling the immune system to trigger appropriate elimination processes (via effector function) once a foreign invader has been detected (via antigen binding). The effector mechanisms are mediated, in the main, through interaction of the IgA Fc region with a variety of specific receptors present on a range of different cell types. These include receptors that bind to the Fc region of IgA such as FcαRI on phagocytes, Fcα/μR present on some follicular dendritic cells, macrophages, plasma cells and Paneth cells, and pIgR on epithelial cells (discussed in detail in Sects. 2.3.5 and 2.7).

The X-ray crystal structure of the Fc region of human IgA1 has been determined in complex with the extracellular part of human FcαRI (Herr et al. 2003) or with SSL7, an IgA-binding protein produced by *Staphylococcus aureus* (Ramsland et al. 2007). Overall, the structure is very similar to the equivalent regions of IgG or IgE. There are some key differences, however. These include the arrangement of the oligosaccharides, which is explored in Sect. 2.3.3, and the location of disulphide bridges between the heavy chains. In IgG, the two heavy chains are anchored to each other by disulphide bonds in the hinge region, but IgA differs in this regard. Instead, the bonds lie at the "top" of the Cα2 domains, tying the upper reaches of the two domains together (Fig. 2.1). Two such bonds link Cys242 with Cys299 in the opposite chain. Additional bonds, possibly between the two Cys241 residues or between Cys241 on one chain and Cys301 on the other, are presumed to exist although the truncated constructs used to obtain the X-ray crystal structures did not extend far enough to resolve these. There are no disulphide links between the hinge regions of IgA, which marks a sharp contrast to the situation in the IgG hinge region. The lack of disulphides in the IgA hinge means that on emerging from the globular Cα2 domains, the heavy chains are presumably relatively free to flex independently of each other. Thus, these different modes of anchoring the two heavy chains together likely have implications in terms of overall flexibility, reach of the Fab arms to distantly spaced antigens, and susceptibility of the hinge peptide to proteolysis.

2.3.2 IgA Subclasses

The number of IgA genes varies from species to species as will be discussed later (Sect. 2.5). In humans, two subclasses of IgA exist. Known as IgA1 and IgA2, they differ in their α-heavy chain constant regions which are encoded by separate Cα genes. Although the differences between the subclasses relate to several specific amino acid differences throughout the length of their heavy chains, the most major differences are seen within the hinge region. The hinge region of IgA1 is much more extended than that of IgA2 due to an insertion event which has introduced two eight amino acid repeats, rich in proline, serine and threonine residues (Figs. 2.1 and 2.5). The IgA1 hinge is usually decorated with three to five O-linked oligosaccharides (Mattu et al. 1998; Royle et al. 2003) attached to particular serine and threonine residues. A small proportion (5–10 %) of molecules are reported to carry six O-linked sugars (Tarelli et al. 2004).

Two IgA2 allotypes, named IgA2m(1) and IgA2m(2), have been well characterised in humans (Fig. 2.2). Another IgA2 variant, presumed to be an allelic form, has also been described (Chintalacharuvu et al. 1994). Although the IgA2m(1) and IgA2m(2) allotypes differ at just six amino acid positions in their heavy chain constant regions, there is a significant difference in their arrangements of their interchain disulphide bridges. While in IgA2m(2), the usual disulphide links between heavy and light chains are present, these are generally lacking in IgA2m(1). In the latter allotype, the association of the heavy chains is stabilised via non-covalent interactions, and a disulphide bridge links the light chains to one another.

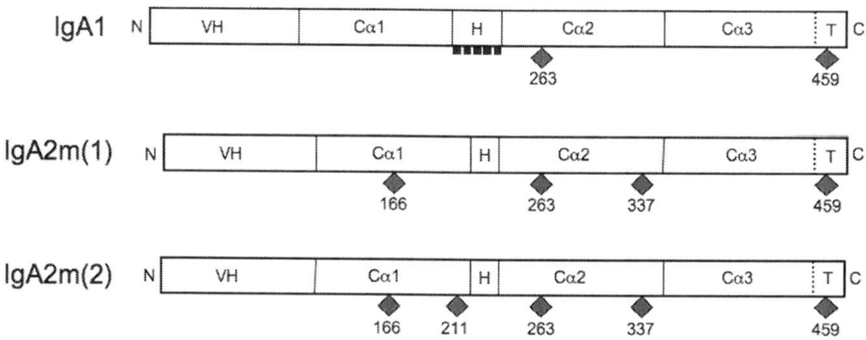

Fig. 2.2 *Glycosylation of the heavy chains of human IgA1, IgA2m(1) and IgA2m(2).* N-linked and O-linked oligosaccharides are depicted as *grey diamonds* and *black squares*, respectively. Attachment sites for N-linked oligosaccharides are numbered using Bur IgA1 numbering

The relative proportions of the human IgA subclasses vary between body compartments. In serum, IgA1 predominates with about 90 % of the IgA being of the IgA1 subclass. In the mucosal secretions, there tends to be a more even distribution of the two subclasses. Typical proportions are around 60 % IgA2 and 40 % IgA1, but there is considerable variability at different mucosal sites.

2.3.3 Glycosylation

N-linked oligosaccharides contribute quite a significant part of the total mass of IgA. In human IgA1, they account for 6–7 % of the mass, while in human IgA2, they make up 8–10 % of the mass (Tomana et al. 1976). Two N-linked sugars, attached at residue Asn263 in the Cα2 domain and Asn459 in the tailpiece, decorate each heavy chain of IgA1 (Fig. 2.2). Human IgA2 molecules possess these same oligosaccharides as well as additional ones, the number varying with allotype. Thus, IgA2m(1) has additional sugars attached to residues Asn166 in the Cα1 domain and Asn337 in the Cα2 domain, while IgA2m(2) has a further Cα1 domain sugar moiety attached at Asn 211 (Fig. 2.2).

Compositional analysis of the N-linked oligosaccharides of serum IgA1 and SIgA revealed the presence of a family of structures based principally around a biantennary mannosyl chitobiose core (Field et al. 1994; Mattu et al. 1998; Royle et al. 2003). A small proportion of the sugars are more branched, mainly as triantennary structures. Within this family, there is considerable variation in the level of fucosylation and the number and type of sugars (galactose and sialic acid) attached at the termini of the branches.

The Cα2 domains of IgA share a common feature with the equivalent Cγ2 domains of IgG and the Cε3 domains of IgE in that they are not closely paired and have N-linked oligosaccharides attached. However, there are important differences between IgA and these other isotypes in the attachment sites and positions occupied by the glycans. In IgA, the glycans attached at Asn 263 lie over the outer surfaces of the Cα2 domain and, in the process, bury some 930 $Å^2$ of Cα2 domain surface area per Fc from solvent contact. They make contacts with the Cα3 domain too, burying a further 914 $Å^2$ of domain surface per Fc from solvent. In contrast, in IgG and IgE, the N-linked oligosaccharides are attached at a different point within the Cγ2 and Cε3 domains and, in each case, occupy the interior of the Fc. Their contribution to stability through burying surface from solvent is therefore less, at around 1044 $Å^2$ per Fc for IgG and 900 $Å^2$ per Fc for IgE. Moreover, their oligosaccharides make no contacts with the Cγ3 or Cε4 domains (the equivalents of the Cα3 domain).

As mentioned in Sect. 2.3.2, the IgA1 hinge carries a number (three to five, or in some cases six) of O-linked oligosaccharides attached at serine and threonine residues (Fig. 2.2). These sugars are much smaller and less complex than the N-linked glycans, comprising mainly N-acetyl galactosamine, galactose and sialic acid. The presence of the O-linked sugars may adjust the conformation adopted by the IgA1 hinge (Narimatsu et al. 2010). The impact of N- and O-linked glycosylation on the functions of IgA will be discussed further in later sections.

2.3.4 Dimeric IgA

As mentioned in Sect. 2.2, IgA has the capacity to polymerise, primarily into dimers. These comprise not just the two IgA monomer units but also one additional molecule of joining chain or J chain (Koshland 1985). This 15 kDa glycoprotein is highly conserved between species and is also involved in the polymerisation of IgM into pentamers. Seemingly unrelated to other known proteins, J chain contains eight cysteine residues and is rich in acidic amino acids. While the three-dimensional structure of J chain remains unsolved, folding models have been proposed. These include a two-domain structure, one of β-sheets followed by one of α-helices (Cann et al. 1982), and a single-domain model featuring a β-barrel structure similar to an immunoglobulin VL domain (Zikan et al. 1985). Following assignment of intra-chain disulphides (see below), another two-domain model emerged that predicted an N-terminal β-barrel domain linked to a second domain of both α-helices and β-strands (Frutiger et al. 1992).

Six of the J chain cysteines form intrachain disulphide bridges, while the remaining two, Cys14 and Cys68, form disulphide bridges to the monomer units of IgA. Specifically, Cys14 links to Cys471, the penultimate tailpiece residue in one of the heavy chains of one of the monomers, while Cys68 links to Cys471 in the tailpiece of one of the heavy chains of the other monomer (Bastian et al. 1992). Mutagenesis experiments showed that these cysteine residues on J chain and the IgA tailpiece were essential for dimer formation (Atkin et al. 1996; Krugmann et al. 1997). These linkages mediate an end-to-end arrangement of the two Fc regions that is consistent with the dimensions of the joined Fc regions as determined from electron microscopy images of dimeric IgA (Fig. 2.1).

The presence of other elements in J chain and IgA has also been shown to contribute to efficient dimer formation. These include the N-linked oligosaccharide attached to residue Asn459 of the tailpiece and the N-linked oligosaccharide which is attached to Asn48 of J chain (Atkin et al. 1996; Krugmann et al. 1997). The domains of the IgA Fc region also appear to contribute to polymerisation, with the presence of the Cα2 domain aiding the efficiency of dimerisation and that of the Cα3 domain impacting on the size of the polymer formed (Yoo et al. 1999).

2.3.5 Secretory IgA

As indicated in Sect. 2.2, the transport of IgA into the mucosal secretions to form SIgA is mediated by a specific receptor, the polymeric immunoglobulin receptor (pIgR). pIgR is expressed on the basolateral surface of epithelial cells that form the mucosal surface. A number of host and microbial factors contribute to the regulation of its expression (Johansen and Kaetzel 2011).

pIgR is composed of a single polypeptide with a large extracellular region (approximately 620 amino acids), a 23-residue transmembrane segment and an intracellular region of just over 100 amino acids (Kaetzel 2005). The extracellular portion is arranged into five immunoglobulin-like domains that show homology to immunoglobulin variable domains and are each stabilised by one or more internal disulphide bridges. The domains are designated 1–5 (D1–D5) from the N-terminus. There are seven N-linked oligosaccharides attached to the domains – two each to D1, D2 and D5 and one to D4. C-terminal to these domains, leading to the membrane, there lies a stretch of non-immunoglobulin-like polypeptide.

The transport process involves binding of dimeric IgA to the receptor at the basolateral surface (Fig. 2.3). The complex is then internalised and transported via vesicular compartments to the apical surface of the epithelial cell. The pIgR is cleaved at a site located between D5 and the cell membrane to release the major portion of the receptor as a fragment known as secretory component (SC). During transport, a disulphide bridge forms between SC and dimeric IgA so that upon release at the apical surface, SC remains as an integral part of the released IgA, now referred to as SIgA (Fig. 2.1).

pIgR binds only to polymeric immunoglobulins, i.e. to dimers (or larger polymers) of IgA and IgM, and not to IgG or IgE. Under experimental conditions, the receptor is able to transport both IgA and IgM polymers at similar rates. However, in the body, access of IgM to pIgR is likely hindered by slow diffusion rates of the pentamer from the circulation through the extracellular matrix and basement membrane. As a result, the transport of IgA dimers across the mucosal surfaces is favoured over that of the larger IgM (Natvig et al. 1997).

Binding between dimeric IgA and pIgR has been studied in some detail, and the elements critical for interaction have been defined. Domains D4 and D5 of pIgR appear to make some contribution to the affinity of the interaction, but it is D1–D3 that play the key roles (Norderhaug et al. 1999). D1, in particular, has a central role

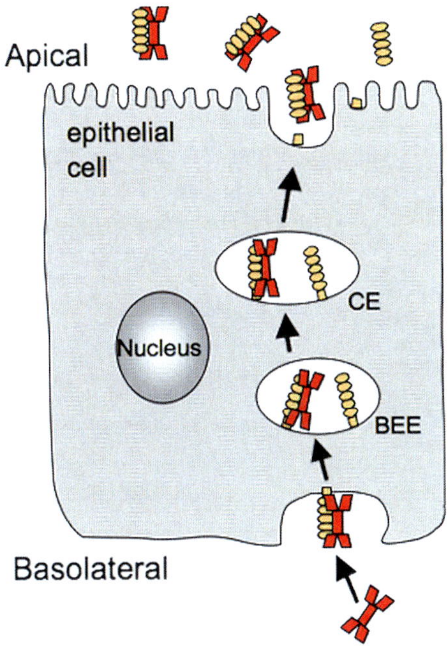

Fig. 2.3 *Receptor-mediated transport of IgA across the epithelial cell layer at mucosal surfaces.* Locally produced dimeric IgA binds to pIgR expressed on the basolateral surface of polarised epithelial cells. pIgR, either unoccupied or bound to dimeric IgA, undergoes endocytosis and is transported across the cell to the apical surface via various intracellular vesicles, including additional vesicular compartments not shown in this simplified diagram. pIgR is cleaved to yield SC which has become covalently linked to the IgA dimer, and the complex (*SIgA*) is released at the mucosal surface. Unoccupied receptor is also cleaved releasing free SC. *BEE* basolateral early endosome, *CE* common endosome

in binding to dimeric IgA, and loops at the end of this domain that occupy positions analogous to the complementarity determining regions (CDRs) that comprise the antigen-binding site in antibody variable domains have been shown to be critical for binding (Bakos et al. 1993; Coyne et al. 1994). The solved X-ray crystal structure for pIgR D1 confirmed that this domain adopts a structure similar to that of immunoglobulin variable domains (Hamburger et al. 2004). Residues in CDR1 (Thr27–Thr33) and CDR2 (Glu53 and Gly54) of D1 that had been implicated in dimeric IgA binding in mutagenesis studies were confirmed to lie close to each other in three-dimensional space.

D1 of pIgR interacts via non-covalent interactions with the Fc region of dimeric IgA. The Cα3 domain is central to this initial binding event (Hexham et al. 1999; Braathen et al. 2002). An exposed loop at the upper surface of the domain comprising residues 402–410 has been implicated (Hexham et al. 1999; White and Capra 2002). Close-lying residues Phe411, Val413, Thr414 and Lys377 appear also to be involved, as well as a stretch lying at the interface with the Cα2 domain consisting of residues Pro440–Ph443 (Lewis et al. 2005). From the available evidence, one can conclude that, together, these motifs form a binding site for pIgR D1 lying essentially across the Cα2-proximal surface of the Cα3 domain. Once D1 has bound the Cα3 domain in the Fc region of dimeric IgA, a disulphide bond is formed between Cys467 of D5 of pIgR and a cysteine residue (Cys311) that lies on the exterior of the Cα2 domain of IgA (Fallgren-Gebauer et al. 1995; Underdown et al. 1977). In addition to these IgA-pIgR interactions, direct interactions between J chain and pIgR also contribute (Johansen et al. 2001).

2.4 Induction of IgA

2.4.1 *Sites of IgA Production*

The biosynthesis of mucosal IgA is completely distinct from that of circulating IgA. Serum IgA is produced in the bone marrow, principally in the form of monomeric IgA1. In contrast, SIgA is produced locally by plasma cells situated at mucosal surfaces. These cells are present in huge numbers, comprising about 80–90 % of all plasma cells in the body. For example, there are calculated to be approximately 10^{10} plasma cells per metre of adult human intestine (Brandtzaeg et al. 1999). The IgA produced by these cells is primarily polymeric as a result of the co-expression of IgA light and heavy chains and J chain, with neither IgA subclass predominating.

The mucosal plasma cells comprise a key part of a highly specialised mucosal immune system that functions largely independently of the systemic immune system. Unsurprisingly, as a result of these very different sites of production, different methods of immunisation can induce serum or secretory IgA responses or a combination of the two. For example, oral immunisation with microbial antigens induces SIgA responses in external secretions but only low serum IgA responses. Conversely, systemic immunisation with antigens that induce dominant IgA responses in plasma does not induce strong IgA responses in mucosal secretions.

The tissue compartments involved with induction of a mucosal IgA response consist of mucosa-associated lymphoid tissue (MALT). MALT is subdivided based on anatomical region, with gut-associated mucosal tissue (GALT) being the largest component. Such sites are specially adapted to generate precursors programmed for IgA production and are exposed to antigens arriving directly from mucosal surfaces. In short, they comprise B cell follicles, made up of a germinal centre of rapidly dividing B cells within a mantle zone of naive B cells, with intervening T cell zones and an array of antigen-presenting cells. These underlie specialised follicle-associated epithelium that contains microfold (M) cells which, aided by dendritic cells, transport exogenous antigenic material from the mucosal lumen to the immune cells below. Questions surrounding whether or not events initiating a human intestinal IgA mucosal response might occur outside GALT have persisted, but it seems likely that this is not the case (Spencer et al. 2012).

Following priming and expansion of mucosal B and T cells in MALT, these cells migrate to local and more distal mucosal effector sites, such as the lamina propria and epithelium of the intestine and respiratory tract. These homing responses are guided through coordinated expression of adhesion molecules, such as the integrin α4β7 which binds to its ligand MAdCAM-1 on endothelial cells, and certain chemokines and their receptors, such as CCL25 and its receptor CCR9 (Brandtzaeg and Johansen 2005; Macpherson et al. 2012).

2.4.2 Class Switch Recombination to Generate IgA

Induction of pathogen-specific mucosal IgA must occur against a background of constant antigenic challenge from food, environmental antigens and a multitude of commensal microorganisms (Macpherson et al. 2012). In fact, IgA induction is highly sensitive to the presence of commensal bacteria. In their absence, laboratory animals have very low levels of IgA, but IgA expression can be switched on by their introduction. This process requires limited sampling of commensal bacteria by intestinal dendritic cells which drives induction of IgA (Macpherson and Uhr 2004), a point which will be returned to below.

In order to produce IgA antibodies, B cells must undergo class switch recombination (CSR), a process involving DNA splicing such that the VDJ segment encoding the variable domain of the heavy chain becomes positioned just upstream of the Cα constant region sequence. CSR is regulated by the specific cytokine milieu. In the case of CSR to IgA, transforming growth factor-β (TGFβ) and interleukin 4 (IL-4) act to promote the switch, with IL-2 and IL-10 having synergistic effects and IL-5 and IL-6 enhancing secretion of IgA (Coffman et al 1989). The mechanism entails engagement of the TGFβ receptor, which signals via SMAD2, SMAD3 and SMAD4 transcription factors in concert with RUNX3, another DNA binding factor. These bind specifically to promoter regions upstream of the initiation (I) α-exon (an initiation point specific to the IgA constant region), initiating transcription and opening the DNA structure. Thereafter, the DNA-editing enzyme activation-induced cytidine deaminase (AID) is able to initiate CSR specifically to IgA.

CSR is also regulated by nuclear factor κB (NFκB), through involvement at the Iα locus and through AID induction. NFκB can be activated by events following from the engagement of CD40 on B cells by CD40L on T cells or in a T cell-independent manner by the cytokines BAFF and APRIL which can be produced by GALT dendritic cells and trigger through specific receptors on B cells (Cerutti 2008). These two alternatives explain how induction of IgA CSR can potentially occur through both T-dependent and T-independent mechanisms.

Returning to the induction of protective IgA antibodies by commensal bacteria mentioned earlier, there are likely a number of mechanisms at play. One possibility is that dendritic cells present bacterial products derived from commensal sampling to B cells, thereby activating Toll-like receptors (TLRs). Through signalling cascades in the B cell, AID expression may then be induced. Microbial products may also stimulate the release of BAFF and APRIL by dendritic cells, which will drive IgA class switching. Another mechanism that has been described involves the triggering of intestinal epithelial cells to release APRIL after they sense bacteria through TLRs (He et al. 2007). These various mechanisms undoubtedly have parts to play in the flexible and complex homeostasis of the mucosal immune system that is necessary to prevent excessive responsiveness to harmless commensals yet allow adequate responsiveness to more harmful microorganisms (Slack et al. 2012).

2.5 IgA in Other Species

2.5.1 Evolutionary Considerations

Studies by Vaerman and colleagues first demonstrated the presence of IgA in several mammals such as dog, cat, cow, sheep, goat, horse and pig (Vaerman et al 1969). It is now known by virtue of genetic sequence analysis and functional comparisons that IgA is present in all categories of mammals (placental, marsupials and monotremes) and also in birds.

The vast majority of mammals have a single IgA isotype. In other words, they possess a single Cα gene encoding an IgA heavy chain constant region. The exceptions are humans, related primates and rabbits. Rabbits and related species (Lagomorpha) have a remarkable 13 Cα genes (Burnett et al. 1989). Eleven of these are apparently expressed, although the level of each varies between different tissues (Spieker-Polet et al. 1993). The role of such a large number of subclasses remains an enigma but perhaps relates to the fact that rabbits have only a single IgG isotype, whereas many mammals that have a single IgA have several IgG subclasses.

Chimpanzees, gorillas and gibbons are like humans in that they have two Cα genes encoding IgA1 and IgA2 subclasses (Kawamura et al. 1992). An equivalent of IgA1 is also present in orangutans, but they have lost their IgA2 equivalent. The dual Cα genes have arisen through duplication of the entire immunoglobulin heavy chain constant region locus in the ancestor common to the apes after the point of divergence from the monkeys. In some species, subsequent deletion or silencing of some of the duplicated genes has occurred (Kawamura and Ueda 1992). The elongated hinge seen in all IgA1 molecules has seemingly evolved only recently in evolutionary terms due to an insertion event.

Allotypic variation in IgA is exemplified by the allotypes of human IgA2 (see Sect. 2.3.2), but it is clear that such diversification is exhibited in various mammalian species. For example, IgA in rhesus macaques shows allotypic variation, with notable allelic polymorphisms in the hinge region (Scinicariello et al. 2004). Allotypic variation has also been observed in the IgA of sooty mangabeys, pig-tailed macaques and baboons (Scinicariello et al. 2006; Rogers et al. 2008) with a high degree of hinge heterogeneity being a prominent feature. Interestingly, the six IgA allotypes in mice also have major differences in their hinge regions (Phillips-Quagliata 2002), while hinge variation is also evident in pig IgA allotypes (Navarro et al. 2000).

Based on analysis in a limited number of avian orders (those represented by chickens, ducks and pigeons), IgA in birds is somewhat unusual in that it lacks a hinge region and has four constant regions in its heavy chain (Mansikka 1992). The level of homology between chicken IgA and mammalian IgAs is around 35–37 %, consistent with evolution of IgA before the divergence of mammals and birds.

In lower vertebrate species that lack IgA, other immunoglobulin isotypes appear to undertake the functions of a mucosal antibody. In some, immunoglobulin analogous to IgM will fulfil this role, while in others, alternative isotypes have evolved independently for this purpose. For example, the external surfaces of amphibians seem to be protected by IgX (Du et al. 2012), while IgT serves as a mucosally active immunoglobulin in teleost fishes (Zhang et al. 2010).

2.5.2 *Species-Dependent Variability in Mucosal Immunoglobulins*

IgA plays a common role in protection of external surfaces in many species (Snoeck et al. 2006). However, there are some important differences between species in terms of the levels and distribution of immunoglobulin isotypes in mucosal secretions. For example, while IgA is the predominant antibody isotype in human colostrum and milk, the same is not true for all mammals. In cows, sheep, goats and horses, the main immunoglobulin isotype in milk and colostrum is IgG. These species lack the transplacental transport of IgG that occurs in humans, and hence, ingestion of colostral-derived IgG by newborn animals is essential for survival. After ingestion, the IgG is transported across the gut epithelium in an active process that occurs for the first 2–3 days of life.

Another distinction is seen between the IgA systems of humans and animals commonly used for experimental purposes (i.e. mice, rats and rabbits). In these species, the serum IgA, as well as that in secretions, is mainly in the form of J chain-containing dimers. Moreover, unlike humans, these species express significant levels of pIgR on their hepatocytes, which results in highly efficient transport of polymeric IgA from the circulation into the bile and thereafter into the intestinal secretions. As a result, the main source of SIgA in the gut lumen of these species is the bile.

2.6 Role in Immune Defence Against Infections

2.6.1 *Neutralisation and Blocking Activity and Synergy with Innate Humoral Factors*

Specific IgA antibodies have been demonstrated to provide effective protection against a range of invading pathogens including viruses, bacteria and protozoa and their products such as toxins (Russell and Kilian 2005; Mantis et al. 2011) (see Fig. 2.4). By direct interaction through their antigen-binding sites, IgA molecules can neutralise the activity of pathogens such as viruses and prevent attachment to host cells. Biologically active toxins derived from plants or pathogens can similarly be neutralised. In a process sometimes referred to as immune exclusion, SIgA can diminish the absorption of antigens to mucosal surfaces. The polymeric nature of SIgA means that it is capable of binding with high avidity to antigenic substances because each molecule has multiple antigen-binding sites (four for dimeric SIgA, eight for tetrameric SIgA).

As well as this general protective role at mucosal surfaces, SIgA has a specific role to play in protection of the newborn. There is overwhelming evidence that SIgA in human colostrum and milk is important in this capacity, and it is established that breastfeeding decreases infant death from both respiratory and gastrointestinal infections (Hanson and Korotkova 2002). SIgA is present at high concentrations in human colostrum and milk (up to 12 mg/ml). It displays specificity for a wide variety of microbial pathogens and can neutralise viruses and toxins. The SIgA in milk most likely acts as a molecular "paint", neutralising the effect of microorganisms by preventing adherence and other invasion processes.

Innate defence factors present in mucosal secretions may interact with SIgA to enhance immune protection. For example, salivary mucins have been reported to undergo non-covalent interactions with SIgA (Fig. 2.4). This is thought to help to concentrate the factors at the tissue-environment interface and enhance the ability of the components to agglutinate bacteria (Biesbrock et al. 1991). Similarly, synergy between SIgA and the antimicrobial effects of the lactoferrin and lactoperoxidase systems has been described (Stephens et al. 1980; Tenovuo et al. 1982; Watanabe et al. 1984).

2.6.2 Intracellular Neutralisation

In addition to the protective effects that SIgA can mediate after release into mucosal secretions, it has become apparent that it may fulfil defensive functions during its passage through the mucosal epithelium. Based on in vitro studies using models of polarised epithelial cells, specific dimeric IgA molecules undergoing pIgR-mediated transcytosis through the cells have been shown to be able to neutralise endocytosed bacterial lipopolysaccharide (LPS) that had been taken up at the apical surface. The LPS and dIgA colocalised within the apical recycling compartment of the cells, preventing the proinflammatory events that would normally be triggered by the LPS (Fernandez et al. 2003) (Fig. 2.4).

Using similar in vitro systems, dimeric IgA molecules directed against viral envelope undergoing epithelial transcytosis have been shown to block growth of Sendai virus following infection at the apical surface (Mazanec et al. 1992) (Fig. 2.4). Using analogous systems, similar inhibitory effects were found for influenza virus (Mazanec et al. 1995), measles virus (Yan et al. 2002), rotavirus (Feng et al. 2002; Corthésy et al. 2006) and human immunodeficiency virus (HIV) (Huang et al. 2005; Wright et al. 2006). It appears that the apical recycling endosomes are the key point of intersection of the IgA and viral proteins.

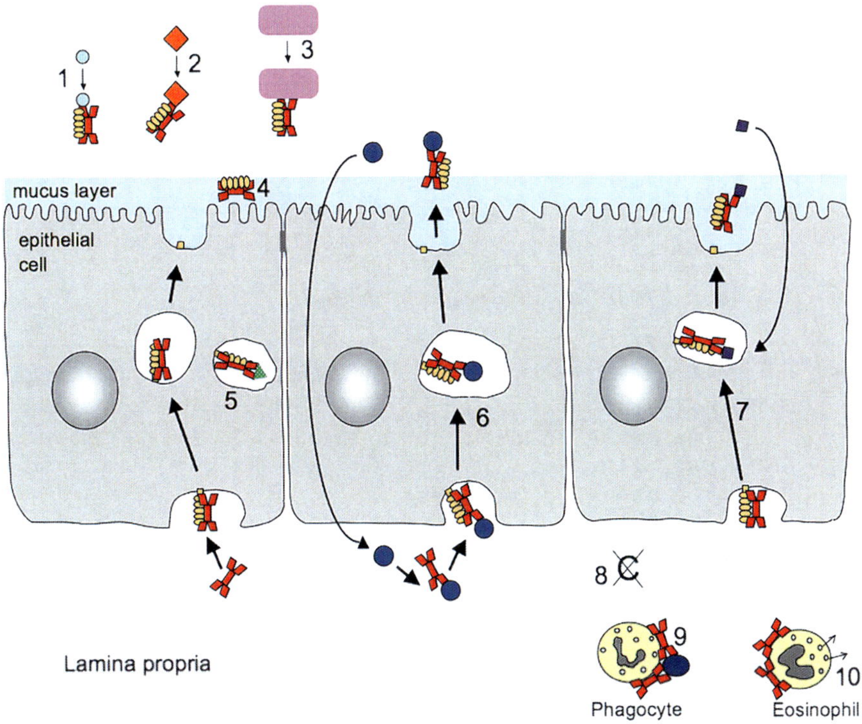

Fig. 2.4 *The roles of IgA in protection at mucosal surfaces. 1* Inhibition of uptake of soluble or particulate antigens, *2* Neutralisation of biologically active toxins and viruses, *3* Inhibition of microbial adherence, *4* Enhancement of activities of innate humoral factors (e.g. mucin, lactoferrin, peroxidase system), *5* Intracellular neutralisation, *6* Elimination of immune complexes formed within the lamina propria, *7* Elimination of immune complexes formed within epithelial cells, *8* Inhibition of complement activation, *9* Antibody-dependent cell-mediated cytotoxicity and opsonisation, *10* Degranulation of eosinophils

Evidence that these in vitro observations may reflect in vivo capabilities has been derived from experiments in which systemically delivered IgA antibodies directed against an inner core protein of rotavirus were shown to prevent infection and cure existing infection in mice (Burns et al. 1996; Feng et al. 2002; Schwartz-Cornil et al. 2002).

2.6.3 Excretion of Antigenic Material Across Mucosal Epithelium

Remarkably, pIgR can transport dimeric IgA alone or in complex with antigen. Consequently, pIgR-mediated transport can deliver antigen-IgA immune complexes from tissues underlying the epithelium (such as the lamina propria) to the lumenal surface of the epithelial cell barrier. This mechanism allows for removal and excretion of soluble antigens, whether derived from microorganisms, diet or the environment (Kaetzel et al. 1991; Robinson et al. 2001) (Fig. 2.4). Virus particles that encounter specific dimeric IgA in the lamina propria, as well as immune complexes formed within epithelial vesicles, can be excreted via this same mechanism, thereby preventing the spread of infectious agents beyond the mucosal surfaces (Yan et al. 2002; Bomsel et al. 1998).

2.6.4 Glycan-Mediated Protective Activity

The N-linked oligosaccharides attached to IgA and SC impart to SIgA the capability for additional non-specific, innate responses to bacterial pathogens and certain of their products. These sugar moieties are able to bind to sugar-dependent receptors or fimbriae on bacterial surfaces and inhibit attachment to host mucosal surfaces (Wold et al. 1990; Ruhl et al. 1996). The demonstration of binding of SC to *Clostridium difficile* toxin A suggests that this is an additional means by which SIgA can recognise certain bacterial components (Dallas and Rolfe 1998) providing another way for SIgA to limit the effects of infection.

2.6.5 Interaction with Complement

IgA lacks the C1q binding motif necessary for binding of the first component of the classical complement pathway and thus is unable to activate this pathway. While this point is clear-cut, the ability of IgA to activate complement via the alternative pathway has been a subject of debate over the years. The consensus view that has emerged is that the reported activation of this pathway by IgA was most likely artefactual. Finally, SIgA has been reported to bind mannose-binding lectin and activate complement via the lectin pathway of complement activation (Roos et al. 2001). Further insights into the physiological significance of such activation are awaited since they may help resolve the controversy that has surrounded complement activation by IgA.

2.7 IgA Receptors

2.7.1 Host Receptors for IgA

In concert with the functions described in Sect. 2.6, IgA interacts with a range of host receptors present on different cell types to generate a variety of outcomes. Transport of IgA into mucosal secretions following interaction of dimeric IgA with pIgR on epithelial cells has already been discussed (Sect. 2.3.5). Another important example is the IgA receptor FcαRI, which mediates a variety of functions to help eliminate pathogenic invaders. The precise functions of other receptors, some characterised to a greater degree than others, are less clear. All demonstrate specificity for the α-heavy chain of IgA, although some, e.g. pIgR and Fcα/μR, are seen to also interact with other immunoglobulin isotypes.

2.7.2 FcαRI

FcαRI, or CD89, is a member of a family of Fc receptors that falls within the much broader immunoglobulin gene superfamily (Bakema and van Egmond 2011a). However, it has key distinctions from the other Fc receptors in this group. Its gene lies in the leukocyte receptor cluster close to killer cell immunoglobulin-like receptors (KIR) and leukocyte immunoglobulin-like receptors on chromosome 19, quite unlike other Fc receptors such as the IgG-specific FcγR and the IgE-specific FcεRI, which are clustered on chromosome 1. Indeed, FcαRI has closer amino acid similarity to the LRC members than to other immunoglobulin gene superfamily Fc receptors. It is a transmembrane glycoprotein with two immunoglobulin-like extracellular domains and a short cytoplasmic region that lacks any known signalling motifs. It associates via its transmembrane region with a dimer of the FcR-γ chain, a signalling chain that possesses two immunoreceptor tyrosine-based activation (ITAM) motifs within its cytoplasmic region. The globular extracellular domains of FcαRI are positioned at an angle of approximately 90° to each other (Herr et al. 2003). Their relative orientation is very different from that of the extracellular domains of other Fc receptors (Herr et al. 2003; Woof and Burton 2004).

FcαRI is expressed on neutrophils, monocytes, eosinophils, platelets and some macrophages and dendritic cells. It is also expressed on Kupffer cells in the liver where it is believed to play a role in mediating phagocytosis of pathogens that have entered the circulation and been opsonised by serum IgA (van Egmond et al. 2000).

Human FcαRI binds equally well to human IgA1 and IgA2 and can engage both serum IgA and secretory IgA, although the outcomes may differ as discussed below. The sites of interaction on both FcαRI and the Fc region of IgA have been defined, initially through mutagenesis analysis and subsequently through solving of the X-ray crystal structure of the complex of IgA1 Fc and the extracellular portion of the receptor (Wines et al. 1999; Wines et al 2001; Carayannopoulos et al. 1996; Pleass et al. 1999; Herr et al. 2003). The hydrophobic core of the interaction essentially involves a region on the membrane distal domain of the receptor comprising Tyr53, Leu54, Phe56, Gly84 and His85 with contributions from Lys55, which interacts with the interface between the two domains of the IgA Fc region, comprising Cα2 residues Leu257 and Leu258 and Cα3 residues Met433, Leu441, Ala442, Phe443, and the aliphatic portion of Arg382. This mode of interaction is completely different from that of IgG with FcγR and IgE with FcεRI, despite considerable homology between both receptors and ligands (Woof and Burton 2004). The latter receptors bind to regions lying at the "top" of the respective Fc regions, well away from this interdomain region.

Engagement of FcαRI by IgA molecules clustered on the antigenic surface of a virus, bacterium or other foreign material can trigger an array of potent killing mechanisms. In the case of serum IgA, these processes include phagocytosis, release of activated oxygen species, cytokine release, degranulation and antibody-dependent cell-mediated cytotoxicity. While SIgA appears unable to trigger phagocytosis via FcαRI, it can elicit a respiratory burst through the receptor on neutrophils in a manner dependent on the presence of the complement receptor CR3 (Mac-1, CD11b/CD18) (Stewart and Kerr 1990; van Spriel et al. 2002).

When FcαRI molecules are effectively cross-linked by binding of IgA immune complexes or by engaging with IgA aggregated on an antigenic surface, they redistribute into detergent-insoluble lipid domains (Lang et al. 1999, 2001), where the Src kinase Lyn phosphorylates the ITAM motifs of the associated FcR-γ dimer. Subsequently, Syk, B lymphocyte kinase (Blk), phospholipase C-γ, Shc and growth factor receptor-bound protein 2 (Grb2) are recruited, and activation of multiple targets, such as PI3K, and a Grb2-containing multi-component adapter protein complex ensues. The GTPase Sos is then engaged and converts GDP-RAS to GTP-RAS, leading to activation of the Raf-1/MEK/MAP kinase and PI3-kinase pathways (reviewed by Bakema and van Egmond 2011a). Ultimately, this interlacing signalling cascade coordinates activation of transcription factors, gene expression and the archetypal mechanisms of pathogen elimination mentioned above – phagocytosis, respiratory burst and ADCC. Interestingly, some FcαRI do not associate with FcR-γ chain dimer but are still able to bind IgA and endocytose IgA immune complexes, although the fate appears to be recycling rather than entry into the degradative pathway. Signalling via FcR-γ chain therefore does not seem essential for endocytosis (Launay et al. 1999). In another twist to the story, binding of monomeric ligand to FcαRI has been demonstrated to trigger inhibitory rather than activatory signals via the FcR-γ chain ITAM leading to downregulation of events such as FcγR-mediated phagocytosis (Pasquier et al. 2005; Blank et al. 2009). This inhibitory signalling via ITAM, referred to as ITAMi, is suggested to have a role in the dampening of excessive immune complex-triggered responses.

2.7.3 Fcα/μR

The expression profiles of Fcα/μR, a receptor that binds both IgA and IgM, differ between humans and mice, suggesting that they play distinct roles in each species (Shibuya and Honda 2006). Little is known, as yet, about expression in other species. In humans, the receptor is found on follicular dendritic cells in tonsil (Kikuno et al. 2007) and on plasma cells, macrophages and Paneth cells in the lamina propria and intestinal germinal centres (Wang et al. 2009), locations that hint at a role in coordinating the immune response in mucosal tissues. However, the functional repertoire of the receptor remains to be elucidated.

The Fcα/μR gene lies close to that of pIgR on human chromosome 1. There are other similarities with pIgR too. The N-terminal immunoglobulin-like domain of Fcα/μR is homologous to D1 of pIgR, and they have been suggested to share structural similarities based on the presence of certain conserved residues (Hamburger et al. 2004). Of particular note, their CDR-like loops are similar. Since these regions in pIgR are believed to interact with dimeric IgA, Fcα/μR may share similar ligand binding characteristics. In common with pIgR, Fcα/μR interacts only with polymeric forms of IgA and IgM (Ghumra et al. 2009). However, unlike pIgR, the presence of J chain does not appear to be essential for binding (Yoo et al. 2011). A site at the Cα2-Cα3 domain interface of the IgA heavy chain, overlapping with those of FcαRI and pIgR, has been shown to be critical for interaction with Fcα/μR (Ghumra et al. 2009).

2.7.4 Other Receptors for IgA

Information on other receptors that bind IgA is more limited. These include the transferrin receptor (TfR or CD71), named for its involvement in iron metabolism, which appears to bind only to monomeric IgA1 possibly via the O-linked sugars of the hinge region (Moura et al. 2001). Another receptor that interacts with the oligosaccharide moieties of IgA, in this case the exposed terminal sugars on desialylated IgA, is the asialoglycoprotein receptor (ASGP-R) (Stockert et al 1982). ASGP-R removes IgA molecules lacking sialic acid from the circulation and hence plays an important role in catabolism and regulation of IgA serum levels.

Another IgA-specific receptor has been described on intestinal M cells in both mice and humans (Mantis et al. 2002). Over the years, IgA has also been reported to bind to B cells, T cells and NK cells, but the precise identities of the receptors involved remain unresolved. Finally, one might mention the putative receptor for SC on eosinophils that may account in part for SIgA's particular ability to trigger degranulation of these cells (Lamkhioued et al. 1995).

2.8 Strategies Used by Pathogens to Perturb IgA Function

2.8.1 Advantages to Microorganisms of Circumventing IgA Function

Numerous pathogens including many that can cause serious disease have evolved mechanisms to specifically subvert the protection provided by IgA. Available evidence indicates that these mechanisms have frequently arisen independently in different organisms, suggesting that there is considerable evolutionary pressure to develop effective systems to attack IgA. Obviously, there are considerable benefits for the organisms in question in terms of enhanced capability for colonisation and spread, particularly via mucosal invasion. Important examples include the production of proteases that cleave IgA into inactive fragments, the utilisation of IgA-binding proteins to competitively block interaction sites on IgA used by host Fc receptors and the use of proteins that can interact with pIgR to facilitate adherence and invasion at mucosal sites.

2.8.2 IgA1 Proteases

Certain bacterial pathogens produce proteases that specifically cleave IgA within the hinge region. The impact of the resulting release of the Fc region from the Fab portions has a dramatic effect on function. The Fab arms are still able to bind to bacterial antigens, but having done so, they lack the means to recruit the effector mechanisms that are normally evoked through the Fc region and hence are unable to trigger killing of the recognised bacteria. Moreover, their attachment prevents access of intact antibodies. Hinge cleavage is therefore an effective strategy to evade the protective functions of IgA.

The majority of IgA proteases cleave only IgA1. These so-called IgA1 proteases are produced by some important human pathogens that as associated with life-threatening infections (Kilian et al. 1996). For example, the principal causes of bacterial meningitis (*Neisseria meningitidis*, *Haemophilus influenzae* and *Streptococcus pneumoniae*) all secrete IgA1 proteases. Closely related but non-pathogenic Neisseria and Haemophilus species do not produce IgA1 proteases, underlining the noted association of these proteases with virulence (Polissi et al. 1998). IgA1 proteases are also produced by bacteria associated with vaginal and urinary tract infections, such as *Neisseria gonorrhoeae* and *Ureaplasma urealyticum*. Certain bacteria associated with disease in the oral cavity, notably *Streptococcus sanguis* and *Streptococcus mitis* known for their role in initiation of plaque formation, secrete IgA1 proteases also. The picture that emerges is that the microorganisms involved tend to colonise and even invade mucosal surfaces. Their production of IgA1 proteases presumably helps them establish a foothold within mucosal tissues.

All IgA1 proteases cleave C-terminal to one of the proline residues of the IgA1 hinge. The particular peptide bond cleaved in each case is indicated in Fig. 2.5. IgA2 lacks the susceptible hinge sequence and, therefore, is resistant to cleavage. The IgA1 proteases are remarkably specific and are known to cleave human IgA1 and the equivalents from gorillas and chimpanzees, but very few other substrates appear to exist. The structural requirements for effective substrate recognition and cleavage have been probed by mutagenesis (Senior et al. 2000; Batten et al. 2003; Senior and Woof 2005a). The susceptible bond must be suitably positioned relative to the Fc region for optimal cleavage (Senior and Woof 2005b). For some enzymes, structural elements within the Cα3 domain must be present for hinge cleavage to occur (Chintalacharuvu et al. 2003; Senior and Woof 2006). This finding, though

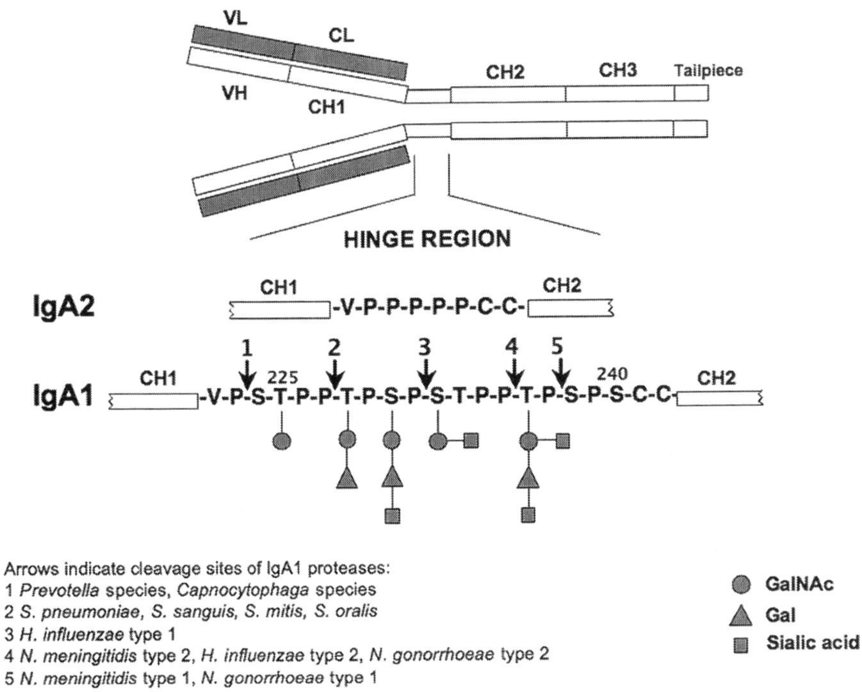

Fig. 2.5 *Differences between the hinge regions of human IgA1 and IgA2*. The amino acid sequences of the hinge regions are shown. Arrows above the IgA1 hinge sequence indicate the cleavage sites of particular IgA1 proteases. The O-linked oligosaccharides carried by the IgA1 hinge are indicated below the sequence

surprising, fits well with a model for IgA1-IgA1 protease interaction proposed on the basis of the solved crystal structure for an IgA1 protease from *H. influenzae* (Johnson et al. 2009). The model proposes that interactions between the IgA1 Fc region and a critical loop region of the protease hold the loop away from the active site, allowing access of the hinge peptide and subsequent cleavage. Therefore, IgA1 proteases appear to identify a substrate for cleavage through a combination of hinge sequence recognition and contextual recognition, at least in some cases.

2.8.3 IgA-Binding Proteins

IgA-binding proteins (IgA-BP) are known to be expressed by some important bacterial pathogens, including some streptococci and *Staphylococcus aureus*. Examples of these proteins in streptococci are Arp4 and Sir22 expressed on group A *streptococcus* (Frithz et al. 1989; Stenberg et al. 1994) and an unrelated protein known as β-protein that is expressed by group B *streptococcus* (Héden et al. 1991). Mutagenesis experiments showed that these IgA-BP all bind to the interdomain region of IgA Fc, at a site overlapping with that for FcαRI (Pleass et al. 2001). Importantly, the streptococcal IgA-BP inhibit both binding of IgA to FcαRI and the subsequent triggering of killing mechanisms (Pleass et al. 2001). Simple blockade at this site therefore allows the bacteria to avoid those elimination processes normally triggered through FcαRI.

In *Staphylococcus aureus*, a similar effect has been described, albeit mediated through a completely different IgA-BP called SSL7. The SSL7 toxin also binds to the interface between Cα2 and Cα3 and competitively inhibits FcαRI binding (Wines et al. 2006; Ramsland et al. 2007). It appears that the evasion strategy of blocking immunoglobulin function by binding to the Fc interdomain region is a common one, used by a variety of bacterial species that infect different mammalian hosts (Lewis et al. 2008; Wines et al. 2012).

2.8.4 Pneumococcal CbpA Binding to pIgR

A surface protein of *Streptococcus pneumoniae* known as CbpA, PspC or SpsA has been shown to bind via a hexapeptide motif to domains D3–D4 of pIgR (Hammerschmidt et al. 2000; Lu et al. 2003; Elm et al. 2004). In vitro, in the absence of free SC or SIgA, the CbpA-pIgR interaction facilitates adherence of *S. pneumoniae* to epithelial cells and subsequent internalisation (Zhang et al. 2000). This co-opting of pIgR may therefore aid pneumococcal adherence to nasopharyngeal cells in vivo, leading to colonisation.

2.9 Potential of Therapeutic IgA Monoclonal Antibodies

Although monoclonal antibodies (mAbs) currently licenced for therapeutic use are all IgG-based, mAbs based on an IgA backbone may offer novel possibilities, given the unique structural and functional properties of this isotype (Dechant and Valerius 2001; Corthésy 2003; Bakema and van Egmond 2011b). A number of studies, mostly early-stage investigations, have indicated that recombinant IgA mAbs directed against bacterial pathogens may offer a means to control or limit certain infections. For example, an IgA-based mAb targeting an adhesin protein on *Streptococcus mutans* as a means to control caries was found to limit oral recolonisation (Ma et al. 1998), and other IgA-based mAbs have shown promise against pathogens such as *S. pneumoniae*, *N. meningitidis*, *Bordetella pertussis* and *Mycobacterium tuberculosis* (van der Pol et al. 2000; Vidarsson et al. 2001; Hellwig et al. 2001; Balu et al. 2011). IgA-based or FcαRI-directed approaches are also showing potential as novel cancer therapeutics (Stockmeyer et al. 2000; van Egmond et al. 2001; Dechant et al. 2002; Otten et al. 2005; Zhao et al. 2008; Lohse et al. 2011; Bakema et al. 2011). Future studies may reveal whether IgA mAbs can offer advantages in the physiological setting.

2.10 Concluding Remarks

IgA is marked out from the other antibody classes by its heterogeneity of form, its location within the body, its novel functions and the vast quantities that are produced each day. Historically, understanding of this complex antibody has lagged behind that of some of the other immunoglobulin isotypes, probably due to difficulties in isolation, compounded by species differences. However, there has been a resurgence of interest, and a clearer picture of IgA structure and function has emerged in recent years, spearheaded by advances in antibody engineering. Further research into IgA promises to open up new avenues for therapy and possibly also for improved vaccination strategies targeted to mucosal surfaces.

References

Atkin JD, Pleass RJ, Owens RJ, Woof JM (1996) Mutagenesis of the human IgA1 heavy chain tailpiece that prevents dimer assembly. J Immunol 157:156–159

Bakema JE, van Egmond M (2011a) The human immunoglobulin A Fc receptor FcαRI: a multifaceted regulator of mucosal immunity. Mucosal Immunol 4:612–624

Bakema JE, van Egmond M (2011b) Immunoglobulin A: a next generation of therapeutic antibodies? MAbs 3:352–361

Bakema JE, Ganzevles SH, Fluitsma DM, Schilham MW, Beelen RH, Valerius T, Lohse S, Glennie MJ, Medema JP, van Egmond M (2011) Targeting FcαRI on polymorphonuclear cells induces tumor cell killing through autophagy. J Immunol 187:726–732

Bakos MA, Kurosky A, Cwerwinski EW, Goldblum RM (1993) A conserved binding site on the receptor for polymeric Ig is homologous to CDR1 of Ig V kappa domains. J Immunol 151:1346–1352

Balu S, Reljic R, Lewis MJ, Pleass RJ, McIntosh R, van Kooten C, van Egmond M, Challacombe S, Woof JM, Ivanyi J (2011) A novel human IgA monoclonal antibody protects against tuberculosis. J Immunol 186:3113–3119

Bastian A, Kratzin H, Eckart K, Hilschmann N (1992) Intra- and inter-chain disulphide bridges of the human J chain in secretory immunoglobulin A. Biol Chem Hoppe Seyler 373:1255–1263

Batten MR, Senior BW, Kilian M, Woof JM (2003) Amino acid sequence requirements in the hinge of human immunoglobulin A1 (IgA1) for cleavage by streptococcal IgA1 proteases. Infect Immun 71:1462–1469

Biesbrock AR, Reddy MS, Levine MJ (1991) Interaction of a salivary mucin-secretory immunoglobulin A complex with mucosal pathogens. Infect Immun 59:3492–3497

Blank U, Launay P, Benhamou M, Monteiro RC (2009) Inhibitory ITAMs as novel regulators of immunity. Immunol Rev 232:59–71

Bomsel M, Heyman M, Hocini H, Lagaye S, Belec L, Dupont C, Desgranges C (1998) Intracellular neutralization of HIV transcytosis across tight epithelial barriers by anti-HIV envelope protein dIgA or IgM. Immunity 9:277–287

Braathen R, Sorensen V, Brandtzaeg P, Sandlie I, Johansen FE (2002) The carboxyl-terminal domains of IgA and IgM direct isotype-specific polymerization and interaction with the polymeric immunoglobulin receptor. J Biol Chem 277:42755–42762

Brandtzaeg P, Johansen FE (2005) Mucosal B cells: phenotypic characteristics, transcriptional regulation, and homing properties. Immunol Rev 206:32–63

Brandtzaeg P, Farstad IN, Johansen FE, Morton HC, Norderhaug IN, Yamanaka T (1999) The B-cell system of human mucosae and exocrine glands. Immunol Rev 171:45–87

Burnett RC, Hanly WC, Zhai SK, Knight KL (1989) The IgA heavy-chain gene family in rabbit: cloning and sequence analysis of 13 Cα genes. EMBO J 8:4041–4047
Burns JW, Siadat-Pajouh M, Krishnaney AA, Greenberg HB (1996) Protective effect of rotavirus VP6-specific IgA monoclonal antibodies that lack neutralizing activity. Science 272:104–107
Cann GM, Zaritsky A, Koshland ME (1982) Primary structure of the immunoglobulin J chain from the mouse. Proc Natl Acad Sci USA 79:6656–6660
Carayannopoulos L, Hexham JM, Capra JD (1996) Localization of the binding site for the monocyte immunoglobulin (Ig) A-Fc receptor (CD89) to the domain boundary between Cα2 and Cα3 in human IgA1. J Exp Med 183:1579–1586
Cerutti A (2008) The regulation of IgA class switching. Nat Rev Immunol 8:421–434
Childers NK, Bruce MG, McGhee JR (1989) Molecular mechanisms of immunoglobulin A defense. Annu Rev Immunol 43:503–536
Chintalacharuvu KR, Raines M, Morrison SL (1994) Divergence of human alpha-chain constant region gene sequences. A novel recombinant alpha2 gene. J Immunol 152:5299–5304
Chintalacharuvu KR, Chuang PD, Dragoman A, Fernandez CZ, Qiu J, Plaut AG, Trinh KR, Gala FA, Morrison SL (2003) Cleavage of the human immunoglobulin A1 (IgA1) hinge region by IgA1 proteases requires structures in the Fc region of IgA. Infect Immun 71:2563–2570
Coffman RL, Lebman DA, Shrader B (1989) Transforming growth factor beta specifically enhances IgA production by lipopolysaccharide-stimulated murine B lymphocytes. J Exp Med 170:1039–1044
Conley ME, Delacroix DL (1987) Intravascular and mucosal immunoglobulin A: two separate but related systems of immune defense? Ann Intern Med 106:892–899
Corthésy B (2003) Recombinant secretory immunoglobulin A in passive immunotherapy: linking immunology and biotechnology. Curr Pharm Biotechnol 4:51–67
Corthésy B, Benureau Y, Perrier C, Fourgeux C, Parez N, Greenberg H, Schwartz-Cornil I (2006) Rotavirus anti-VP6 secretory immunoglobulin A contributes to protection via intracellular neutralization but not via immune exclusion. J Virol 80:10692–10699
Coyne RS, Siebrecht M, Peitsch MC, Casanova JE (1994) Mutational analysis of polymeric immunoglobulin receptor/ligand interactions. Evidence for the involvement of multiple complementarity determining region (CDR)-like loops in receptor domain I. J Biol Chem 269:31620–31625
Dallas SD, Rolfe RD (1998) Binding of Clostridium difficile toxin A to human milk secretory component. J Med Microbiol 47:879–888
Dechant M, Valerius T (2001) IgA antibodies for cancer therapy. Crit Rev Oncol Hematol 39:69–77
Dechant M, Vidarsson G, Stockmeyer B, Repp R, Glennie MJ, Gramatzki M, van de Winkel JGJ, Valerius T (2002) Chimeric IgA antibodies against HLA class II effectively trigger lymphoma cell killing. Blood 100:4574–4580
Du CC, Mashoof SM, Criscitiello MF (2012) Oral immunization of the African clawed frog (Xenopus laevis) upregulates the mucosal immunoglobulin IgX. Vet Immunol Immunopathol 145:493–498
Elm C, Braathen R, Bergmann S, Frank R, Vaerman J-P, Kaetzel CS, Chhatwal GS, Johansen F-E, Hammerschmidt S (2004) Ectodomains 3 and 4 of human polymeric immunoglobulin receptor (hpIgR) mediate invasion of *Streptococcus pneumoniae* into the epithelium. J Biol Chem 279:6296–6304
Fallgren-Gebauer E, Gebauer W, Bastian A, Kratzin H, Eiffert H, Zimmerman B, Karas M, Hilschmann N (1995) The covalent linkage of the secretory component to IgA. Adv Exp Med Biol 371A:625–628
Feng N, Lawton JA, Gilbert J, Kuklin N, Vo P, Prasad BV, Greenberg HB (2002) Inhibition of rotavirus replication by a non-neutralizing, rotavirus VP6-specific IgA mAb. J Clin Invest 109:1203–1213
Fernandez MI, Pedron T, Tournebize R, Olivo-Marin JC, Sansonetti PJ, Phalipon A (2003) Anti-inflammatory role for intracellular dimeric immunoglobulin A by neutralization of lipopolysaccharide in epithelial cells. Immunity 18:739–749

Field MC, Amatayakul-Chantler S, Rademacher TW, Rudd PM, Dwek RA (1994) Structural analysis of the N-glycans from human immunoglobulin A1: comparison of normal human serum immunoglobulin A1 from that isolated from patients with rheumatoid arthritis. Biochem J 299:261–275
Frithz E, Héden L-O, Lindahl G (1989) Extensive sequence homology between IgA receptor and M proteins in *Streptococcus pyogenes*. Mol Microbiol 3:1111–1119
Frutiger S, Hughes GJ, Paquet N, Luthy R, Jaton JC (1992) Disulfide bond assignment in human J chain and its covalent pairing with immunoglobulin M. Biochemistry 31:12643–12647
Ghumra A, Shi J, Mcintosh RS, Rasmussen IB, Braathen R, Johansen FE, Sandlie I, Mongini PK, Areschoug T, Lindahl G, Lewis MJ, Woof JM, Pleass RJ (2009) Structural requirements for the interaction of human IgM and IgA with the human Fcα/μ receptor. Eur J Immunol 39:1147–1156
Hamburger AE, West AP, Bjorkman PJ (2004) Crystal structure of a polymeric immunoglobulin binding fragment of the human polymeric immunoglobulin receptor. Structure 12:1925–1935
Hammerschmidt S, Tillig MP, Wolff S, Vaerman J-P, Chhatwal GS (2000) Species-specific binding of human secretory component to SpsA protein of *Streptococcus pneumoniae* via a hexapeptide motif. Mol Microbiol 36:726–736
Hanson LA, Korotkova M (2002) The role of breastfeeding in prevention of neonatal infection. Semin Neonatol 7:275–281
He B, Xu W, Santini PA, Polydorides AD, Chiu A, Estrella J, Shan M, Chadburn A, Villanacci V, Plebani A, Knowles DM, Rescigno M, Cerutti A (2007) Intestinal bacteria trigger T cell-independent immunoglobulin A(2) class switching by inducing epithelial-cell secretion of the cytokine APRIL. Immunity 26:812–826
Héden L-O, Frithz E, Lindahl G (1991) Molecular characterization of an IgA receptor from group B streptococci: sequence of the gene, identification of a proline-rich region with unique structure and isolation of N-terminal fragments with IgA-binding capacity. Eur J Immunol 21:1481–1490
Hellwig SMM, van Spriel AB, Schellekens JFP, Mooi FR, van de Winkel JGJ (2001) Immunoglobulin A-mediated protection against *Bordetella pertussis* infection. Infect Immun 69:4846–4850
Herr AB, Ballister ER, Bjorkman PJ (2003) Insights into IgA-mediated immune responses from the crystal structures of human FcαRI and its complex with IgA1-Fc. Nature 423:614–620
Hexham JM, White KD, Carayannopoulos LN, Mandecki W, Brisette R, Yang YS, Capra JD (1999) A human immunoglobulin (Ig)A Cα3 domain motif directs polymeric Ig receptor-mediated secretion. J Exp Med 189:747–752
Huang YT, Wright A, Gao X, Kulick L, Yan H, Lamm ME (2005) Intraepithelial cell neutralization of HIV-1 replication by IgA. J Immunol 174:4828–4835
Johansen FE, Kaetzel CS (2011) Regulation of the polymeric immunoglobulin receptor and IgA transport: new advances in environmental factors that stimulate pIgR expression and its role in mucosal immunity. Mucosal Immunol 4:598–602
Johansen FE, Braathen R, Brandtzaeg P (2001) The J chain is essential for polymeric Ig receptor-mediated epithelial transport of IgA. J Immunol 167:5185–5192
Johnson TA, Qiu J, Plaut AG, Holyoak T (2009) Active-site gating regulates substrate selectivity in a chymotrypsin-like serine protease the structure of *Haemophilus influenzae* immunoglobulin A1 protease. J Mol Biol 389:559–574
Kaetzel CS (2005) The polymeric immunoglobulin receptor: bridging innate and adaptive immune responses at mucosal surfaces. Immunol Rev 206:83–99
Kaetzel CS, Robinson JK, Chintalacharuvu KR, Vaerman JP, Lamm ME (1991) The polymeric immunoglobulin receptor (secretory component) mediates transport of immune complexes across epithelial cells: a local defense function for IgA. Proc Natl Acad Sci USA 88:8796–8800
Kawamura S, Ueda S (1992) Immunoglobulin C_H gene family in hominoids and its evolutionary history. Genomics 13:194–200

Kawamura S, Saitou N, Ueda S (1992) Concerted evolution of the primate immunoglobulin α-gene through gene conversion. J Biol Chem 267:7359–7367
Kikuno K, Kang DW, Tahara K, Torii I, Kubagawa HM, Ho KJ, Baudino L, Nishizaki N, Shibuya A, Kubagawa H (2007) Unusual biochemical features and follicular dendritic cell expression of human Fcα/μ receptor. Eur J Immunol 37:3540–3550
Kilian M, Reinholdt J, Lomholt H, Poulsen K, Frandsen EV (1996) Biological significance of IgA1 proteases in bacterial colonization and pathogenesis: critical evaluation of experimental evidence. APMIS 104:321–338
Koshland ME (1985) The coming of age of the immunoglobulin J chain. Annu Rev Immunol 3:425–453
Krugmann S, Pleass RJ, Atkin JD, Woof JM (1997) Structural requirements for assembly of dimeric IgA probed by site-directed mutagenesis of J chain and a cysteine residue of the α chain CH2 domain. J Immunol 159:244–249
Lamkhioued B, Gounni AS, Gruart V, Pierce A, Capron A, Capron M (1995) Human eosinophils express a receptor for secretory component. Role in secretory IgA-dependent activation. Eur J Immunol 25:117–125
Lang ML, Shen L, Wade WF (1999) Gamma-chain dependent recruitment of tyrosine kinases to membrane rafts by the human IgA receptor FcαR. J Immunol 163:5391–5398
Lang ML, Shen L, Gao H, Cusack WF, Lang GA, Wade WF (2001) Fcα receptor cross-linking causes translocation of phosphatidylinositol-dependent protein kinase 1 and protein kinase Bα to MHC class II peptide-loading-like compartments. J Immunol 166:5585–5593
Launay P, Patry C, Lehuen A, Pasquier B, Blank U, Monteiro RC (1999) Alternative endocytic pathway for immunoglobulin A Fc receptors (CD89) depends on the lack of FcRγ association and protects against degradation of bound ligand. J Biol Chem 274:7216–7225
Lewis MJ, Pleass RJ, Batten MR, Atkin JD, Woof JM (2005) Structural requirements for the interaction of human IgA with the human polymeric Ig receptor. J Immunol 175:6694–6701
Lewis MJ, Meehan M, Owen P, Woof JM (2008) A common theme in interaction of bacterial immunoglobulin-binding proteins with immunoglobulins illustrated in the equine system. J Biol Chem 283:17615–17623
Lohse S, Derer S, Beyer T, Klausz K, Peipp M, Leusen JH, van de Winkel JG, Dechant M, Valerius T (2011) Recombinant dimeric IgA antibodies against the epidermal growth factor receptor mediate effective tumor cell killing. J Immunol 186:3770–3778
Lu L, Lamm ME, Li H, Corthésy B, Zhang JR (2003) The human polymeric immunoglobulin receptor binds to *Streptococcus pneumoniae* via domains 3 and 4. J Biol Chem 278:48178–48187
Ma JK, Hikmat BY, Wycoff K, Vine ND, Chargelegue D, Yu L, Hein MB, Lehner T (1998) Characterization of a recombinant plant monoclonal secretory antibody and preventive immunotherapy in humans. Nat Med 4:601–606
Macpherson AJ, Uhr T (2004) Induction of protective IgA by intestinal dendritic cells carrying commensal bacteria. Science 328:1705–1709
Macpherson AJ, Geuking MB, McCoy KD (2012) Homeland security: IgA immunity at the frontiers of the body. Trends Immunol 33:160–167
Mansikka A (1992) Chicken IgA H chains. Implications concerning the evolution of H chain genes. J Immunol 149:855–861
Mantis NJ, Cheung MC, Chintalacharuvu KR, Rey J, Corthésy B, Neutra MR (2002) Selective adherence of IgA to murine Peyer's patch M cells: evidence for a novel IgA receptor. J Immunol 169:1844–1851
Mantis NJ, Rol N, Corthesy B (2011) Secretory IgA's roles in immunity and mucosal homeostasis in the gut. Mucosal Immunol 4:603–611
Mattu TS, Pleass RP, Willis AC, Kilian M, Wormald MR, Lellouch AC, Rudd PM, Woof JM, Dwek RA (1998) The glycosylation and structure of human serum IgA1, Fab and Fc regions and the role of N-glycosylation on Fcα receptor interactions. J Biol Chem 273:2260–2272

Mazanec MB, Kaetzel CS, Lamm ME, Fletcher D, Nedrud JG (1992) Intracellular neutralization of virus by immunoglobulin A antibodies. Proc Natl Acad Sci USA 89:6901–6905
Mazanec MB, Coudret CL, Fletcher DR (1995) Intracellular neutralization of influenza virus by immunoglobulin A anti-hemagglutinin monoclonal antibodies. J Virol 69:1339–1343
Mestecky J, Russell MW, Jackson S, Brown TA (1986) The human IgA system: a reassessment. Clin Immunol Immunopathol 40:105–114
Mestecky J, Lue C, Russell MW (1991) Selective transport of IgA: cellular and molecular aspects. Gastroenterol Clin North Am 20:441–471
Mestecky J, Bienenstock J, McGhee JR, Lamm ME, Strober W, Cebra JJ, Mayer L, Ogra PL (2005) Historical aspects of mucosal immunology. In: Mestecky JR, Bienenstock J, Lamm ME, Mayer L, McGhee JR, Strober W (eds) Mucosal immunology, 3rd edn. Elsevier Academic, London
Mostov KE, Kraehenbuhl JP, Blobel G (1980) Receptor-mediated transcellular transport of immunoglobulin: synthesis of secretory component as multiple and larger molecular forms. Proc Natl Acad Sci USA 77:7257–7261
Moura IC, Centelles MN, Arcos-Fajardo M, Malheiros DM, Collawn JF, Cooper MD, Monteiro RC (2001) Identification of the transferrin receptor as a novel immunoglobulin (Ig)A1 receptor and its enhanced expression on mesangial cells in IgA nephropathy. J Exp Med 194:417–426
Narimatsu Y, Kubota T, Furukawa S, Morii H, Narimatsu H, Yamasaki K (2010) Effect of glycosylation on the cis/trans isomerization of prolines in IgA1-hinge peptide. J Am Chem Soc 132:5548–5549
Natvig IB, Johansen FE, Nordeng TW, Haraldsen G, Brandtzaeg P (1997) Mechanism for enhanced external transfer of dimeric IgA over pentameric IgM: studies of diffusion, binding to the human polymeric Ig receptor, and epithelial transcytosis. J Immunol 159:4330–4340
Navarro P, Christenson RK, Ekhardt G, Bosworth B, Lunney JK, Rothschild M, Lemke J, Butler JE (2000) Genetic differences in the frequency of the hinge variants of porcine IgA is breed dependent. Vet Immunol Immunopathol 73:287–295
Norderhaug IN, Johansen FE, Krajci P, Brandtzaeg P (1999) Domain deletions in the human polymeric Ig receptor disclose differences between its dimeric and pentameric IgM interaction. Eur J Immunol 29:3401–3409
Otten MA, Rudolph E, Dechant M, Tuk CW, Reijmers RM, Beelen RH, van de Winkel JGJ, van Egmond M (2005) Immature neutrophils mediate tumor cell killing via IgA but not IgG Fc receptors. J Immunol 174:5472–5480
Pasquier B, Launay P, Kanamaru Y, Moura IC, Pfirsch S, Ruffié C, Hénin D, Benhamou M, Pretolani M, Blank U, Monteiro RC (2005) Identification of FcαRI as an inhibitory receptor that controls inflammation: dual role of FcR γ ITAM. Immunity 22:31–42
Phillips-Quagliata JM (2002) Mouse IgA allotypes have major differences in their hinge regions. Immunogenetics 53:1033–1038
Pleass RJ, Dunlop JI, Anderson CM, Woof JM (1999) Identification of residues in the CH2/CH3 domain interface of IgA essential for interaction with the human Fcα receptor (FcαR) CD89. J Biol Chem 274:23508–23514
Pleass RJ, Areschoug T, Lindahl G, Woof JM (2001) Streptococcal IgA-binding proteins bind in the Cα2-Cα3 interdomain region and inhibit binding of IgA to human CD89. J Biol Chem 276:8197–8204
Polissi A, Pontiggia A, Feger G, Altieri M, Mottl H, Ferrari L, Simon D (1998) Large-scale identification of virulence genes from *Streptococcus pneumoniae*. Infect Immun 66:5620–5629
Ramsland PA, Willoughby N, Trist HM, Farrugia W, Hogarth PM, Fraser JD, Wines BD (2007) Structural basis for evasion of IgA immunity by Staphylococcus aureus revealed in the complex of SSL7 with Fc of human IgA1. Proc Natl Acad Sci USA 104:15051–15056
Robinson JK, Blanchard TG, Levine AD, Emancipator SN, Lamm ME (2001) A mucosal IgA-mediated excretory immune system in vivo. J Immunol 166:3688–3692
Rogers KA, Jayashankar L, Scinicariello F, Attanasio R (2008) Nonhuman primate IgA: genetic heterogeneity and interaction with CD89. J Immunol 180:4816–4824

Roos A, Bouwman LH, van Gijlswijk-Janssen DJ, Faber-Krol MC, Stahl GL, Daha MR (2001) Human IgA activates the complement system via the mannan-binding lectin pathway. J Immunol 167:2861–2868
Royle L, Roos A, Harvey DJ, Wormald MR, van Gijlswijk-Jannsen D, Redwan E-RM, Wilson IA, Daha MR, Dwek RA, Rudd PM (2003) Secretory IgA N- and O-linked glycans provide a link between the innate and adaptive immune systems. J Biol Chem 278:20140–20153
Ruhl S, Sandberg AL, Cole MF, Cisar JO (1996) Recognition of immunoglobulin A1 by oral actinomyces and streptococcal lectins. Infect Immun 64:5421–5424
Russell MW, Kilian M (2005) Biological activities of IgA. In: Mestecky JR, Bienenstock J, Lamm ME, Mayer L, McGhee JR, Strober W (eds) Mucosal immunology, 3rd edn. Elsevier Academic, Amsterdam
Schwartz-Cornil I, Benureau Y, Greenberg H, Hendrickson BA, Cohen J (2002) Heterologous protection induced by the inner capsid proteins of rotavirus requires transcytosis of mucosal immunoglobulins. J Virol 76:8110–8117
Scinicariello F, Engleman CN, Jayashankar L, McClure HM, Attanasio R (2004) Rhesus macaque antibody molecules: sequences and heterogeneity of alpha and gamma constant regions. Immunology 111:66–74
Scinicariello F, Masseoud F, Jayashankar L, Attanasio R (2006) Sooty mangabey (Cercocebus torquatus atys) IGHG and IGHA genes. Immunogenetics 58:955–965
Senior BW, Woof JM (2005a) Effect of mutations in the human immunoglobulin A1 (IgA1) hinge on its susceptibility to cleavage by diverse bacterial IgA1 proteases. Infect Immun 73:1515–1522
Senior BW, Woof JM (2005b) The influences of hinge length and composition on the susceptibility of human IgA to cleavage by diverse bacterial IgA1 proteases. J Immunol 174:7792–7799
Senior BW, Woof JM (2006) Sites in the CH3 domain of human IgA1 that influence sensitivity to bacterial IgA1 proteases. J Immunol 177:3913–3919
Senior BW, Dunlop JI, Batten MR, Kilian M, Woof JM (2000) Cleavage of a recombinant human immunoglobulin A2 (IgA2)-IgA1 hybrid antibody by certain bacterial IgA1 proteases. Infect Immun 68:463–469
Shibuya A, Honda SI (2006) Molecular and functional characteristics of the Fcα/μR, a novel Fc receptor for IgM and IgA. Springer Semin Immunopathol 28:377–382
Slack E, Balmer ML, Fritz JH, Hapfelmeier S (2012) Functional flexibility of intestinal IgA – broadening the fine line. Front Immunol 3(article 100):1–10
Snoeck V, Peters IR, Cox E (2006) The IgA system: a comparison of structure and function in different species. Vet Res 37:455–467
Spencer J, Klavinskis LS, Fraser LD (2012) The human intestinal IgA response: burning questions. Front Immunol 3(article 108):1–10
Spieker-Polet H, Yam PC, Knight KL (1993) Differential expression of 13 IgA-heavy chain genes in rabbit lymphoid tissues. J Immunol 61:5457–5465
Stenberg L, O'Toole PW, Mestecky J, Lindahl G (1994) Molecular characterization of protein Sir, a streptococcal cell surface protein that binds both immunoglobulin A and immunoglobulin G. J Biol Chem 269:13458–13464
Stephens S, Dolby JM, Montreuil J, Spik G (1980) Differences in inhibition of the growth of commensal and enteropathogenic strains of Escherichia coli by lactotransferrin and secretory immunoglobulin A isolated from human milk. Immunology 41:597–603
Stewart WW, Kerr MA (1990) The specificity of the human neutrophil IgA receptor (FcαR) determined by measurement of chemiluminescence induced by serum or secretory IgA1 or IgA2. Immunology 71:328–334
Stockert RJ, Kressner MS, Collins JC, Sternlieb I, Morell AG (1982) IgA interaction with the asialoglycoprotein receptor. Proc Natl Acad Sci USA 79:6229–6231
Stockmeyer B, Dechant M, van Egmond M, Tutt AL, Sundarapandiyan K, Graziano RF, Repp R, Kalden JR, Gramatzki M, Glennie MJ, van de Winkel JGJ, Valerius T (2000) Triggering Fcα-

receptor I (CD89) recruits neutrophils as effector cells for CD20-directed antibody therapy. J Immunol 165:5954–5961

Tarelli E, Smith AC, Hendry BM, Challacombe SJ, Pouria S (2004) Human serum IgA1 is substituted with up to six O-glycans as shown by matrix assisted laser desorption ionisation time-of-flight mass spectrometry. Carbohydr Res 339:2329–2335

Tenovuo J, Moldoveanu Z, Mestecky J, Pruitt KM, Rahemtulla BM (1982) Interaction of specific and innate factors of immunity: IgA enhances the antimicrobial effect of the lactoperoxidase system against Streptococcus mutans. J Immunol 128:726–731

Tomana M, Niedermeier W, Mestecky J, Skvaril F (1976) The differences in carbohydrate composition between the subclasses of IgA immunoglobulins. Immunochemistry 13:325–328

Underdown BJ, DeRose J, Plaut A (1977) Disulfide bonding of secretory component to a single monomer subunit in human secretory IgA. J Immunol 118:1816–1821

Vaerman JP, Heremans JF, van Kerckhoven G (1969) Identification of IgA in several mammalian species. J Immunol 103:1421–1423

van der Pol W, Vidarsson G, Vile HA, van de Winkel JG, Rodriguez ME (2000) Pneumococcal capsular polysaccharide-specific IgA triggers efficient neutrophil effector functions via FcαRI (CD89). J Infect Dis 182:1139–1145

van Egmond M, van Garderen E, van Spriel AB, Damen CA, van Amersfoort ES, van Zandbergen G, van Hattum J, Kuiper J, van de Winkel JG (2000) FcαRI-positive liver Kupffer cells: reappraisal of the function of immunoglobulin A in immunity. Nat Med 6:680–685

van Egmond M, van Spriel AB, Vermeulen H, Huls G, Van Garderen E, van de Winkel JGJ (2001) Enhancement of polymorphonuclear cell-mediated tumor cell killing on simultaneous engagement of FcγRI (CD64) and FcαRI (CD89). Cancer Res 61:4055–4060

van Spriel AB, Leusen JH, Vilé H, van de Winkel JG (2002) Mac-1 (CD11b/CD18) as accessory molecule for FcαR (CD89) binding of IgA. J Immunol 169:3831–3836

Vidarsson G, van der Pol WL, van den Elsen JM, Vilé H, Jansen M, Duijs J, Morton HC, Boel E, Daha MR, Corthésy B, van de Winkel JG (2001) Activity of human IgG and IgA subclasses in immune defense against *Neisseria meningitidis* serogroup B. J Immunol 166:6250–6256

Wang R, Fu Y, Zhao Q, Pan L, Zhang W (2009) Human Fcα/μR and pIgR distribute differently in intestinal tissues. Biochem Biophys Res Commun 381:148–152

Watanabe T, Nagura H, Watanabe K, Brown WR (1984) The binding of human milk lactoferrin to immunoglobulin A. FEBS Lett 168:203–207

White KD, Capra JD (2002) Targeting mucosal sites by polymeric immunoglobulin receptor-directed peptides. J Exp Med 196:551–555

Wines BD, Hulett MD, Jamieson GP, Trist HM, Spratt JM, Hogarth PM (1999) Identification of residues in the first domain of human Fcα receptor essential for interaction with IgA. J Immunol 162:2146–2153

Wines BD, Sardjono CT, Trist HH, Lay CS, Hogarth PM (2001) The interaction of FcαRI with IgA and its implications for ligand binding by immunoreceptors of the leukocyte receptor cluster. J Immunol 166:1781–1789

Wines BD, Willoughby N, Fraser JD, Hogarth PM (2006) A competitive mechanism for staphylococcal toxin SSL7 inhibiting the leukocyte IgA receptor, FcαRI, is revealed by SSL7 binding at the Cα2/Cα3 interface of IgA. J Biol Chem 281:1389–1393

Wines BD, Trist HM, Farrugia W, Ngo C, Trowsdale J, Areschoug T, Lindahl G, Fraser JD, Ramsland PA (2012) A conserved host and pathogen recognition site on immunoglobulins: structural and functional aspects. Adv Exp Med Biol 946:87–112

Wold A, Mestecky J, Tomana M, Kobata A, Ohbayashi H, Endo T, Edén CS (1990) Secretory immunoglobulin A carries oligosaccharide receptors for *Escherichia coli* type 1 fimbrial lectin. Infect Immun 58:3073–3077

Woof JM, Burton DR (2004) Human antibody-Fc receptor interactions illuminated by crystal structures. Nat Rev Immunol 4:89–99

Woof JM, Russell MW (2011) Structure and function relationships in IgA. Mucosal Immunol 4:590–597

Wright A, Yan H, Lamm ME, Huang YT (2006) Immunoglobulin A antibodies against internal HIV-1 proteins neutralize HIV-1 replication inside epithelial cells. Virology 356:165–170
Yan H, Lamm ME, Björling E, Huang YT (2002) Multiple functions of immunoglobulin A in mucosal defense against viruses: an in vitro measles virus model. J Virol 76:10972–10979
Yoo EM, Coloma MJ, Trinh KR, Nguyen TQ, Vuong LU, Morrison SL, Chintalacharuvu KR (1999) Structural requirements for polymeric immunoglobulin assembly and association with J chain. J Biol Chem 274:33771–33777
Yoo EM, Trinh KR, Lim H, Wims LA, Morrison SL (2011) Characterization of IgA and IgM binding and internalization by surface-expressed human Fcα/μ receptor. Mol Immunol 48:1818–1826
Zhang JR, Mostov KE, Lamm ME, Nanno M, Shimida S, Ohwaki M, Tuomanen E (2000) The polymeric immunoglobulin receptor translocates pneumococci across human nasopharyngeal epithelial cells. Cell 102:827–837
Zhang YA, Salinas I, Li J, Parra D, Bjork S, Xu Z, LaPatra SE, Bartholomew J, Sunyer JO (2010) IgT, a primitive immunoglobulin class specialized in mucosal immunity. Nat Immunol 11:827–835
Zhao J, Kuroki M, Shibaguchi H, Wang L, Huo Q, Takami N, Tanaka T, Kinugasa T, Kuroki M (2008) Recombinant human monoclonal IgA antibody against CEA to recruit neutrophils to CEA-expressing cells. Oncol Res 17:217–222
Zikan J, Novotny J, Trapane TL, Koshland ME, Urry DW, Bennett JC, Mestecky J (1985) Secondary structure of the immunoglobulin J chain. Proc Natl Acad Sci USA 82:5905–5909

Chapter 3
Crystal Structures of Human IgG-Fc Fragments and Their Complexes with Fcγ Receptors

Peter Sondermann

Abstract In the second half of the last century, protein crystallography emerged as a revolutionary technique in life sciences that enabled the visualization of proteins on an atomic level. Crystal structures boosted the understanding of the function of proteins and their complexes and allow in combination with biochemical data a thorough understanding of the underlying biology. This is in particular true for immunology where crystallography still provides invaluable information in the field.

This chapter focuses on the crystallographic analysis of the Fc part of IgG and its binding to Fcγ receptors, an interaction that is crucial for a proper functioning of the immune system.

Keywords Fc receptors • Crystal structure • Glycosylation • Fucosylation • Immunoglobulins

3.1 Introduction

Antibodies consist of two Fab (from "fragment antigen binding") and one Fc (from "fragment crystallizable") domain. The two Fab moieties are formed each by a heterodimer of a light (LC) and the N-terminal half of the heavy chain (HC), while the C-terminal parts of two heavy chains dimerize to form the Fc fragment resulting in an overall $(LC\text{-}HC)_2$ arrangement within the antibody (Fig. 3.1, inset). Both light and heavy chains fold into two and four immunoglobulin domains (Bork et al. 1994;

P. Sondermann Ph.D (✉)
CSO, SuppreMol GmbH, Am Klopferspitz 19a, 82152 Martinsried, Germany
e-mail: sondermann@suppremol.com

F. Nimmerjahn (ed.), *Molecular and Cellular Mechanisms of Antibody Activity*,
DOI 10.1007/978-1-4614-7107-3_3, © Springer Science+Business Media New York 2013

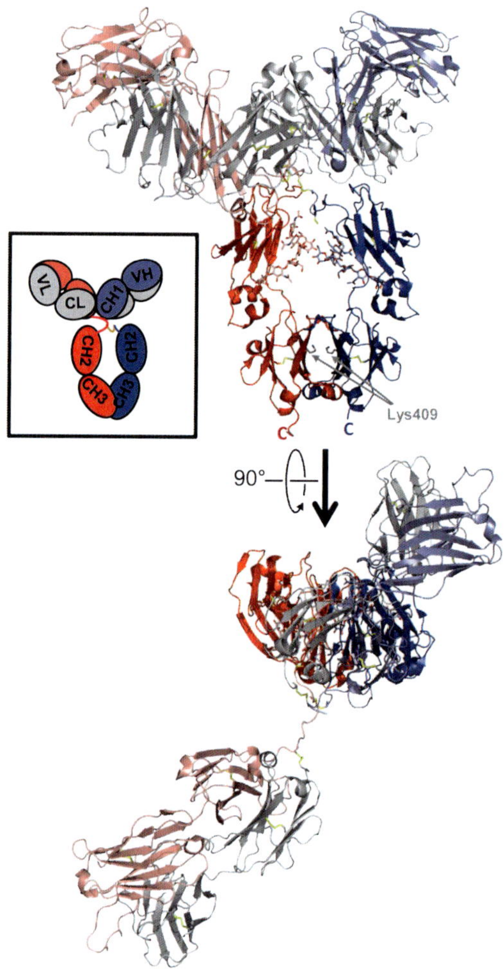

Fig. 3.1 *The IgG molecule.* Crystal structure of an intact human IgG1 (PDB-code 1HZH (Saphire et al. 2001)). Heavy chains are colored *red* and *blue,* while the light chains are depicted in *gray.* The N-glycosylation linked to N297 at the CH2 domains, disulphide bridges (*yellow*), and the Lys409 residues responsible for proper CH3 homodimer formation are shown in ball-and-stick representation. The trace of the blue hinge peptide is only partially visible due to its flexibility in the crystal structure. The lower part of the figure was obtained by rotation of the upper image around the *x*-axis as indicated by the arrow. The inset depicts domain structure and their designation within the antibody as orientated in the upper part of the figure. All structure images were produced with the program PYMOL (http://www.pymol.org)

Chothia et al. 1997), respectively, that exhibit structural homology despite a rather low sequence identity. Immunoglobulin domains are formed of two beta-sheets each consisting of aligned beta-strands that are packed face-to-face to form a beta-sandwich that is in antibodies always stabilized by a disulphide bridge connecting both sheets. Each of these domains has a size of 100–110 amino acids yielding an overall molecular weight of the antibody of around 150 kDa.

The different functions mediated by antibodies are clearly separated within the molecule. In the Fab fragments the variable domains (VL and VH) located at the N-terminus are responsible for antigen recognition, while humoral and cellular effector functions are mediated via the Fc fragment formed by the C-terminal part of the heavy chains. In addition, the Fc part confers an enhanced half-life to the antibody mediated by the interaction to FcRn (Olafsen 2012). The Fabs are loosely connected with the Fc by the hinge peptide that allows the spatial adjustment of the three fragments for an optimum binding to antigen and effector proteins.

Due to the critical role of antibodies in the immune system, the interest in understanding their function has always been a lively field of research. Especially interaction with FcγRs which connects the humoral with the cellular immune response has attracted significant attention during the last decade.

This chapter describes the structure of the Fc fragment and its interaction with human Fc receptors.

3.2 The Structure of Whole IgG Molecules

The importance of antibodies for the immune system has been early recognized, and the investigation of their structure was facilitated by their high abundance in the serum (10–20 mg/ml) which allowed the preparation of the required large amounts for crystallographic studies. However, crystallizability relies on homogenous material which is not present within the antibody populations found in healthy individuals. Heterogeneity is not only caused by the various IgG subclasses but also by the different specificities encoded within the CDRs of the variable regions. Therefore, the successful crystallization of IgGs can only be performed with monoclonal antibodies that are either derived from lymphoma patients or that can nowadays be produced using recombinant techniques. In addition to the problem of the sequence variation, the flexibility of the antibody molecule impedes the buildup of a rigid crystal lattice which limits the examples of full antibody structures (Saphire et al. 2002).

Although the principal crystallizability of antibodies has been already observed in 1938 (von Bonsdorff et al. 1938), the first immunoglobulin x-ray image was obtained in 1968 from a monoclonal cryoglobulin termed "DOB" which is characterized by a hinge deletion (Terry et al. 1968) by which the flexibility between the Fab and Fc fragments is significantly reduced. It took another 8 years before the first full antibody was analyzed crystallographically (Huber et al. 1976). In this study only the electron density for the Fab fragments of the antibody termed "KOL" could be traced, while the Fc fragment's structure was disordered. Nevertheless, the crystal structure of the isolated Fc fragment was solved separately and modeled into the structure derived from the crystals obtained from the entire IgG molecule revealing for the first time the quaternary structure of an entire though hinge-deleted IgG molecule.

Before the first entire structure of a non-mutated human IgG1 without hinge deletion produced by recombinant techniques was described in 2001 (Saphire et al. 2001), the structures of two murine full IgG molecules were reported (Harris et al. 1992, 1998). All three structures confirmed the overall arrangement proposed by Huber and colleagues. As predicted the hinge peptides that connect the Fab fragments with the Fc domain are partially unstructured and confer high flexibility to the three substructures which is a prerequisite for the proper antibody function.

As a result of this flexibility, the different contacts in the crystal, and the different variable regions, the antibodies differ in the orientation of the fragments to each other which therefore represents only one conformation of a large set of possible structures (Fig. 3.1, Sandin et al. 2004).

3.3 The Protein Structure of Fc Fragments

In contrast to the sequence and structure heterogeneity observed for the Fab fragments contributed by the variable domains, the Fc region is apart from a few sequence variations between the four subclasses conserved within IgG. Fc can easily be prepared by IgG cleavage within the hinge peptide into Fab and Fc fragments using various proteases (e.g., papain or plasmin). The first crystal structure of an Fc fragment which usually spontaneously crystallizes, e.g., by dialysis against water, was derived from hIgG1 (hFc1) and was determined by Huber et al. (1976), revealing also its principal carbohydrate structure. The overall structure of the Fc is horseshoe like with the carbohydrate tree covalently attached at Asn297 located at the tips of the CH2 domains. The α1,6-arm termed according to its linkage to the central mannose residue is tightly associated to the CH2 domains, while the α1,3-arm protrudes into the central cavity (Fig. 3.2). In its uncomplexed form electron density for the Fc protein moiety is visible at the first residue within the CH2 domains (Pro238 according to the Eu protein numbering system, (Rutishauser et al. 1970)), while the lower hinge peptide though present in the protein is due to its flexibility usually disordered in crystals of the free Fc. Nevertheless, both heavy chains in hIgG1 are covalently but loosely tethered to each other by two disulphide bridges formed by Cys226 and Cys229 residues with the equivalent cysteine residues of the other heavy chain. Apart from the covalent dimerization by the two disulphide bridges, the Fc fragment is also non-covalently dimerized by intense CH3/CH3-domain contacts. The resulting rigid CH3 dimer exhibits a low intrinsic flexibility (judged by the B factors which indicate true static or dynamic mobility of an atom within a crystal structure). The CH2 domains are connected to the CH3 domains via the "elbow hinge" and exhibit B factors that increase in relation to the distance from the CH3 domain which indicates some flexibility within this linkage.

The hinge peptide in hIgG1 exhibits characteristics of a poly-proline helix that supports the efficient formation of the interchain disulphide bridges. Though not mandatory for dimer arrangement, disulphide bridges do not quantitatively form in IgG4 molecules which may result under distinct conditions in a "Fab-arm exchange" (van der Neut et al. 2007). In this reaction an hIgG4 half molecule consisting of an HC-LC heterodimer exchanges with that of another hIgG4 half molecule. In case of a therapeutic IgG4 molecule, a format which has been frequently used and that is merely free of effector functions, it was observed that this exchange also occurs in vivo with body-own IgG4 molecules resulting in antibodies being bispecific with the second specificity contributed by the exchanged LC-HC monomer. This exchange of half molecules is promoted by the CH3/CH3 interface residue Arg409

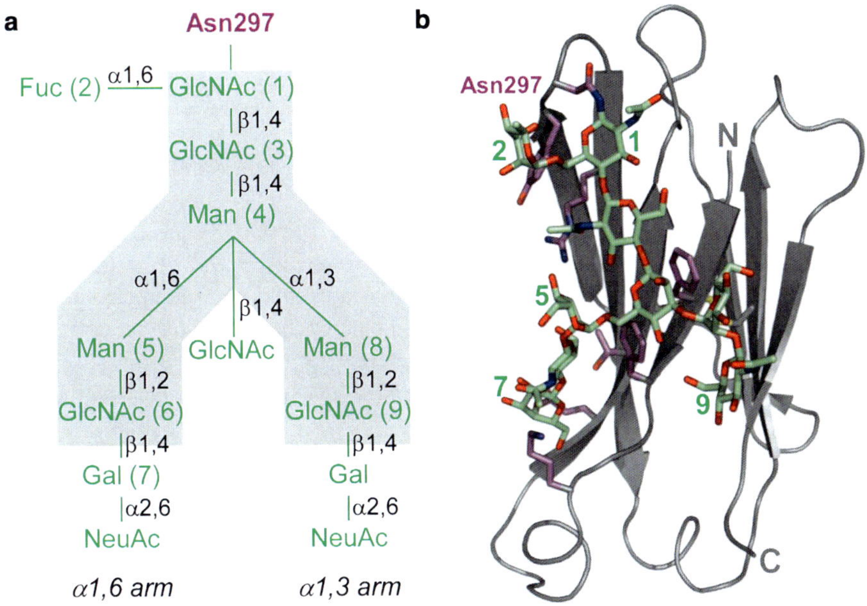

Fig. 3.2 *The N-glycosylation on the Fc fragment*. (**a**) Scheme of the glycosylation structure that is found on the CH2 domain attached at Asn297. The numbering shown behind the carbohydrate residues is the frequently used one from Deisenhofer (Deisenhofer 1981). The conserved oligosaccharide core is shaded gray. The used abbreviations are *GlcNAc* N-Acetylglucosamine (*NAG*), *Fuc* Fucose, *Man* Mannose, *Gal* Galactose, *NeuAc* N-Acetylneuraminic acid (sialic acid). (**b**) N-glycosylated CH2 domain of the Fc fragment human IgG1 (PDB-code 1FC1 (Deisenhofer 1981). The glycosylation is oriented and numbered as in the scheme in (**a**). Amino acid side chains that perform major contacts to the carbohydrate moiety are colored magenta and depicted as sticks. Note that the glycosylation structure might not be traceable as in its entirety the terminal residues may not appear in the electron density due to their flexibility

which destabilizes the interaction between both CH3 domains, while the respective Lys residue present in the other subclasses results in a rigid dimerization (Fig. 3.1) (Rose et al. 2011). Interestingly, the exchange of IgG4 half molecules seems to be conserved also in other mammalian species (Labrijn et al. 2011) and is likely to represent a relevant mechanism intended to modulate humoral immunity as also body-own IgG4 molecules are undergoing this process.

From the biotechnical point of view, this exchange reaction clearly represents safety and efficacy issues for therapeutic antibodies (Labrijn et al. 2009). As a consequence the use of the IgG4 subclass in its native form is now mostly avoided by drug developers. However, the Fab-arm exchange reaction can be minimized either by restoration of the poly-proline-like sequence in the hinge region by a Ser228Pro mutation which allows a more quantitative hinge disulphide bridge formation or by exchange of Arg409 to Lys that leads to a stronger CH3/CH3 dimer interaction (van der Neut et al. 2007).

3.4 The N-Glycosylation of Fc Fragments

One mandatorily used N-glycosylation site is present in the constant region of the heavy chain of IgG that is located at the tips of the CH2 domains at position Asn297. At least in serum IgG, this site is quantitatively occupied by biantennary carbohydrate structures (Fig. 3.2a). Although glycosylation of IgG is crucial for eliciting effector functions mediated by the Fc part (Lund et al. 1990), it was only recently found that even considerable minor changes within the glycosylation structure can modulate effector functions in a rather dramatic way. Although the molecular bases of these effects are only partially understood, they may represent a further option for the immune system to influence the extent of IgG-mediated effector functions in addition to the amount, subclass, and antigen affinity of the generated antibodies (see Sect. 3.6.4).

Despite the vast possible variety of different glycoforms that could theoretically be attached to the Fc fragment, only a limited number of structures are found (Butler et al. 2003; Jefferis et al. 1990a). The observed glycosylation pattern is of the biantennary complex type and consists of a central conserved core heptasaccharide that might be modified by terminal galactose and sialic acid residues, fucose attached at the first N-acetylglucosamine (GlcNAc), "bisecting" GlcNAc linked to the central mannose residue, or combinations thereof (Fig. 3.2a). The limited number of glycostructures is likely caused by the limited accessibility for the various transferases and glycosidases of the carbohydrate tree within the Fc's central cavity. Nevertheless, further modifications/variations might be observed although to a minor extent in case the mAbs are recombinantly expressed but may constitute substantial fractions if nonmammalian expression systems are used where a different set of enzymes involved in the glycosylation pathway is present (Wright and Morrison 1997; Cabanes-Macheteau et al. 1999). A further remarkable feature of the Fc glycosylation is that in contrast to most other glycoproteins, the half-life of nonsialylated IgG is not compromised (Qian et al. 2007; Huang et al. 2006).

In a unique arrangement the stem and the α1,6-arm of carbohydrate moiety is tightly associated to the surface of the CH2 domain mainly via hydrophobic interactions (Fig. 3.2b) making it in the majority of Fc structures fully traceable. The α1,3-arm protrudes into the open cavity coming potentially into contact with the α1,3-arm associated to the other CH2 domain. The carbohydrates in the α1,3 chain exhibit therefore a higher flexibility and are often only partially traceable within crystal structures.

The described arrangement leads to a dual role in achieving and maintaining structural integrity of the Fc domain. On the one hand, the glycostructure influences the quaternary structure of the Fc. The localization of the α1,3-arm on both chains results in their contact within the central cavity of the Fc holding the CH2 domains apart resulting in an "open" conformation of the Fc fragment. Secondly, the tertiary structure within the CH2 domain is rigidified by the close interaction of the carbohydrate tree with predominantly hydrophobic side chains of the CH2 domain (Fig. 3.2b).

This unique tight association of carbohydrate and protein is essential for Fc-mediated effector functions as unglycosylated IgG is devoid of FcγR and complement binding (Tao and Morrison 1989; Jefferis et al. 1998; Mimura et al. 2000). Stepwise truncation of the carbohydrate structure correlates with a reduced affinity for FcγRIIB and a lower thermal stability of the CH2 domains (Mimura et al. 2001). The structure determination of these glycovariants revealed that the CH2 domains approach each other in relation to the degree of carbohydrate tree truncation (Krapp et al. 2003). At the same time a reorientation within the CH2 domain is observed. These structural changes accompanied by the observed "softening" of the structure may explain the affinity loss upon successive deglycosylation.

In fact, the crystal structure of unglycosylated Fc shows a substantially different tertiary and quaternary structure which can be ascribed to the structural changes solely concerning the CH2 domains, while the CH3 domains' monomeric and dimeric structure remains unaffected by the presence of glycosylation (Feige et al. 2009). This unglycosylated Fc exhibits a pronounced "closed" conformation in which the protein moieties of both CH2 domains approach each other (Fig. 3.3a). The observed structural rearrangement is mediated by a rotation around the elbow hinge located between the CH2 and CH3 domains. As already observed in the structures of the above mentioned glycosylation variants, the lack of glycosylation does also affect the structure of the CH2 domain itself. The major change in its structure is characterized by a reorientation of the C'/E loop that contains Asn297 to which the glycosylation is usually attached (Fig. 3.3b). This loop is significantly involved in FcγR binding, and both effects seem to be responsible for the modulation of FcγR binding. In addition, the unglycosylated CH2 domains are less rigid compared to their glycosylated counterpart indicated by higher B factors and by a reduced thermal stability (Mimura et al. 2001). Although these crystal structures of the glycosylated and unglycosylated Fc fragment are derived from murine IgG1 Fc fragments, it can be anticipated that human Fc will show a comparable reorientation upon deglycosylation. This can be concluded from the glycosylated structures of murine and human IgG1 that exhibit a high similarity which is a result of the sequence homology of both proteins (Fig. 3.4).

The influence of terminal sialylated or nonfucosylated Fc carbohydrates on FcγR affinity will be discussed in Sect. 3.6.4.

3.5 The Structure of FcγRs

Due to its importance within the immune system, the interest in the knowledge of the interaction site of FcγRs with immunoglobulins has been intensely investigated. Before the use of crystallography, efforts mainly concentrated on mapping of the interaction site using IgG subclasses, IgGs from different species, and mutants derived thereof. The majority of the results obtained in these studies pointed to a prominent role of the hinge peptide (Jefferis et al. 1990b; Sarmay et al. 1992), but

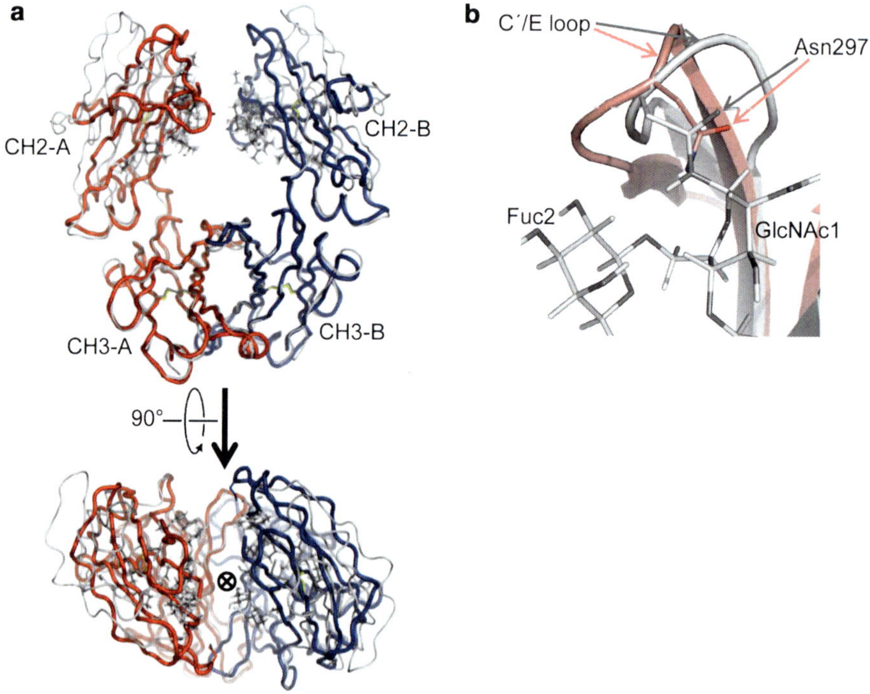

Fig. 3.3 *Influence of glycosylation on the Fc structure.* Superposition of a glycosylated and a nonglycosylated Fc fragment. (**a**) Front and top view of a murine IgG1 Fc fragment with its associated carbohydrates (all *gray*, PDB-code 1IGY (Harris et al. 1998)) in comparison with that of a nonglycosylated murine Fc (*red* and *blue*, PDB-code 3HKF (Feige et al. 2009)). Carbohydrates, the Fc glycosylation site, and disulphide bridges are depicted as sticks. For the structural alignment only the Cα atoms of the CH3 domains were used using a least square fit algorithm. The lower part of the figure was obtained by rotation of the top image around the *x*-axis as indicated by the arrow resulting in a view along the C2 symmetry axis which is indicated by ⊗. (**b**) Comparison of the CH2 domains highlighting the influence of glycosylation on the structure of the C′/E loop. For the figure the same structures and color scheme as in (**a**) were used. The carbohydrate residues are numbered according to Fig. 3.2

also the other parts of the Fc fragment were identified to be involved (Canfield and Morrison 1991). For the FcγRs the results pointed to the second domain and more specific to the loops in the domain's N-terminal region (Tamm et al. 1996; Hogarth et al. 1992; Hulett et al. 1994, 1995).

The first crystal structure of the extracellular and principally monomeric region of an FcγR (FcγRIIB, (Sondermann et al. 1999)) could not resolve the question on the interaction with IgG but disclosed already some interesting structural features of this molecule. From sequence analysis it is known that the low-affinity FcγRs (FcγRII and FcγRIII) consist in their extracellular region of solely β-strands that fold into two Ig domains (D1 and D2) which mediate a micromolar affinity to IgG.

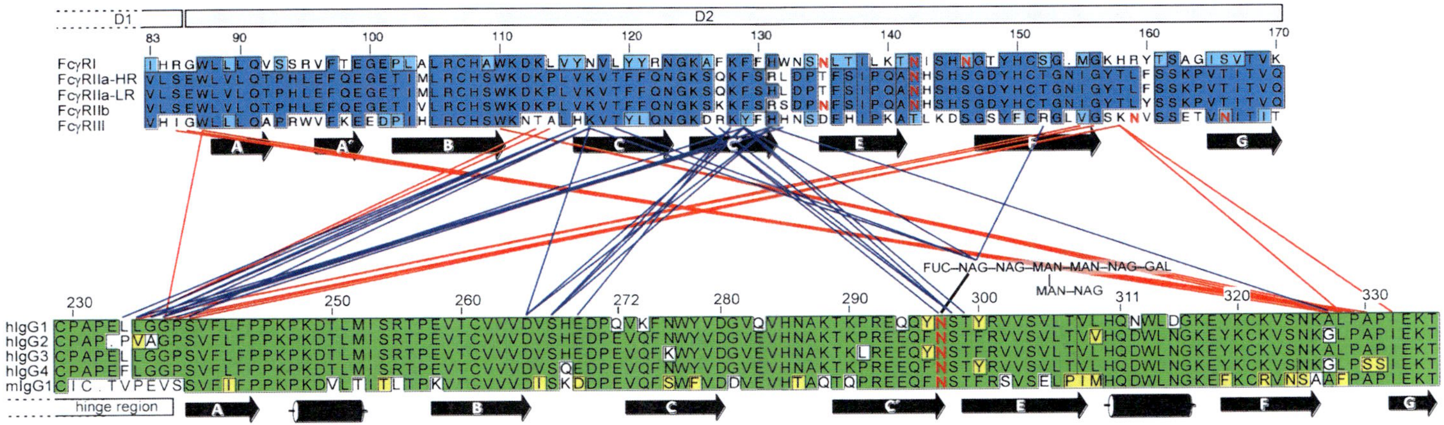

Fig. 3.4 *Contacts between FcγRIIIB and IgG1-Fc in their complex.* Structure-based sequence alignment of FcγRIII and hIgG1 with their respective homologues. Amino acid residues of FcγRIIIB (*upper part*) that contact amino acid residues of hFc1 (*lower part*) in the FcγRIIIB/hFc1 structure (Sondermann et al. 2000) are connected with red or blue lines for contacts of FcγRIII with CH2-A or CH2-B, respectively. The homologue FcγRs and IgGs are respectively aligned to both complex components. The degree of homology correlates with the depth of the coloring

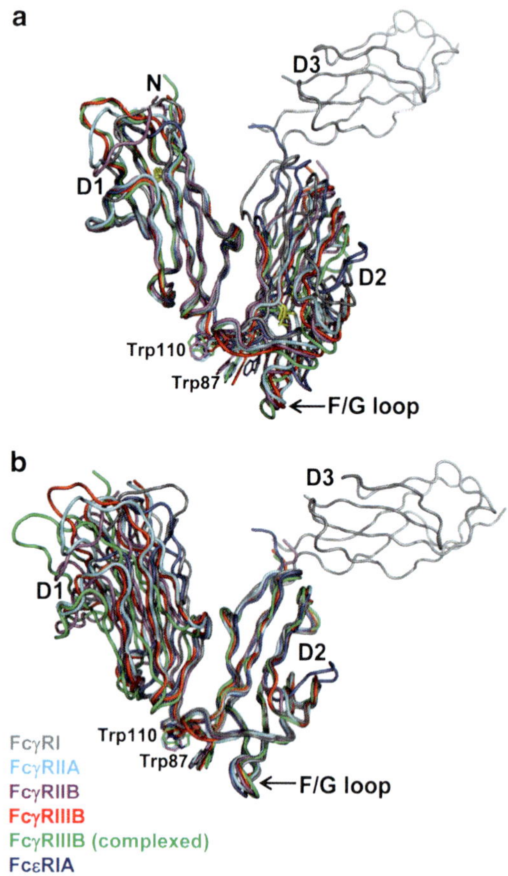

Fig. 3.5 *Superposition of FcγRs/FcεRIα*. The structures of the presented FcγRs (PDB-codes: FcγRI, 3RJD (Lu et al. 2011); FcγRIIA, 1FCG (Maxwell et al. 1999); FcγRIIB, 2FCB (Sondermann et al. 1999); FcγRIIIB, 1E4J (Sondermann et al. 2000); FcγRIIIB/Fc complex, 1E4K (Sondermann et al. 2000)) and FcεRIα (PDB-code 1F2Q (Garman et al. 1998)) were superimposed with a least square fit algorithm using either the Cα atoms of D1 (**a**) or the Cα atoms of D2 (**b**). The molecules are colored as indicated with the intradomain disulphide bridges of all proteins shown in yellow. The two tryptophan residues forming the proline sandwich in the IgG complex are depicted as sticks. For consistency reasons the numbering scheme of FcγRII has been used throughout this chapter which may differ in other publications from that of FcγRIII

Interestingly, the structure of FcγRIIB exhibits a unique intense D1/D2 interaction that is mediated by residues which interactions are conserved between the FcγRs (Sondermann et al. 1999) including the homologue FcεRIα and are representing an extremely close and steep Ig domain arrangement within one polypeptide chain. This overall heart-shaped structure was also found irrespectively of the used expression system for the other FcγRs including FcεRIα. The fold of the isolated domains is highly conserved within the receptor family as demonstrated by structural alignment of either D1 or D2; therefore, their overall structures merely differ by different interdomain angles (Fig. 3.5).

FcγRI possesses an additional third C-terminal domain (D3) that confers its nanomolar affinity towards IgG1 (Hulett and Hogarth 1998). The crystal structure shows as predicted from sequence homology that the first two domains have an identical fold and orientation to each other like that of the low-affinity FcγRs, while D3 is positioned approximately 30 Å away from the IgG binding site (Lu et al. 2011). This suggests that D3 of FcγRI is probably not directly contacting the Fc fragment but may primarily stabilize the conformation of the receptor (Fig. 3.6).

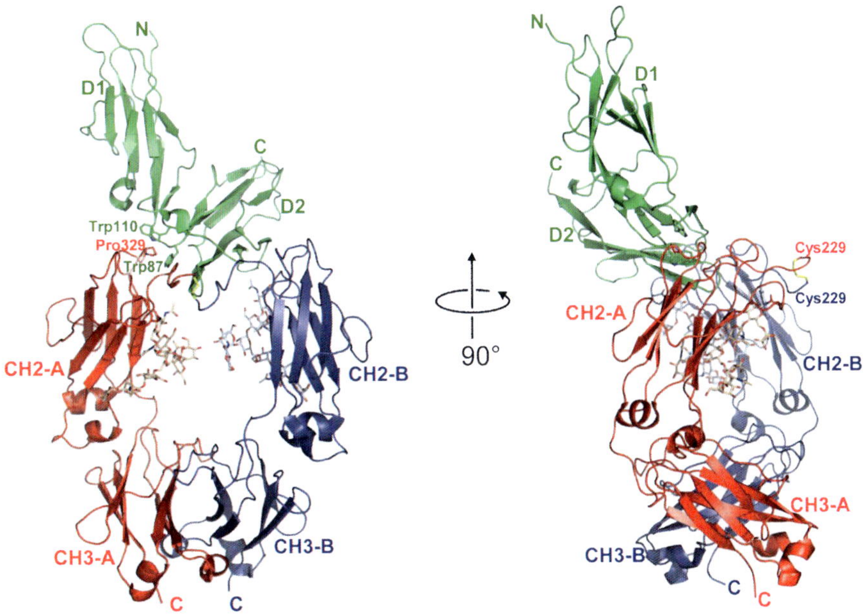

Fig. 3.6 *Structure of FcγRIII in complex with Fc (front and side view). Left*: ribbon representation with the dimer axis of hFc1 (*red* and *blue*) on the y-axis (PDB-code 1E4K (Sondermann et al. 2000)). The "proline sandwich" consisting of Pro 329 of the CH2-A domain and Trp 87 and Trp 110 of sFcγRIII (*green*) is shown in stick representation together with the glycosylation of the Fc fragment and the interchain disulphide bridge (*yellow*) of the Cys 229 residues. *Right*: side view of the complex obtained by a rotation of 90 ° around the *y*-axis as indicated by the arrow (The figure was modified from Sondermann et al. (2000))

3.6 The Structure of FcγR/Fc Complexes

3.6.1 The Structure of the FcγRIIIB/hIgG1-Fc Complex

The first available FcγR/IgG-Fc complex structure was that of unglycosylated FcγRIIIB in complex with hFc1 (Sondermann et al. 2000) and could solve the enigma of the 1:1 stoichiometry despite the dimeric character of the Fc fragment. This monomeric arrangement is mandatory to prevent a continuous activation of the immune system just by the abundant amounts of IgG. Due to the high homology within the FcγR family and within the IgG subclasses, the observed tertiary and quaternary structure is paradigmatic for all FcγR/IgG complexes and has been independently confirmed (Radaev et al. 2001) but also by complex structures of the FcεRIα/IgE-Fc (Garman et al. 2000) and the FcγRIIA-HR/hFc1 complex (Ramsland et al. 2011).

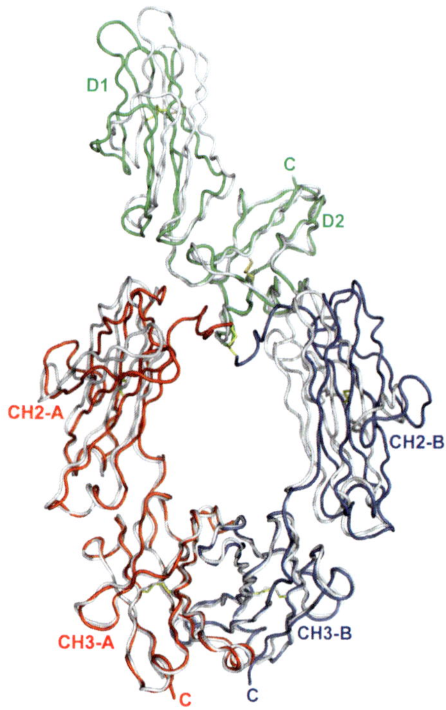

Fig. 3.7 *Overlay bound and unbound Fc*. Superposition of the sFcγRIIIB/hFc1 complex structure with that obtained for the free FcγRIII and hFc1. The alignment of the structures was performed with a least square algorithm using the Cα atoms. For hFc1 the residues of both CH3 domains and for FcγRIIIB only that of D2 were used. Orientation of the complex and color scheme is the same as in Fig. 3.6 with the free uncomplexed proteins depicted in *gray*

In the FcγRIIIB/hFc1 complex the receptor contacts the Fc fragment with the tip of the heart-shaped molecule formed by the C-terminal region of D1 and the N-terminal part of D2. This region with the protruding F/G loop inserts into the opening of the horseshoe-shaped Fc fragment where the top of both CH2 domains is contacted but without directly interfering with the C2 symmetry axis (Fig. 3.3b) of the Fc fragment (Fig. 3.6). Such an arrangement would principally allow a second receptor to bind oppositely to the first one. Consequently, the areas and therefore also the individual residues that are contacted in both CH2 domains are different. The possible 2:1 stoichiometry is however prevented by the hinge peptides that are bent in direction of the potential second binding site thereby breaking the symmetry and blocking the binding of a second receptor to the Fc (Fig. 3.6). At the same time the FcγR binding is accompanied by an asymmetric "opening" of the Fc fragment which is mediated by the elbow hinge located between the CH2 and CH3 domains. Compared to the unbound Fc fragment, the distance between both tips measured at the Pro329 residues widens by 7 Å (Sondermann et al. 2000) indicating the substantial rearrangement which is more pronounced in the B chain within the upper Fc fragment, while the structure of the CH3 dimer is almost not affected by FcγR binding (Fig. 3.7). Also the sFcγRIII molecule experiences an opening of its interdomain angle from 70° to 80° (Fig. 3.7 (Sondermann et al. 2000)) which is facilitated by the presence of only two hydrogen bonds between D1 and D2, while for the interdomain contacts in the FcγRII isoforms, seven hydrogen

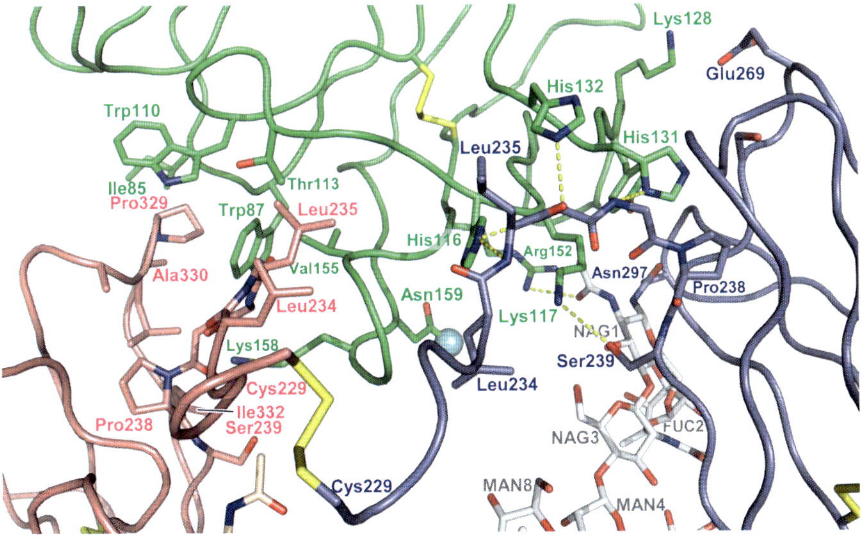

Fig. 3.8 *The binding site of the FcγRIIIB/IgG1-Fc complex.* A detailed view of the residues mediating the sFcγRIIIB/hFc1 interaction in an orientation as the previous figure using the same coloring scheme. The backbone residues of sFcγRIIIB are colored green, while that of CH2-A and CH2-B are red and blue, respectively. Potential hydrogen bonds between sFcγRIII and hFc1 are shown as dashed lines. The N-glycosylation site at Asn159 that is not occupied in this complex due to use of an unglycosylated FcγRIIIB is depicted as a cyan-colored ball. The interchain disulphide bridge between both Cys229 (*yellow*) and residues relevant for the interaction are labeled and depicted as sticks. NAG1: N-Acetylglucosamine bound to Asn297 (see Fig. 3.2) (The figure was modified from Sondermann et al. (2000))

bonds are found (Sondermann et al. 1999). The weak interdomain interaction in FcγRIII leads to high unfavorable entropic changes upon Fc binding which is different from the more rigid FcγRII forms (Maenaka et al. 2001).

Within the complex, three interaction sites of the FcγR with the Fc fragment can be defined (Fig. 3.8). Interaction site 1 is formed by contacts of both receptor domains with the F/G loop of the one of the CH2 domains. Central to this interaction is the "proline sandwich," a motif conserved in all FcγRs (including FcεRIα) and IgG molecules consisting of Trp87 and Trp110 of the receptor that wedges Pro329 of the Fc fragment. This motif seems to be the primary docking site of the complex, and change of the loop's substructure results in a substantial affinity loss (Canfield and Morrison 1991).

Binding site 2 has a significantly larger interaction area that is formed mainly by FcγR's side chains of residues from the C' strand that contact the second CH2 domain's B/C and C'/E loop which also contains the Fc glycosylation site Asn297. In addition, marginal contacts to the carbohydrates exist. As these are fully solvent exposed, they most likely do not contribute significantly to the affinity of the complex.

Interaction site 3 involves all contacts of the hinge peptides that are in the majority mediated by the B/C and the F/G loop and the C' strand of the receptors' second domain. Both hinge peptides have to bend sharply to allow a close contact with the receptor. Only glycine residues present at position 236 and 237 can adopt phi/psi angles that allow such an orientation which is prohibited for other residues. The affinity between the hinge peptides and the receptor is mainly mediated by Leu235 of both hinge peptides with an inferior contribution of Leu234. An exchange of Leu234 for Phe as in IgG4 (Spiegelberg et al. 1976) or that of both Leu residues for Ala (Woodle et al. 1999) results in a significant affinity reduction to FcγRs.

3.6.2 *Modeling of Other FcγR/IgG-Fc Complexes*

Due to the high degree of conservation of FcγRs and the IgG subclasses, the modeling of FcγR/Fc complexes is possible that allows an explanation of the resulting affinities for several of the complexes (Sondermann et al. 2001). For example, in principle low affinity of human IgG2 to FcγRs can be deduced from the complex structure. In hIgG2 the Gly-Gly motif of human IgG1 at position 236/237 is exchanged for Ala-Gly, and the preceding Leu-Leu motif at position 234/235 is exchanged for Pro-Val (Fig. 3.4). Both exchanges contribute to the observed low affinity of human IgG2 for all FcγRs as discussed in Sect. 3.6.1. An exemption represents the low-responder (LR) allele of FcγRIIA which carries a histidine residue at position 131 and that binds hIgG2 significantly better than the high-responder (HR) allele which carries an Arg at that position. As the histidine is smaller than arginine, the less bent hinge peptide of hIgG2 can be better accommodated leading to less sterical hindrance (Fig. 3.8). On the other hand the HR allele designated after its stronger interaction with murine IgG1 (Tax et al. 1984) can form a productive interaction to this immunoglobulin. In such an arrangement the turn of the hinge peptide is more relaxed due to the exchange of the rigid proline at position 238 for a serine that may even allow a productive interaction by the formation of a salt bridge between Glu237 from mIgG1 and Arg131 of the FcγRIIA-HR allele (Fig. 3.8). The described LR/HR dimorphism of FcγRIIA may have implications for disease susceptibility when the hIgG2 isotype predominates the immune response as observed for some viral infections. The occurrence of H. influenza infections is statistically lower in the Japanese population (Musser et al. 1990), who carry a higher frequency of the LR-allele (Clark et al. 1991).

A further clinical relevant dimorphic allelic variation of an FcγR is that of FcγRIII at position 155 (which is in a different but frequently used numbering scheme position 158). Recipients of monoclonal antibody therapies that rely on ADCC which is mainly mediated by FcγRIIIA (Bowles et al. 2006) exhibit a significant lower response in case they are homo- or heterozygous carrier of the FcγRIIIA-Phe155 allele (Cartron et al. 2002; Musolino et al. 2008; Louis et al. 2004). A phenylalanine at this position lowers the affinity for IgG1 to a tenth of the respective valine variant (FcγRIIIA-V155, (Ferrara et al. 2006)) which suggests a correlation of the treatment response with the FcγRIIIA affinity.

The siginificant lower affinity of FcγRIIIA-Phe155 to hIgG1 can not easily be explained. In the FcγRIII/IgG1-Fc complex, Val155 contacts Leu235 of the A chain (Fig. 3.8). Due to its location on the rim of the binding site, phenylalanine residue at this position could in principle be accommodated without sterical clashes. However, Phe155 is also closely located to Trp87 which is part of the "proline sandwich" where it could destabilize this important structural motif which could also be a reason for the observed reduced affinity of this variant to IgG1.

3.6.3 *Design of Immunoglobulin Mutants with Enhanced Affinity to FcγRs*

Using the FcγRIII/Fc complex also immunoglobulins can be designed that exhibit a different binding pattern for FcγRs. Such IgG variants might be advantageous for therapeutic use, e.g., by mediating enhanced ADCC via FcγRIIIA. One example encompasses the IgG1 mutant Ser239Asp/Ala330Leu/Ile332Glu that is described to have a more than 100-fold higher affinity to both allelic variants of FcγRIIIA at position 155 (Val/Phe) than an unmodified antibody (Lazar et al. 2006). The aspartate at position 239 on both chains would form salt bridges with Lys117 and Lys158 (Fig. 3.8). The other two mutations only affect the interaction of the A chain with the receptor, while the corresponding positions on the B chain are distant from the receptor. The leucine at position 330 and a glutamate at position 332 would consequently interact productively with the receptor's residues Lys158 and Ile85, respectively (Fig. 3.8).

In a similar approach antibodies with mutated Fc regions have been generated that preferentially bind to FcγRIIB (Chu et al. 2008). Such variants might be useful for the treatment of immune disorders. In addition, nonglycosylated IgG molecules have been described that are able to rescue the FcγR affinity (Sazinsky et al. 2008; Jung et al. 2010). Such antibodies might be of relevance for biotechnological applications due to their facilitated prokaryotic expression and the reduced number of glycovariants that in general impose a regulatory hurdle for human use.

3.6.4 *The Influence of Fc Carbohydrates on FcγR affinity Affinity*

The pharmacological effect of intravenous IgG therapy (IVIg) was demonstrated to depend on the fully sialylated fraction of IgG (Kaneko et al. 2006). In preclinical studies the therapeutic efficacy was correlating to the content of sialic acid bound α2,6 to the preceding galactose, while α2,3-linked sialic acid showed no effect in these models. At the same time the affinity of terminally α2,6-sialylated IgGs to all FcγRs is strongly reduced (Kaneko et al. 2006), but such sialylated Fc fragments gain affinity for the anti-inflammatory C-type lectin DC-SIGN (Anthony et al. 2008a).

The identification of the molecular basis of the affinity drop for FcγRs is difficult as the availability of Fc structures that contain sialic acid residues is very limited.

The only crystal structure of an uncomplexed Fc with traceable sialic acid residues was obtained from the hinge-deleted human IgG1 "Mcg" (Deutsch and Suzuki 1971; Guddat et al. 1993). In this structure (PDB-code 1MCO), the sialic acid is only traceable on the α1,6-arm and indicates a similar CH2 carbohydrate interaction as observed for nonsialylated IgG, but the sialic acid was built into the model as α2,3 linked to the preceding galactose residue. This linkage is underrepresented in pooled human IgG where the sialic acid residues are merely α2,6 linked ((Anthony et al. 2008b), Fig. 3.2a).

A second structure of a sialylated Fc fragment exists, but it is present in a complexed form with FcγRIIA (PDB-code 3RY6, (Ramsland et al. 2011)). The Fc fragment found in this structure of the FcγRIIA-HR/hFc1 complex contains an entirely traceable fully α2,6-sialylated carbohydrate structure on the CH2 domain that interacts with binding site 2 of the receptor (Sect. 3.6.1). The interaction of this carbohydrate with the protein does not interfere with the FcγR binding site but exhibits a significantly different association to the CH2 domain including a reorientation of the C'/E loop. The structure of the other CH2 domain that interacts with FcγRIIA in the "proline sandwich" does not contain a traceable sialylation on the α1,6-arm. Therefore, it remains elusive if the observed structural reorientation is induced by the FcγRIIA-HR binding or if it is a result of the sialylation itself which could explain the generally lowered FcγR affinity of sialylated IgG.

Also the presence or absence of other single monosaccharide units on the Fc glycosylation structure can dramatically impact affinity and activity of the Fc region. It is established that in some diseases, the glycosylation pattern of IgG is changing. For example, patients suffering from autoimmune diseases exhibit a reduced level of galactosylated antibodies (Holland et al. 2002; van Zeben et al. 1994; Parekh et al. 1985; Bond et al. 1997) which exhibit an enhanced affinity for C1q, while the affinity to FcγRs remains unchanged (Hodoniczky et al. 2005). The level of galactosylated antibodies is elevated in patients suffering from rheumatoid arthritis during remission and also in affected persons during pregnancy (Arnold et al. 2007; Rook et al. 1991) which correlates with a reduction of disease symptoms.

A further modification of the N-glycosylation pattern in the Fc fragment relates to the fucosylation on the first carbohydrate residue (Fig. 3.2a) which is present in around 70 % of the serum IgGs (Butler et al. 2003). The lack of the core fucose residue in IgG1 molecules leads to about 50-fold affinity increase for FcγRIIIA (Okazaki et al. 2004), while the affinity to the other FcγRs remains unaltered. This higher affinity to FcγRIIIA translates also into enhanced ADCC (Umaña et al. 1999). Surface plasmon resonance measurements identified that this high-affinity binding requires in addition to the nonfucosylated Fc also at Asn159 (in a different numbering scheme Asn162) glycosylated FcγRIIIA (Ferrara et al. 2006).

The crystal structure of the isolated nonfucosylated Fc turned out to be very similar to that of fucosylated Fc except for the hydration mode around Tyr296 preceding the glycosylation site Asn297 (Matsumiya et al. 2007). The crystal structure

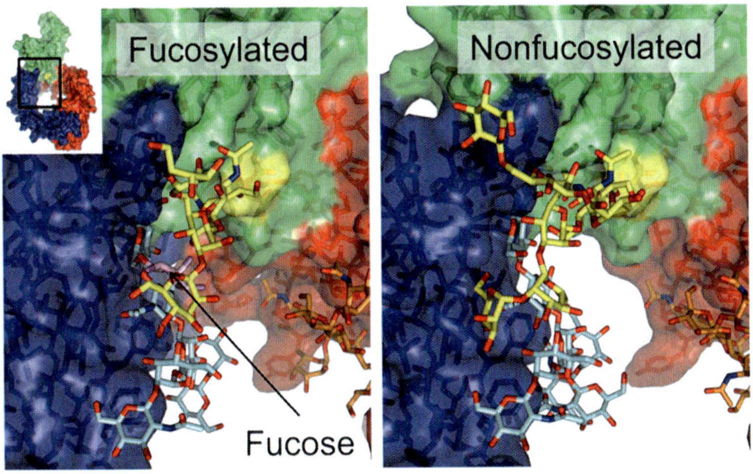

Fig. 3.9 *Complex of glycosylated FcγRIIIA with afucosylated Fc.* Clipping of the binding site of glycosylated FcγRIIIA/IgG1-Fc complex as shown in the inset marked by a rectangle. The protein moieties are depicted as semitransparent surface with sFcγRIIIA in green and CH2/CH3-A and CH2/CH3-B in red and blue, respectively, maintaining the coloring scheme from the previous figures. Note that the orientation is different, and the complex is seen from the "back". *Left*: the interaction of the FcγRIII glycosylation structure attached at Asn159 with native (fucosylated) IgG is impaired by the fucose residue (*magenta*) that prevents the optimum interaction with the Fc (PDB-code 3SGJ (Ferrara et al. 2011)). *Right*: in the complex of glycosylated FcγRIIIA with non-fucosylated hFc1 (PDB-code 3SGK (Ferrara et al. 2011)), the interaction of the receptor's carbohydrate moiety with the CH2 domain is more intense leading to a 50-fold higher affinity of the two proteins

obtained for the complex of nonfucosylated IgG-Fc and in position Asn159-glycosylated FcγRIIIA demonstrated that the carbohydrate attached at Asn159 interacts in fact with the Fc in the region where the fucose sticks out of the continuous surface of the CH2 domain (Fig. 3.9, (Ferrara et al. 2011)). If the fucose is present on the Fc, the interaction of the FcγRIII-carbohydrate with the Fc is impaired. Consequently, the affinity of Asn159-glycosylated FcγRIIIA to fucosylated Fc is reduced compared to the nonglycosylated receptor. This reduction is likely to be caused by the required displacement of the carbohydrate structure out of the contact interface during complex formation (Ferrara et al. 2006; Drescher et al. 2003) that is obviously not compensated by the minor interaction of the receptor's carbohydrate with the CH2 domain.

FcγRIIIA found in other species carries like the human alleles several N-glycosylation sites that are distributed over the protein. Even between rather related species, the position of the glycosylation sites varies with the exemption of that located at Asn159 which is conserved in all species that express FcγRIII that corresponds in mouse to FcγRIV (Fig. 3.10).

	Amino acid position in FcγRIIIA 132 140 150 159 169
Man (Val155)	HNSDFYIPKATLKDSGSYFCRGLVGSKNVSSETVNITI
Macaca	QNSDFYIPKATLKDSGSYFCRGLIGSKNVSSETVNITI
Marmoset	QNSDFFIPKATLKDNGSYFCRGLVGGKNVSSESVNITV
Dog	NNSEYHIPAATSEHNGSYFCRGIIGKKNESSEAVNIII
Cat	KNSDFYIPKATSKHSGSYFCRGLIGNKNESSEAVNITV
Ferret	GNSEFYIPAARSEHNGSYFCRGLIGKRNESSEAVDIII
Rabbit	QNSDLHIPEATRNHSGSYFCRGLIGHHNMSSETVTITV
Mouse (FcγRIV)	ENSELLIPKATHNDSGSYFCRGLIGHNNKSSASFRISL
Hamster	KNSELHIPNATQNHSGSYFCRGIIGRNNKSSETLRITV
Pig	QNFEYHIPNATLKDGGSYFCRGIIKNYNLSSEAVKVTV
Cattle	GNSDFHIPEAKLEHSGSYFCRGIIGSKNESSESVQITV
Rat	RNSELSISKATHADSGSYFCRGIIGRNNISSASLQISI
Panda	TDSEFHIPAAASKHSGSYFCRGLIGKKNESSEAVDIII

Fig. 3.10 *Sequence alignment of FcγRIII of different species*. The FcγRIII proteins from various mammals were taken from the protein database, and the stretch of amino acid 132–169 was aligned. The cysteine residue involved in disulphide bridge formation in D2 is colored red, while potential N-glycosylation sites are colored green. The N-glycosylation site present at position159 (blue) is conserved in all FcγRIII-like proteins identified up to date

This surprising finding leads to the intriguing hypothesis that the glycosylation at Asn159 is used by the immune system to tune the affinity of the FcγRIII/IgG interaction by a modulation of the fucosylation content of the produced IgG. Antibodies against human platelet antigens (HPAs) can be formed during pregnancy and cross the placenta where they cause fetal-maternal alloimmune thrombocytopenia (FMAIT). Affinity-purified IgG1 anti-HPA alloantibodies were found to have markedly less fucose and slightly higher galactose which enhances their pathogenicity, compared with normal serum IgG1 from the same donors (Wuhrer et al. 2009). This strongly suggests that the level of pathogenic nonfucosylated antibodies is actively modulated. This enhanced antibody activity may have evolved for the protection of pregnant women from pathogens while their immune systems are in a state of mild systemic immunoregulation (Kumpel and Manoussaka 2012).

Similarly, FcγRIII has been found to be diversely glycosylated on different cell types resulting in differential ligand binding (Edberg and Kimberly 1997). This finding reveals that the immune system possesses in principle the capability also to modulate the affinity to IgG on the receptor side. In summary, the conservation of the glycosylation site Asn159 in FcγRIII, its impact on IgG binding and the enhanced affinity to afucosylated IgG may provide an additional modulator for tuning the antibody mediated immune response.

3.7 Conclusions

The binding to FcγRs represents a central event in the antibody-mediated immune response. Our current knowledge suggests that the immune system has developed a toolbox that allows the tuning of the immune response by modulation of the interaction of the FcγR/IgG complex in terms of specificity and strength. In addition to allelic variation on the FcγR side in distinct populations, the variation of the IgG glycosylation pattern seems to be a versatile tool. Addition of sialic acid results in a general reduction of the affinity towards FcγRs accompanied by converting cytotoxic potential into anti-inflammatory activity. On the other hand, the lack of fucose generates antibodies that possess an extremely high potency, a principle that is started also to be exploited therapeutically (Mössner et al. 2010). Although the primary sequences of immunoglobulins and their receptors are known since decades, their impact on structure and consequently onto function has been identified only in the last few years which suggests that antibodies as one of the best investigated proteins still have the potential to surprise us also in the future with new and so far unknown functionalities.

References

Anthony RM, Wermeling F, Karlsson MC, Ravetch JV (2008a) Identification of a receptor required for the anti-inflammatory activity of IVIG. Proc Natl Acad Sci USA 105(50):19571–19578

Anthony RM, Nimmerjahn F, Ashline DJ, Reinhold VN, Paulson JC, Ravetch JV (2008b) Recapitulation of IVIG anti-inflammatory activity with a recombinant IgG Fc. Science 320(5874):373–376

Arnold JN, Wormald MR, Sim RB, Rudd PM, Dwek RA (2007) The impact of glycosylation on the biological function and structure of human immunoglobulins. Annu Rev Immunol 25:21–50

Bond A, Alavi A, Axford JS, Bourke BE, Bruckner FE, Kerr MA et al (1997) A detailed lectin analysis of IgG glycosylation, demonstrating disease specific changes in terminal galactose and N-acetylglucosamine. J Autoimmun 10(1):77–85

Bork P, Holm L, Sander C (1994) The immunoglobulin fold. Structural classification, sequence patterns and common core. J Mol Biol 242(4):309–320

Bowles JA, Wang SY, Link BK, Allan B, Beuerlein G, Campbell MA et al (2006) Anti-CD20 monoclonal antibody with enhanced affinity for CD16 activates NK cells at lower concentrations and more effectively than rituximab. Blood 108(8):2648–2654

Butler M, Quelhas D, Critchley AJ, Carchon H, Hebestreit HF, Hibbert RG et al (2003) Detailed glycan analysis of serum glycoproteins of patients with congenital disorders of glycosylation indicates the specific defective glycan processing step and provides an insight into pathogenesis. Glycobiology 13(9):601–622

Cabanes-Macheteau M, Fitchette-Lainé AC, Loutelier-Bourhis C, Lange C, Vine ND, Ma JK et al (1999) N-Glycosylation of a mouse IgG expressed in transgenic tobacco plants. Glycobiology 9(4):365–372

Canfield SM, Morrison SL (1991) The binding affinity of human IgG for its high affinity Fc receptor is determined by multiple amino acids in the CH2 domain and is modulated by the hinge region. J Exp Med 173(6):1483–1491

Cartron G, Dacheux L, Salles G, Solal-Celigny P, Bardos P, Colombat P et al (2002) Therapeutic activity of humanized anti-CD20 monoclonal antibody and polymorphism in IgG Fc receptor FcγRIIIa gene. Blood 99(3):754–758

Chothia C, Hubbard T, Brenner S, Barns H, Murzin A (1997) Protein folds in the all-beta and all-alpha classes. Annu Rev Biophys Biomol Struct 26:597–627

Chu SY, Vostiar I, Karki S, Moore GL, Lazar GA, Pong E (2008) Inhibition of B cell receptor-mediated activation of primary human B cells by coengagement of CD19 and FcγRIIb with Fc-engineered antibodies. Mol Immunol 45(15):3926–3933

Clark MR, Stuart SG, Kimberly RP, Ory PA, Goldstein IM (1991) A single amino acid distinguishes the high-responder from the low-responder form of Fc receptor II on human monocytes. Eur J Immunol 21(8):1911–1916

Deisenhofer J (1981) Crystallographic refinement and atomic models of a human Fc fragment and its complex with fragment B of protein A from Staphylococcus aureus at 2.9- and 2.8-Å resolution. Biochemistry 20(9):2361–70

Deutsch HF, Suzuki T (1971) A crystalline G1 human monoclonal protein with an excessive H chain deletion. Ann N Y Acad Sci 190:472–486

Drescher B, Witte T, Schmidt RE (2003) Glycosylation of FcγRIII in N163 as mechanism of regulating receptor affinity. Immunology 110(3):335–340

Edberg JC, Kimberly RP (1997) Cell type-specific glycoforms of FcγRIIIa (CD16): differential ligand binding. J Immunol 159(8):3849–3857

Feige MJ, Nath S, Catharino SR, Weinfurtner D, Steinbacher S, Buchner J (2009) Structure of the murine unglycosylated IgG1 Fc fragment. J Mol Biol 391(3):599–608

Ferrara C, Stuart F, Sondermann P, Brünker P, Umaña P (2006) The carbohydrate at FcγRIIIa Asn-162. An element required for high affinity binding to non-fucosylated IgG glycoforms. J Biol Chem 281(8):5032–5036

Ferrara C, Grau S, Jäger C, Sondermann P, Brünker P, Waldhauer I et al (2011) Unique carbohydrate-carbohydrate interactions are required for high affinity binding between FcγRIII and antibodies lacking core fucose. Proc Natl Acad Sci USA 108(31):12669–12674

Garman SC, Kinet JP, Jardetzky TS (1998) Crystal structure of the human high-affinity IgE receptor. Cell 95(7):951–61

Garman SC, Wurzburg BA, Tarchevskaya SS, Kinet JP, Jardetzky TS (2000) Structure of the Fc fragment of human IgE bound to its high-affinity receptor FcεRIα. Nature 406(6793):259–266

Guddat LW, Herron JN, Edmundson AB (1993) Three-dimensional structure of a human immunoglobulin with a hinge deletion. Proc Natl Acad Sci USA 90(9):4271–4275

Harris LJ, Larson SB, Hasel KW, Day J, Greenwood A, McPherson A (1992) The three-dimensional structure of an intact monoclonal antibody for canine lymphoma. Nature 360(6402):369–372

Harris LJ, Skaletsky E, McPherson A (1998) Crystallographic structure of an intact IgG1 monoclonal antibody. J Mol Biol 275(5):861–872

Hodoniczky J, Zheng YZ, James DC (2005) Control of recombinant monoclonal antibody effector functions by Fc N-glycan remodeling *in vitro*. Biotechnol Prog 21(6):1644–1652

Hogarth PM, Hulett MD, Ierino FL, Tate B, Powell MS, Brinkworth RI (1992) Identification of the immunoglobulin binding regions (IBR) of FcγRII and FcεRI. Immunol Rev 125:21–35

Holland M, Takada K, Okumoto T, Takahashi N, Kato K, Adu D et al (2002) Hypogalactosylation of serum IgG in patients with ANCA-associated systemic vasculitis. Clin Exp Immunol 129(1):183–190

Huang L, Biolsi S, Bales KR, Kuchibhotla U (2006) Impact of variable domain glycosylation on antibody clearance: an LC/MS characterization. Anal Biochem 349(2):197–207

Huber R, Deisenhofer J, Colman PM, Matsushima M, Palm W (1976) Crystallographic structure studies of an IgG molecule and an Fc fragment. Nature 264(5585):415–420

Hulett MD, Hogarth PM (1998) The second and third extracellular domains of FcγRI (CD64) confer the unique high affinity binding of IgG2a. Mol Immunol 35(14–15):989–996

Hulett MD, Witort E, Brinkworth RI, McKenzie IF, Hogarth PM (1994) Identification of the IgG binding site of the human low affinity receptor for IgG FcγRII. Enhancement and ablation of binding by site-directed mutagenesis. J Biol Chem 269(21):15287–15293

Hulett MD, Witort E, Brinkworth RI, McKenzie IF, Hogarth PM (1995) Multiple regions of human FcγRII (CD32) contribute to the binding of IgG. J Biol Chem 270(36):21188–21194

Jefferis R, Lund J, Mizutani H, Nakagawa H, Kawazoe Y, Arata Y et al (1990a) A comparative study of the N-linked oligosaccharide structures of human IgG subclass proteins. Biochem J 268(3):529–537

Jefferis R, Lund J, Pound J (1990b) Molecular definition of interaction sites on human IgG for Fc receptors (huFcγR). Mol Immunol 27(12):1237–1240

Jefferis R, Lund J, Pound JD (1998) IgG-Fc-mediated effector functions: molecular definition of interaction sites for effector ligands and the role of glycosylation. Immunol Rev 163:59–76

Jung ST, Reddy ST, Kang TH, Borrok MJ, Sandlie I, Tucker PW (2010) Aglycosylated IgG variants expressed in bacteria that selectively bind FcγRI potentiate tumor cell killing by monocyte-dendritic cells. Proc Natl Acad Sci USA 107(2):604–609

Kaneko Y, Nimmerjahn F, Ravetch JV (2006) Anti-inflammatory activity of immunoglobulin G resulting from Fc sialylation. Science 313(5787):670–673

Krapp S, Mimura Y, Jefferis R, Huber R, Sondermann P (2003) Structural analysis of human IgG-Fc glycoforms reveals a correlation between glycosylation and structural integrity. J Mol Biol 325(5):979–989

Kumpel BM, Manoussaka MS (2012) Placental immunology and maternal alloimmune responses. Vox Sang 102(1):2–12

Labrijn AF, Buijsse AO, van den Bremer ET, Verwilligen AY, Bleeker WK, Thorpe SJ et al (2009) Therapeutic IgG4 antibodies engage in Fab-arm exchange with endogenous human IgG4 in vivo. Nat Biotechnol 27(8):767–771

Labrijn AF, Rispens T, Meesters J, Rose RJ, den Bleker TH, Loverix S et al (2011) Species-specific determinants in the IgG CH3 domain enable Fab-arm exchange by affecting the non-covalent CH3-CH3 interaction strength. J Immunol 187(6):3238–3246

Lazar GA, Dang W, Karki S, Vafa O, Peng JS, Hyun L et al (2006) Engineered antibody Fc variants with enhanced effector function. Proc Natl Acad Sci USA 103(11):4005–4010

Louis E, El Ghoul Z, Vermeire S, Dall'Ozzo S, Rutgeerts P, Paintaud G et al (2004) Association between polymorphism in IgG Fc receptor IIIa coding gene and biological response to infliximab in Crohn's disease. Aliment Pharmacol Ther 19(5):511–519

Lu J, Ellsworth JL, Hamacher N, Oak SW, Sun PD (2011) Crystal structure of Fcγ receptor I and its implication in high affinity γ-immunoglobulin binding. J Biol Chem 286(47):40608–40613

Lund J, Takahashi N, Popplewell A, Goodall M, Pound JD, Tyler R et al (1990) Expression and characterization of truncated forms of humanized L243 IgG1. Architectural features can influence synthesis of its oligosaccharide chains and affect superoxide production triggered through human Fcγ receptor I. Eur J Biochem 267(24):7246–7257

Maenaka K, van der Merwe PA, Stuart DI, Jones EY, Sondermann P (2001) The human low affinity Fcγ receptors IIa, IIb, and III bind IgG with fast kinetics and distinct thermodynamic properties. J Biol Chem 276(48):44898–44904

Matsumiya S, Yamaguchi Y, Saito J, Nagano M, Sasakawa H, Otaki S et al (2007) Structural comparison of fucosylated and nonfucosylated Fc fragments of human immunoglobulin G1. J Mol Biol 368(3):767–779

Maxwell KF, Powell MS, Hulett MD, Barton PA, McKenzie IF, Garrett TP et al (1999) Crystal structure of the human leukocyte Fc receptor, FcgRIIa. Nat Struct Biol 6(5):437–42

Mimura Y, Church S, Ghirlando R, Ashton PR, Dong S, Goodall M et al (2000) The influence of glycosylation on the thermal stability and effector function expression of human IgG1-Fc: properties of a series of truncated glycoforms. Mol Immunol 37(12–13):697–706

Mimura Y, Sondermann P, Ghirlando R, Lund J, Young SP, Goodall M (2001) Role of oligosaccharide residues of IgG1-Fc in FcγRIIb binding. J Biol Chem 276(49):45539–45547

Mössner E, Brünker P, Moser S, Püntener U, Schmidt C, Herter S et al (2010) Increasing the efficacy of CD20 antibody therapy through the engineering of a new type II anti-CD20 antibody with enhanced direct and immune effector cell-mediated B-cell cytotoxicity. Blood 115(22):4393–4402

Musolino A, Naldi N, Bortesi B, Pezzuolo D, Capelletti M, Missale G et al (2008) Immunoglobulin G fragment C receptor polymorphisms and clinical efficacy of trastuzumab-based therapy in patients with HER-2/neu-positive metastatic breast cancer. J Clin Oncol 26(11):1789–1796

Musser JM, Kroll JS, Granoff DM, Moxon ER, Brodeur BR, Campos J (1990) Global genetic structure and molecular epidemiology of encapsulated Haemophilus influenzae. Rev Infect Dis 12(1):75–111

Okazaki A, Shoji-Hosaka E, Nakamura K, Wakitani M, Uchida K, Kakita S et al (2004) Fucose depletion from human IgG1 oligosaccharide enhances binding enthalpy and association rate between IgG1 and FcγRIIIa. J Mol Biol 336(5):1239–1249

Olafsen T (2012) Fc engineering: serum half-life modulation through FcRn binding. Methods Mol Biol 907:537–556

Parekh RB, Dwek RA, Sutton BJ, Fernandes DL, Leung A, Stanworth D et al (1985) Association of rheumatoid arthritis and primary osteoarthritis with changes in the glycosylation pattern of total serum IgG. Nature 316(6027):452–457

Qian J, Liu T, Yang L, Daus A, Crowley R, Zhou Q (2007) Structural characterization of N-linked oligosaccharides on monoclonal antibody cetuximab by the combination of orthogonal matrix-assisted laser desorption/ionization hybrid quadrupole-quadrupole time-of-flight tandem mass spectrometry and sequential enzymatic digestion. Anal Biochem 364(1):8–18

Radaev S, Motyka S, Fridman WH, Sautes-Fridman C, Sun PD (2001) The structure of a human type III Fcγ receptor in complex with Fc. J Biol Chem 276(19):16469–16477

Ramsland PA, Farrugia W, Bradford TM, Sardjono CT, Esparon S, Trist HM (2011) Structural basis for FcγRIIa recognition of human IgG and formation of inflammatory signaling complexes. J Immunol 187(6):3208–3217

Rook GA, Steele J, Brealey R, Whyte A, Isenberg D, Sumar N et al (1991) Changes in IgG glycoform levels are associated with remission of arthritis during pregnancy. J Autoimmun 4:779–794

Rose RJ, Labrijn AF, van den Bremer ET, Loverix S, Lasters I, van Berkel PH et al (2011) Quantitative analysis of the interaction strength and dynamics of human IgG4 half molecules by native mass spectrometry. Structure 19(9):1274–1282

Rutishauser U, Cunningham BA, Bennett C, Konigsberg WH, Edelman GM (1970) The covalent structure of a human gamma G-immunoglobulin. 8. amino acid sequence of heavy-chain cyanogen bromide fragments H5–H7. Biochemistry 9(16):3171–3181

Sandin S, Ofverstedt LG, Wikström AC, Wrange O, Skoglund U (2004) Structure and flexibility of individual immunoglobulin G molecules in solution. Structure 12(3):409–415

Saphire EO, Parren PW, Pantophlet R, Zwick MB, Morris GM, Rudd PM, Dwek RA, Stanfield RL, Burton DR, Wilson IA (2001) Crystal structure of a neutralizing human IgG against HIV-1: a template for vaccine design. Science 293(5532):1155–1159

Saphire EO, Stanfield RL, Crispin MD, Parren PW, Rudd PM, Dwek RA et al (2002) Contrasting IgG structures reveal extreme asymmetry and flexibility. J Mol Biol 319(1):9–18

Sarmay G, Lund J, Rozsnyay Z, Gergely J, Jefferis R (1992) Mapping and comparison of the interaction sites on the Fc region of IgG responsible for triggering antibody dependent cellular cytotoxicity (ADCC) through different types of human Fcγ receptor. Mol Immunol 29(5):633–639

Sazinsky SL, Ott RG, Silver NW, Tidor B, Ravetch JV, Wittrup KD (2008) Aglycosylated immunoglobulin G1 variants productively engage activating Fc receptors. Proc Natl Acad Sci USA 105(51):20167–20172

Sondermann P, Huber R, Jacob U (1999) Crystal structure of the soluble form of the human Fcγ-receptor IIb: a new member of the immunoglobulin superfamily at 1.7 A resolution. EMBO J 18(5):1095–1103

Sondermann P, Huber R, Oosthuizen V, Jacob U (2000) The 3.2-Å crystal structure of the human IgG1 Fc fragment-FcγRIII complex. Nature 406(6793):267–273

Sondermann P, Kaiser J, Jacob U (2001) Molecular basis for immune complex recognition: a comparison of Fc-receptor structures. J Mol Biol 309(3):737–749

Spiegelberg HL, Perlmann H, Perlmann P (1976) Interaction of K lymphocytes with myeloma proteins of different IgG subclasses. J Immunol 117(5 Pt 1):1464–1471

Tamm A, Kister A, Nolte KU, Gessner JE, Schmidt RE (1996) The IgG binding site of human FcγRIIIB receptor involves CC' and FG loops of the membrane-proximal domain. J Biol Chem 271(7):3659–3666

Tao MH, Morrison SL (1989) Studies of aglycosylated chimeric mouse-human IgG. Role of carbohydrate in the structure and effector functions mediated by the human IgG constant region. J Immunol 143(8):2595–2601

Tax WJ, Hermes FF, Willems RW, Capel PJ, Koene RA (1984) Fc receptors for mouse IgG1 on human monocytes: polymorphism and role in antibody-induced T cell proliferation. J Immunol 133(3):1185–1189

Terry WD, Matthews BW, Davies DR (1968) Crystallographic studies of a human immunoglobulin. Nature 220(5164):239–241

Umaña P, Jean-Mairet J, Moudry R, Amstutz H, Bailey JE (1999) Engineered glycoforms of an antineuroblastoma IgG1 with optimized antibody-dependent cellular cytotoxic activity. Nat Biotechnol 17(2):176–180

van der Neut KM, Schuurman J, Losen M, Bleeker WK, Martínez-Martínez P, Vermeulen E et al (2007) Anti-inflammatory activity of human IgG4 antibodies by dynamic Fab arm exchange. Science 317(5844):1554–1557

van Zeben D, Rook GA, Hazes JM, Zwinderman AH, Zhang Y, Ghelani S et al (1994) Early agalactosylation of IgG is associated with a more progressive disease course in patients with rheumatoid arthritis: results of a follow-up study. Br J Rheumatol 33(1):36–43

von Bonsdorff B, Groth H, Packalén T (1938) On the presence of a high-molecular crystallizable protein in blood serum in myeloma. Fol haemat (Leipzig) 59:184

Woodle ES, Xu D, Zivin RA, Auger J, Charette J, O'Laughlin R (1999) Phase I trial of a humanized, Fc receptor nonbinding OKT3 antibody, huOKT3γ1(Ala-Ala) in the treatment of acute renal allograft rejection. Transplantation 68(5):608–616

Wright A, Morrison SL (1997) Effect of glycosylation on antibody function: implications for genetic engineering. Trends Biotechnol 15(1):26–32

Wuhrer M, Porcelijn L, Kapur R, Koeleman CA, Deelder A, de Haas M et al (2009) Regulated glycosylation patterns of IgG during alloimmune responses against human platelet antigens. J Proteome Res 8(2):450–456

Chapter 4
The Role of IgG in Immune Responses

Jeanette H.W. Leusen and Falk Nimmerjahn

Abstract The glycoprotein immunoglobulin G (IgG) is a major effector molecule of the humoral immune response in man and accounts for about 75 % of the total immunoglobulins in plasma of healthy individuals. Antibodies of the IgG class express their predominant activity during a secondary antibody response. In comparison to antibodies of the IgM class, IgG antibodies have a relatively high affinity and persist in the circulation for a long time. Whereas all other Ig isotypes contain multiple glycosylation sites in the antibody constant and variable domains, IgG molecules have only one sugar moiety attached to each of the CH2 domains in the IgG constant fragment. This sugar moiety consists of a heptameric core structure of N-acetylglucosamine and mannose residues with variable additions of terminal (galactose and sialic acid) and branching (fucose and N-acetylglucosamine) residues. In the serum more than 30 different IgG glycoforms can be identified, and the pattern of IgG glycosylation can change during inflammatory responses. Removal of this sugar moiety abrogates IgG function, and certain IgG glycoforms have been demonstrated to have enhanced effector functions, suggesting that regulation of IgG glycosylation may be a means of modulating IgG activity.

Keywords Immunoglobulins • Humoral immunity • IgG • IgG subclasses • Opsonisation • Phagocytosis • Autoimmune disease

---

J.H.W. Leusen, Ph.D. (✉)
Department of Immunology KC02.085.2, Laboratory for Immunotherapy, University Medical Centre Utrecht, Lundlaan 6, Utrecht 3584, EA, The Netherlands
e-mail: jleusen@umcutrecht.nl

F. Nimmerjahn
Institute of Genetics, Department of Biology, University of Erlangen-Nuremberg, Erwin-Rommel-Str. 3, Erlangen 91058, Germany
e-mail: fnimmerj@biologie.uni-erlangen.de

F. Nimmerjahn (ed.), *Molecular and Cellular Mechanisms of Antibody Activity*,
DOI 10.1007/978-1-4614-7107-3_4,

4.1 Immunoglobulins and Humoral Immunity

The glycoprotein immunoglobulin G (IgG) is a major effector molecule of the humoral immune response in man and accounts for about 75 % of the total immunoglobulins in plasma of healthy individuals. Antibodies of the IgG class express their predominant activity during a secondary antibody response. In comparison to antibodies of the IgM class, IgG antibodies have a relatively high affinity and persist in the circulation for a long time. Whereas all other Ig isotypes contain multiple glycosylation sites in the antibody constant and variable domains, IgG molecules have only one sugar moiety attached to each of the CH2 domains in the IgG constant fragment. This sugar moiety consists of a heptameric core structure of N-acetylglucosamine and mannose residues with variable additions of terminal (galactose and sialic acid) and branching (fucose and N-acetylglucosamine) residues. In the serum more than 30 different IgG glycoforms can be identified, and the pattern of IgG glycosylation can change during inflammatory responses. Removal of this sugar moiety abrogates IgG function, and certain IgG glycoforms have been demonstrated to have enhanced effector functions, suggesting that regulation of IgG glycosylation may be a means of modulating IgG activity.

4.2 Properties of Human IgG Subclasses

Quantitatively, the relative serum concentrations of the human IgG subclasses are as follows: IgG1 > IgG2 > IgG3 = IgG4. The four subclasses show more than 95 % homology in the amino acid sequences of the constant domains of the heavy chains. The four IgG subclasses show their most conspicuous differences in the amino acid composition and structure of the 'hinge region', which is between the Fab arms (fragment antigen binding) and the two carboxy-terminal domains (C_H2 and C_H3) of both heavy chains and determines the flexibility of the molecule (Burton et al. 1986; Roux et al. 1997); see Fig. 4.1.

The flexibility of the hinge region decreases in the order IgG3 > IgG1 > IgG4 > IgG2. IgG3 is very susceptible to cleavage by proteolytic enzymes, such as papain, plasmin, trypsin and pepsin, whereas IgG2 is relatively resistant. IgG1 and IgG4 exhibit an intermediary sensitivity, depending upon the enzyme used. Since these proteolytic enzymes all cleave IgG molecules near or within the hinge region, it is likely that the high sensitivity of IgG3 to enzyme digestion is related to its accessible hinge.

As a consequence of the structural differences, the four IgG subclasses show differences in some of their biological properties (Table 4.1).

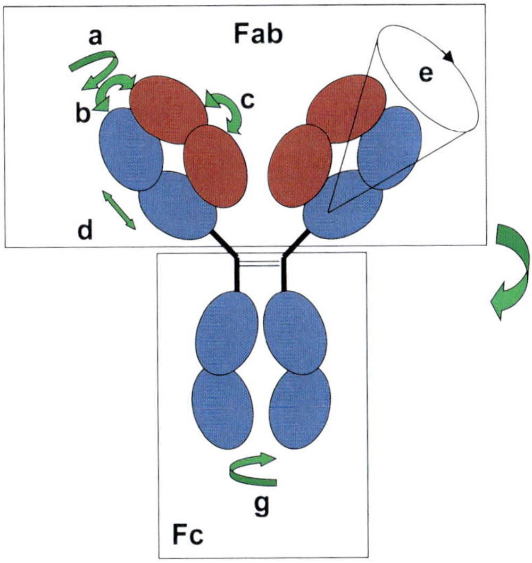

Fig. 4.1 The flexibility of the IgG molecule. The different modes of flexibility include (**a**) rotation about the long axis, (**b**) hinge-like folding or bending, (**c**) elbow bending at the switch region, (**d**) translation (in-and-out motion), (**e**) conical wagging, (**f**) Fab–Fc planar folding and (**g**) Fc rotation

Table 4.1 Biological properties of human IgG subclasses

	IgG1	IgG2	IgG3	IgG4
Proportion of total IgG (%)	43–75	16–48	1.7–7.5	0.8–11.7
Adult serum level range (g/l)	4.9–11.4	1.5–6.4	0.20–1.10	0.08–1.40
(Mean, g/l)	(6.98)	(3.8)	(0.51)	(0.56)
Half-life (days)	21	21	7	21
Placental transfer	+	+	+	+
Antibody response to:				
Proteins	++	+/−	++	+/−
Polysaccharides	+	++	(−)	(−)
Allergens	+	(−)	(−)	++
Complement activation				
C1q binding	++	+	+++	−
C1q binding, high epitope density enhancement alternative pathway	−	+	−	+/−
Binding to Fcγ receptors: (from Bruhns et al. Blood 2009)				
FcγRI (CD64)	++ ++	−	+++ +	+ +++
FcγRIIA-H131	++ +	+++	+++	+
FcγRIIA-R131	++ +	+	++	+
FcγRIIB/C	++	+	++	++
FcγRIIIa (CD16) Y158	++ +	+	++ ++	++
FcγRIIIa (CD16) F158	+++	+	+++	++
FcγRIIIb-NA1	++	−	+++	−
FcγRIIIb-NA2	++	−	++ +	−
Binding to staphylococcal protein A	++	++	*	+
Binding to streptococcal protein G	++	++	++	++

*IgG3 allotype dependent

4.3 Antibody Activity of IgG Subclasses

The IgG subclass distribution in specific antibody responses has been found to vary with the structure of the antigen. Bacterial and viral protein antigens, such as tetanus toxoid or outer-membrane components, are T cell-dependent (thymus-dependent) antigens and require interaction with helper T lymphocytes in order to stimulate B lymphocytes to antibody production. These trigger all four IgG subclasses, but IgG1 is the prevailing isotype, sometimes in combination with IgG3 (Ferrante et al. 1990). Anti-protein antibodies of the IgG2 subclass generally provide only a marginal contribution. On the other hand, IgG antibodies against polysaccharide antigens, which are generally T cell-independent (thymus-independent), show a much more pronounced subclass distribution: immunisation with several encapsulated bacteria leads to an almost exclusive IgG2 anti-polysaccharide response (Siber et al. 1990). An exception is seen in children under the age of 2–3 years, in whom anti-polysaccharide antibodies have been found to occur in the IgG1 subclass (Morell et al. 1990).

Repeated, long-term antigenic stimulation with T cell-dependent antigens may lead to a marked IgG4 antibody response (Aalberse et al. 1983). In general, antiviral IgG antibodies are highly restricted to IgG1 and IgG3, with IgG3 antibodies appearing first in the course of infection. The IgG subclass distribution in an antibacterial response will be more heterogeneous, since bacteria contain many different antigenic epitopes, with considerable variations in their protein and carbohydrate structures.

4.4 Effector Functions of IgG Subclasses

The most important biological activities of antibodies are related to their effector functions, aimed at inactivation or removal of infectious agents and their products (e.g. bacteria, viruses, toxins). Antibodies of the IgG class exert two major effector functions mediated by the (constant) Fcγ fragment: activation of complement and binding to Fcγ receptors.

4.5 Complement Activation

Serum bactericidal activity is induced when the complete complement cascade (C1–C9) is activated. The formation of the membrane attack complex (MAC) is the final result of activation by the classical pathway or the alternative pathway of complement (Fig. 4.2). Unlike the classical pathway, activation of the alternative pathway is not dependent upon antibodies, although many indications exist for an enhancing activity of antibodies in the alternative pathway. The MAC (C5b-9) is

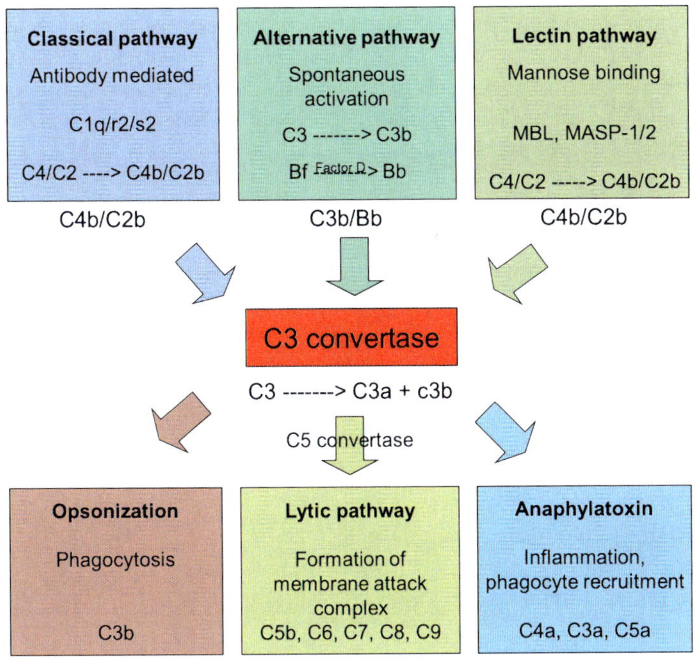

Fig. 4.2 Overview of complement pathways (Adapted from Janeway 2001)

inserted into the membrane of target cells, after which these are killed by a lytic mechanism. Besides, opsonisation of microorganisms by complement components (C3b, iC3b and C3dg) may result in phagocytosis. The structural diversity in the hinge region of IgG subclasses entails differences in their ability to activate complement. The initial event in the activation of the complement cascade by the classical pathway is the binding of C1q to sites present on the C_H2 domains of IgG. This binding occurs when the latter is aggregated, usually through formation of an antigen–antibody complex. Other components required for complement activation include a disulphide bond in the amino-terminal end of C_H2 (normally the hinge region), the presence of carbohydrates and paired C_H3 domains. The capacity of the four human IgG subclasses (in monomeric form) to bind C1q decreases in the order: IgG3 > IgG1 > IgG2 > IgG4 (Flanagan and Rabbitts 1982). IgG4 does not activate complement. Hinge-dependent Fab–Fab and Fab–Fc flexibility determines the accessibility of the complement binding site to C1q. IgG3, the human IgG subclass with the longest hinge, is the most effective complement activator.

Although IgG2 is a poor activator of the classical complement cascade, effective opsonisation and killing of certain bacterial strains and protection in bioassays by purified IgG2 have been reported. This may be explained by a combined effect of both bacterial antigen and IgG2 on the activation of complement. Density and accessibility of antigenic determinants on the bacteria may determine the actual complement activation by IgG subclasses. Alternatively, IgG2 may utilise the alternative pathway in complement activation in an Fcγ-independent way, bypassing C1q binding (Weinberg et al. 1986; Bjornson and Lobel 1987; Kuijpers et al. 1992). Complement activation via the alternative route appears to be especially important in the opsonisation and killing of a variety of bacterial strains.

4.6 Opsonisation and Induction of Phagocytosis

Most microorganisms are by themselves incapable to activate complement or to bind phagocytes. Antibodies act as flexible adapters mediating the adherence of infectious agents to leucocytes via Fcγ receptors. In mouse strains deficient in FcγRs, antibody-mediated effector functions, including antibody-dependent cellular cytotoxicity, phagocytosis of immune complexes (IC) and the release of proinflammatory mediators, are blocked (Takai 2002; Ravetch and Clynes 1998).

Leucocyte receptors for the Fcγ domain of IgG (FcγR) are members of the Ig supergene family, and three classes have been described in humans: FcγRI (CD64), FcγRII (CD32) and FcγRIII (CD16) (Nimmerjahn and Ravetch 2008).

4.6.1 FcγRI

Three genes have been characterised for the class I IgG receptor – Fcγ RIA, IB and IC – and all are mapped to chromosome 1, band q21.1. The prototypic FcγRI, encoded by the FcγRIa1 transcript, has an extracellular region with three Ig-like domains, a transmembrane domain and a cytoplasmic tail. Although several other transcripts have been described, no evidence exists to support their identity as true FcγRI isoforms. FcγRIa represents the sole leucocyte FcγR capable of binding monomeric IgG with high affinity $K_a = 10^8$–10^9 L mol^{-1}. The receptor shows high specific binding for the subclasses IgG1, IgG3 and IgG4 (Table 4.1).

FcγRI is constitutively expressed on monocytes, macrophages and myeloid dendritic cells. IFN-γ can enhance surface expression of FcγRI on these cells, and FcγRI expression can be induced on granulocytes by IFN-γ or G-CSF stimulation. In contrast, cytokines such as IL-4, IL-10 and TGF-β downregulate activating Fcγ receptors, including FcγRI, and enhance expression of inhibitory receptor FcγRIIb (Pricop et al. 2001; Tridandapani et al. 2003). In vitro, IFN-γ treatment leads to increased Fcγ receptor-induced cytokine production (Boruchov et al. 2005).

FcγRIa exists as a heteroligomeric receptor complex with a ligand-binding α-chain and an FcR γ-chain homodimer (Fig. 4.3). The α-chain does not bear known signalling domains but can trigger endocytosis. The FcR γ-chain molecules are critical for stabilisation of FcγRI surface expression, for signalling and for ligand-binding affinity of the receptor. The FcR γ-chain contains a conserved ITAM signalling motif in its cytoplasmic region, consisting of two YxxL boxes separated by seven amino acids. Upon FcγRIa cross-linking, this ITAM is phosphorylated on

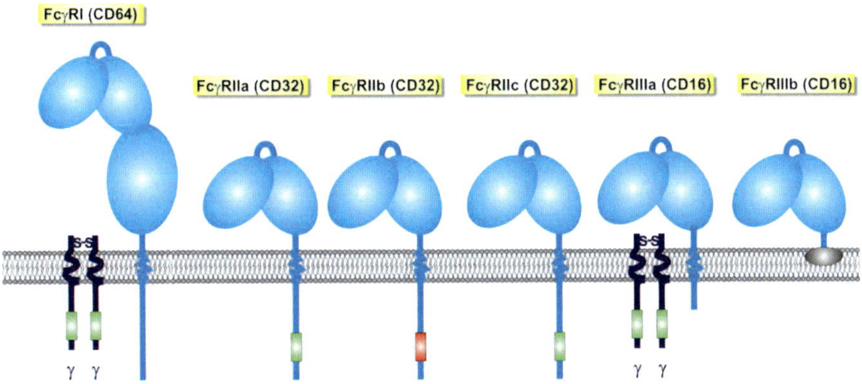

Fig. 4.3 Schematic representation of human Fcγ receptors. The extracellular parts of Fcγ receptors are composed of two or three immunoglobulin domains. Based on their function, FcR can mediate activating signals through ITAM (*green box*), which consist of YXXLXXXXXXXYXXL (Y = tyrosine, L = leucine, X = any amino acid). The ITAM is located on the associated g-chain or on the intracellular part of the receptors (FcγRIIa and IIc). The inhibiting receptor FcγRIIb bears an ITIM (*red box*), which consists of Y/V/S/xYxxL/V (V = valine, S = serine). FcγRIIIb is a glycosylphosphatidylinositol (GPI)-linked receptor that has no intracellular tail

both tyrosines by Src and Syk protein tyrosine family kinases, initiating signalling events. Serine/threonine phosphorylation of the FcR γ–chain is probably important for FcγRIa signal transduction.

Expression of FcγRIa increases in a variety of inflammatory and proinflammatory conditions, such as human immunodeficiency virus (HIV) infection and rheumatoid arthritis, indicating a role for this molecule under such conditions. However, the role of FcγRI in immunity in vivo remains unclear as, due to its high affinity, FcγRI is believed to be saturated with monomeric IgG. Despite this, upon cellular activation, FcγRI can readily bind immune complexes despite pre-engaged monomeric IgG (van der Poel et al. 2010). Also, several in vivo studies documented a role for FcγRI, varying from a contribution during inflammation and autoimmune reactions (Barnes et al. 2002; Ioan-Facsinay et al. 2002; Baudino et al. 2008), or during mAb-based immunotherapy in melanoma and B cell lymphoma models (Bevaart et al. 2006; Minard-Colin et al. 2008), to a malaria model in which transgenic expression of human FcγRI was central for effective mAb treatment (McIntosh et al. 2007). Furthermore, FcγRI can induce potent proinflammatory signalling compared to FcγRIIa (Dai et al. 2009) and can efficiently mediate both MHC-II antigen presentation and cross-presentation (Keler et al. 2000; Mosser and Edwards 2008; Liu et al. 1996; Heijnen et al. 1996).

To date, no polymorphisms have been described for FcγRI that alter receptor affinity or function. However, recently three nonsynonymous single nucleotide polymorphisms (SNPs) were identified that need to be further analysed to understand if and how these potential SNPs affect receptor functioning (van der Poel et al. 2011).

Table 4.2 Cell distribution and function of Fcγ receptors

Expression (induced)	Modulation	Functions	
FcγRI	Monocytes, macrophages, CD34+myeloid progenitor cells, dendritic cells (PMN: IFN-γ, G-CSF)	Upregulation: G-CSF, IFN-γ, IL-10Downregulation:IL-4, IL-13	Internalisation, antigen presentation, superoxide generation, ADCC, phagocytosis, cytokine production
FcγRII	IIa: monocytes, macrophages, PMN, eosinophils, basophils, Langerhans cells, platelets, placental endothelial cells, T cells, subpopulation	Downregulation: IL-4	IIa: internalisation, ADCC, phagocytosis, cytokine production, respiratory burst
	IIb: B cells, basophils, macrophages		IIb2: internalisation
	IIc: NK cells		IIb1/IIb2: downmodulation of B cells, mast cells, macrophages
FcγRIII	IIIa: macrophages, NK cells, (monocytes, subpopulation: TGFβ), T cells, subpopulation	Upregulation: TGFβ	IIIa: phagocytosis, ADCC, superoxide generation, cytokine production, induction of adhesion, apoptosis
		Downregulation: IL-4	
	IIIb: PMN		IIIb: superoxide generation, ADCC, degranulation

4.6.2 *FcγRII*

Three genes have been identified for FcγRII (IIA, IIB, IIC), located on chromosome 1q23–24, all encoding 40-kDa glycoproteins. Six transcripts, resulting from alternative splicing, have been documented, all encoding true isoforms. These molecules are conserved in their extracellular and transmembrane regions but differ markedly in their cytoplasmic domains. In FcγRIIa additional variation is introduced due to a genetic polymorphism at amino acid position 131 within the extracellular region (either histidine or arginine), resulting in different affinities for IgG isotypes (Table 4.1). FcγRII molecules interact only with complexed or polymeric IgG, and family members are expressed on most types of blood leucocytes (Table 4.2).

FcγRIIa is primarily expressed on myeloid cells, whereas FcγRIIb isoforms are found on mast cells, basophils and B cells (Van den Herik-Oudijk et al. 1995). The FcγRIIc isoform has been detected on NK cells (Metes et al. 1998). In about 70 % of the human population, however, the FcγR2c gene is a pseudogene and does not encode for a functional protein.

FcγRII cross-linking may either result in activation or downmodulation of immune responses, depending on the specific FcγRII isoform triggered. FcγRIIa cross-linking triggers functions varying from internalisation, phagocytosis, cytokine production, antigen presentation and superoxide generation to antibody-dependent cell-mediated cytotoxicity (ADCC). FcγRIIa is unique among leucocyte FcRs in that it contains a noncanonical ITAM directly in its cytoplasmic region. FcγRIIa can therefore transmit a phagocytic signal in the absence of other FcR subunits, and receptor cross-linking results in tyrosine phosphorylation of the γ-chain. FcγRIIa can, in addition, associate with the FcR γ–chain on monocytes and macrophages. Both the α-chain and the γ-chain can activate signalling pathways, involving a variety of nonreceptor kinases (Table 4.2). Signalling via FcγRIIa has furthermore been shown to involve membrane molecules such as CD45, CD148 and the integrin complement receptor 3 (CR3) (CD11b/CD18).

The FcγRIIb1 and IIb2 isoforms are unique in that they bear an ITIM within their cytoplasmic tails. An ITIM is a highly conserved (13 amino acids) motif containing one YxxL box, and FcγRIIb may inhibit cell activation triggered by receptors that signal via ITAM (B cell receptor (BCR), FcγRI or FcγRIIa). Inhibition requires ITIM phosphorylation and involves activation of the phosphatases SHP-1 (B cells) and SHIP (mast cells and B cells) and the CD19 molecule on B cells. Furthermore, the FcγRIIb2 isoform proved effective in internalisation of small immune complexes in contrast to FcγRIIb1 (Van de Winkel and Hogarth 1998).

The third gene, FcγRIIc, probably results from an unequal crossover event between genes IIA and IIB and encodes a protein which resembles FcγRIIa intracellularly and FcγRIIb extracellularly. Interestingly, an FcγRIIc isoform has been identified on natural killer (NK) cells from some donors, involved in triggering cytotoxicity (Metes et al. 1998).

FcγRIIa-His131 represents the sole leucocyte FcR capable of interaction with IgG2. Therefore, polymorphisms of FcγRIIa are considered to be heritable risk factors for both infectious diseases (such as meningitidis and periodontitis) and autoimmune diseases (such as systemic lupus erythematosus) (Van der Pol and Van de Winkel 1998a). Polymorphisms of FcγRIIb will be discussed in a separate paragraph.

Soluble FcγRs have been documented as potent inhibitors of immune complex-induced inflammation. Recombinant soluble FcγRII not only prevents binding of immune complexes to membrane-bound FcRs but also alters the behaviour of complexes by preventing their precipitation. Targets for therapy with soluble Fcγ receptors may include rheumatoid arthritis, systemic lupus erythematosus and immune cytopenias (Van de Winkel and Hogarth 1998).

4.6.3 FcγRIII

This receptor class is encoded by two genes (IIIA, IIIB), both localised on chromosome 1q23–24, specifying highly glycosylated molecules of 50 80 kDa. FcγRIIIa encodes a transmembrane receptor with two extracellular Ig-like domains and a 25-amino acid cytoplasmic tail, with medium affinity for IgG ($K_a = 3 \times 10^7$ L mol^{-1}). Expression and function of this receptor is dependent upon association with the FcR γ-chain on monocytes and macrophages or with γ- or ζ-chains on NK cells. The receptor is differentially glycosylated between myeloid and lymphoid cells.

Expression of FcγRIIIa can be upregulated by transforming growth factor β (TGFβ) and downregulated by IL-4. The ligand-binding chain of FcγRIIIa has recently been reported to be active in initiating calcium fluxes. Biological function of this receptor, however, is critically dependent on either the ζ and/or γ accessory chains.

FcγRIIIa has been shown to be polymorphic, and two allotypes have been defined that differ by a single amino acid at position 158 (either valine or phenylalanine). The allotypes exhibit different capacities to bind IgG1, IgG3 and IgG4 (see Table 4.1). hFcγRIIIA$_{F158}$ has been linked to SLE (Wu et al. 1997) and to rheumatoid arthritis (RA) (Nieto et al. 2000). Furthermore, genetic variation of this receptor determines therapeutic outcome upon treatment with rituximab and trastuzumab (Cartron et al. 2002; Weng and Levy 2003; Zhang et al. 2007; Musolino et al. 2008).

The FcγRIIIb subclass exhibits low affinity for IgG and is expressed exclusively on PMNs. Two allotypes and one gene duplication exist that are designated FcγRIIIb-NA1, IIIb-NA2 and IIIb-SH, respectively. FcγRIIIb has two extracellular Ig-like domains and is uniquely anchored to the outer leaflet of the lipid bilayer by a glycosylphosphatidylinositol linkage. FcγRIIIb has been suggested to interact with either FcγRIIa or the integrin CR3 (CD11b/CD18) or accessory signalling molecules localised in specialised membrane domains (called 'rafts') to confer signalling ability (Vidarsson and van de Winkel 1998). FcγRIIIb cross-linking can trigger a variety of activation events including superoxide generation, degranulation and ADCC. FcγRIIIb can be released into serum upon cell activation or apoptosis induction through serine protease cleavage. The level of soluble FcγRIIIb in the circulation is now appreciated to reflect the total body mass of neutrophils (van der Pol and van de Winkel 1998).

FcγRIIIb bears the neutrophil antigen (NA) polymorphism, which has been implicated in alloimmune and autoimmune neutropenias and blood transfusion reactions. The FcγRIIIb-NA1 exhibits higher affinity for immune-complexed IgG1 and IgG3 than IIIb-NA2, and IgG1- and IgG3-opsonised particles are more efficiently phagocytosed via the IIIb-NA1 allotype. In addition, it has been shown that the FcγRIIIb allelic polymorphism is of clinical importance for progression of periodontal disease (van der Pol and van de Winkel 1998).

4.7 FcγR Genetic Polymorphism, IgG Subclasses and Disease

In humans, IgG2 antibodies have been found to play a key role in immunity against infection with encapsulated bacteria. This defence does not only depend upon complement activation, since IgG2 has a poor complement-activating activity, at least when the epitope density is low. Indeed, it has been shown that phagocytosis by neutrophil granulocytes is the major underlying mechanism. Moreover, the FcγRIIa polymorphism is important; neutrophils from FcγRIIa-H131 homozygous individuals phagocytose better than those of FcγRIIa-R131 individuals (Erbe et al. 1991). This polymorphism will be more critical when IgG2 levels are low, as is the case in children. In children surviving meningococcal septic shock, the allotype FcγRIIa-R131 occurred twice as often as in normal donors (Bredius et al. 1994).

With respect to FcγRIIIb polymorphism, it was reported that the combined FcγRIIa-R131 and FcγRIIIb-NA2 phenotype is associated with meningococcal infection (Fijen et al. 1993).

Thus, next to a quantitative defect of IgG subclass levels, effector functions of humoral immunity may also be negatively affected by defects in Fcγ receptors.

4.8 FcγRIIB Polymorphisms and Autoimmune Disease

Several single nucleotide polymorphisms (SNPs) have been identified within the promoter and coding regions of *FCGR2B*. In the promoter, there are two SNPs at nucleotides −386 and −120. These SNPs form two haplotypes, a common −386 G:−120 T haplotype and a less common −386 C:−120 A haplotype. These haplotypes may alter transcription factor binding and expression. Seven nonsynonymous SNPs have been identified in the human *FCGR2B* gene but only one has been studied in detail. This nonsynonymous T-to-C transition (rs1050501) in exon 5 of the gene results in the substitution of a threonine for isoleucine at position 232 within the transmembrane helix of FcγRIIB (FcγRIIB^{T232}) and was first identified in 2002 (Kyogoku et al. 2002). The FcγRIIB^{T232} variant was described to protect against malaria. On cellular level, it is excluded from lipid rafts and has notably impaired inhibitory function. Therefore, this single amino acid polymorphism of FcγRIIB is defined as a 'loss of function' mutation (Floto et al. 2005). The polymorphism that encodes FcγRIIB^{T232} has been associated with increased susceptibility to systemic lupus erythematosus (SLE) (Kono et al. 2005). The frequency of homozygosity of 232 T is at least doubled in individuals with SLE compared with that in healthy controls in three Asian and one Caucasian cohort (reviewed in Smith and Clatworthy 2010).

4.9 Cooperation of FcR and Complement

Inactivation of the complement pathway has no major effect on late antibody functions such as ADCC and phagocytosis but rather seems to modulate IgG function (Nimmerjahn and Ravetch 2005). However, sometimes inactivation of late complement components results in a strong reduction of antibody-mediated inflammation. In a model of autoimmune haemolytic anaemia (AIHA) and IC-induced lung disease, it was shown that the activated complement component C5a modulates expression of FcγRs and that in mice deficient in the C5a-receptor antibody-mediated inflammation was abrogated (Kumar et al. 2006; Shushakova et al. 2002). Although the entire pathway has not yet been elucidated fully, generation of C5a seems to be independent of the classical complement activation pathways and depends upon cellular FcγRs (Schmidt and Gessner 2005). More information on this important pathway can also be found in the complement chapter of the book.

4.10 Insights into the Roles of Fcγ Receptors for IgG Activity In Vivo: FcγRs in Mice

Most of our current insights on how IgG molecules mediate their function in vivo are derived from mouse model systems. The expression of Fcγ receptors in mice is largely comparable with man, with a few exceptions. Mice have four Fcγ receptors: FcγRI, FcγRIIB, FcγRIII and FcγRIV (Fig. 4.4). As in humans, FcγRI is a high-affinity receptor compared to the rest of the molecules which can only interact with IgG in the form of immune complexes. Like in humans, mice have only one inhibitory Fcγ receptor, FcγRIIB, which is broadly expressed on innate immune effector cells and is the only FcgR on B cells. Unlike humans, mice have no FcγRIIa or c and no FcγRIIIb. With respect to mouse IgG, four different subclasses have been

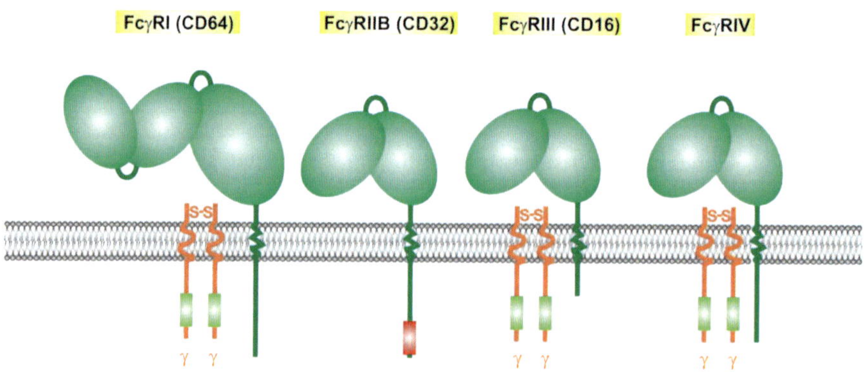

Fig. 4.4 Overview of mouse Fcγ receptors. Mice contain four Fcγ receptors: three activating receptors, associated with the γ-chain, and one inhibiting receptor containing an ITIM

identified and have been termed IgG1, IgG2a, IgG2b and IgG3. Of note, these subclasses behave very differently from the human subclasses and despite of similar names should not be compared to the human IgG molecules. Thus, mouse IgG1 cannot activate the complement pathway and selectively binds to mouse FcγRIII and to the inhibitory FcγRIIB (Table 4.2). IgG2a (depending on the mouse strain, this subclass is sometimes also termed IgG2c) can bind to all activating and the inhibitory FcγR and can activate the classical pathway of complement activation. IgG2b binds to both FcγRIII and FcγRIV and the inhibitory FcγRIIB, in addition to being able to activate the complement pathway. IgG3 does not bind to most of the FcγRs with the exception of some minimal binding to FcγRI but can activate the complement pathway quite efficiently. Of note, IgG2a is the only subclass which can bind with high affinity to FcγRI as a monomeric molecule. In contrast to humans, most mouse FcγRs also have a very low affinity for IgE which may be of relevance for some mouse models as discussed below. Many Fc receptor knockout and transgenic mice have been generated over the past two decades and taught us a lot about the function of these receptors. First, the FcR gamma subunit deficient mice were generated by gene targeting in embryonic stem cells (reviewed by Takai 1996). The FcR γ-chain is required for efficient surface expression of all activatory Fc receptors in mice. The lack of this associated chain leads to multiple defects in FcR-mediated effector cell responses, including absence of phagocytic activity against opsonised red blood cells by activated macrophages and loss of antibody-dependent cell-mediated cytotoxicity (ADCC) capacity. FcR γ-deficient mice are unable to mount the type II and III hypersensitivity reactions.

Later, also FcγR single knockouts were generated: FcγRIIB-/- mice were reported in 1996 (Takai et al. 1996), FcγRIII-/- mice in 1996 (Hazenbos et al. 1996) and 1999 (Ujike et al. 1999), FcγRI-/- mice in 2002 (Barnes et al. 2002; Ioan-Facsinay et al. 2002), FcRn-/- in 2003 (Chaudhury et al. 2003) and, finally, FcγRIV-/- mice in 2010 (Nimmerjahn et al. 2010). This array of mouse strains has been very helpful in delineating the role of the individual FcγRs for IgG activity. In the following paragraphs the most important phenotypes of these mice will be discussed.

4.10.1 FcγRI Knockout Mice

FcγRI shows a rather restricted expression pattern and can be found on monocytes and macrophages. FcγRI-/- mice demonstrated that FcγRI contributes substantially to severity of arthritis, hypersensitivity responses and protection from bacterial infection (Ioan-Facsinay et al. 2002). Also, in the absence of FcγRI, antibody responses were elevated, implying the removal of a control point by the deletion of FcγRI (Barnes et al. 2002). The activity of IgG2a, the only IgG subclass which can bind to FcγRI with high affinity, was reduced in FcγRI knockout mice in some but not all tumour model systems. More recent data suggest that both FcγRI and FcγRIV are required for IgG2a activity in passive therapeutic and autoimmune model systems. The varying requirement of this receptor for IgG2a activity may be explained by the high affinity for IgG2a which results in saturation of FcγRI on cells in the blood where IgG is present in the mg/ml range. Upon effector cell migration into the periphery, this receptor may be more accessible for IgG2a binding.

4.10.2 FcγRIIB Knockout Mice

FcγRIIB is broadly expressed in the innate and adaptive immune system with only very few exceptions including T cells and NK cells. FcγRIIB-/- mice backcrossed to a C57BL/6 background spontaneously develop hypergammaglobulinaemia, autoantibodies and an immune complex-mediated disease resembling SLE (Bolland and Ravetch 2000); this does not occur on mice backcrossed to a BALB/c background, indicating the importance of strain-specific epistatic effects on autoimmune susceptibility. The restoration of FcγRIIB expression levels on bone marrow cells of (NZB × NZW) F1 mice (a mouse model of spontaneous SLE) using an FcγRIIB-expressing retrovirus can prevent autoimmunity (McGaha et al. 2005), and modest transgenic overexpression of FcγRIIB on B cells markedly reduces the development of SLE in MRL–lpr mice (Brownlie et al. 2008). Thus, studies in genetically modified mice implicate FcγRIIB in the pathogenesis of autoimmunity and the maintenance of B cell tolerance. However, a study generating FcγRIIB(-/-) mice by gene targeting in B6-derived ES cells revealed that FcγRIIB deficiency does not cause autoimmunity but amplifies spontaneous autoimmunity determined by autoimmune susceptibility loci again emphasising the role of other modulating genes of the specific mouse background (Boross et al. 2011). Recent studies in humanised mouse models revealed the role of FcγRIIB as a gatekeeper of humoral tolerance also for the human system (Baerenwaldt et al. 2011). Besides setting a threshold for B cell activation, isolated cross-linking of this receptor without any concomitant positive signals was suggested to induce apoptosis in B cells (Fig. 4.5). This situation may be relevant for plasma cells, which downregulate B cell receptor expression but maintain to express high levels of FcγRIIB. This may enable to kill a portion of plasma cells in the bone marrow and allow new plasma cells to occupy niches in the bone marrow which are essential for their long-term survival. In many autoimmune prone strains such as BXSB, NZBW and NOD mice, FcγRIIB expression on plasma cells is reduced, which may protect these autoantibody-producing cells from being eliminated by immune complexes. Thus, increasing FcγRIIB expression on these cells may be a strategy to reduce autoantibody-producing plasma cells which will be discussed in the next chapter.

In addition to the phenotype in the B cell compartment, FcγRIIB knockout animals show an enhanced IgG activity. Lack of FcγRIIB leads to elevated IC-mediated inflammation and phagocytosis as demonstrated by an enhanced Arthus reaction, systemic anaphylaxis, anti-GBM glomerulonephritis, immunothrombocytopenia, haemolytic anaemia, collagen-induced arthritis and IgG-mediated clearance of pathogens and tumour cells. Consistent with the higher affinity of IgG1 for FcγRIIB compared to its only activating FcγRIII, the greatest enhancement was seen for the

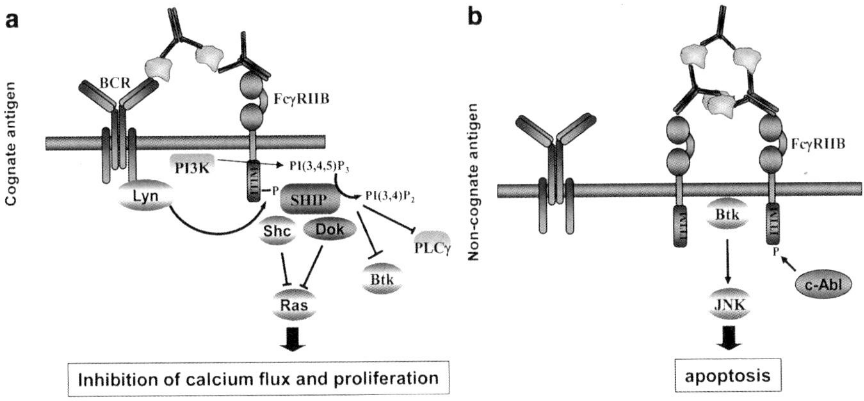

Fig. 4.5 Signalling pathways induced by FcγRIIB cross-linking on B cells. (**a**) Co-cross-linking of FcγRIIB with the B cell receptor (BCR) or activating FcγRs (not depicted) results in a Lyn-dependent phosphorylation of the ITIM motif in the cytoplasmatic tail of the inhibitory FcγRIIB. This recruits SHIP to the ITIM domain, leading to the inhibition of downstream activating signalling pathways, such as the release of calcium. (**b**) Isolated triggering of the inhibitory FcγRIIB on B cells was shown to induce apoptosis in an ITIM-independent but Btk and c-Abl-dependent manner. See text for further details

IgG1 subclass in several model systems. As the inhibitory FcγRIIB is coexpressed with activating FcγRs on the majority of innate immune effector cells including mast cells, basophils, eosinophils, neutrophils, monocytes and macrophages, this suggests that an immune complex can bind to both activating and the inhibitory Fcγ receptor (Fig. 4.6). Upon co-aggregation with its activating counterpart, LYN phosphorylates the ITIM motif which leads to the recruitment of SHIP (SH2-domain containing inositol 5′ phosphatase). SHIP activation leads to enhanced hydrolysis of phosphatidyl inositol intermediates and thereby interferes with the membrane recruitment of BTK and PLCγ, resulting in inhibition of ITAM signalling mediated calcium release and downstream effector functions. Moreover, tyrosine phosphorylated SHIP can bind to SHC and DOK, thereby inhibiting activation of the Ras pathway and ultimately cell proliferation. In effect, the inhibitory FcγR will set a threshold for cell activation as both activatory and counterbalancing inhibitory signalling pathways will be initiated by immune complex binding. The importance of this threshold may be exemplified by the differential impact of FcγRIIB on the individual IgG subclasses. By measuring the affinities of all antibody isotypes for the various FcRs and dividing the affinities of antibody isotypes for activating FcRs by the affinity for the inhibitory receptor, a so-called A/I ratio was established which predicted antibody activity in vivo. Thus, IgG1 (which only binds to the FcγRIII/FcγRIIB pair) has a lower affinity for the activating than for the inhibitory FcR (A/I-ratio « 1) resulting in lower activity. In contrast, IgG2a and IgG2b have 20–40-fold higher affinity for the activating FcγRIV than for FcγRIII/FcγRIIB, which results in

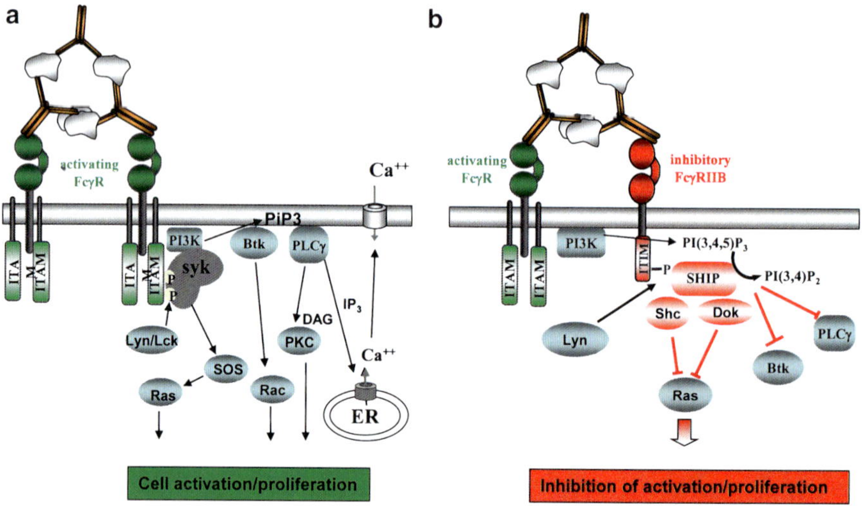

Fig. 4.6 Signalling pathways of activating and inhibitory Fc receptors. (**a**) Immune complexes-triggered cross-linking of activating FcRs induces phosphorylation of the ITAM motif by members of the SRC-kinase family, resulting in activation of downstream signalling events and cell activation. (**b**) Simultaneous triggering of the inhibitory FcR interferes with cell activation by recruitment of the phosphatase SHIP and inhibition of activating signalling pathways. The strength of activating versus inhibitory signals triggered by immune complexes is determined by the affinity of individual antibody isotypes to activating and inhibitory FcRs and other factors such as the cytokine environment (see text for details)

preferential triggering of FcγRIV and a lower degree of negative regulation by the inhibitory Fcγ receptor. In humans, the presence of different activating FcR alleles that have differential affinities for certain human antibody isotypes might result in a similar situation, thus changing the A/I ratio and antibody activity.

There are several factors that can alter this threshold set by the inhibitory Fc receptor (Fig. 4.7). Cytokines, for example, often regulate expression of the associated signalling adaptors (β- and γ-chains) which leads to a concomitant change in α-chain expression, as shown for transforming growth factor β (TGF-β), IL-4 and IL-10. Moreover, inflammatory cytokines/mediators, such as TNF-α, C5a or LPS, tend to upregulate activating receptors, whereas TH-2 cytokines such as TGF-β, L-4 and IL-10 seem to decrease expression. These effects can be cell type specific as IL-4 upregulates the inhibitory FcγRIIB on myeloid cells but has the opposite effect on activated B cells.

The outcome of cytokine-mediated changes in activating and inhibitory expression might vary depending on the differential regulation of IgG isotypes by the inhibitory Fcγ receptor. Thus, IgG2a and to a lesser degree IgG2b antibodies might be insensitive to elevated FcγRIIB expression, relative to the severe impairment of IgG1 activity. During antibody-mediated inflammation, however, the steady-state

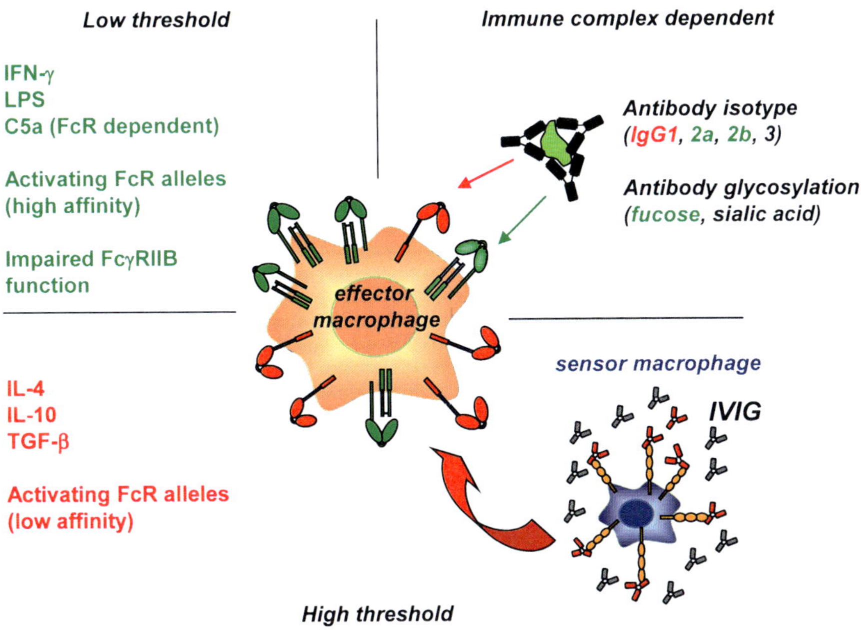

Fig. 4.7 Factors that influence FcR-dependent effector cell activation. Factors that shift the balance towards cell activation or inhibition are shown in *green* or *red*, respectively. Inflammatory cytokines/mediators such as IFN-γ, LPS or C5a upregulate activating FcRs (shown in *green*) resulting in a lower threshold for cell activation. In addition, allelic variants of activating and inhibitory receptors that influence antibody binding or FcR function might have similar effects. A higher threshold for cell activation is induced by anti-inflammatory cytokines or therapeutics such as IVIG that upregulate the inhibitory FcγRIIB (shown in *red*). In addition, activating FcR alleles with low affinity for antibody isotypes will trigger activating signalling pathways less efficiently. Moreover, the composition of ICs with respect to antibody isotype and antibody glycosylation pattern determines if activating or inhibitory signals will dominate

ratios change in favour of the activating Fcγ receptors as the release of inflammatory mediators, such as IFN-γ and C5a, can upregulate activating Fcγ receptors and at the same time reduce FcγRIIB expression levels. Under these circumstances, autoreactive IgG1 antibodies are capable of triggering severe damage. In fact, the regulation of FcR expression by cytokines is coupled to the regulation of isotypes by these same cytokines: TH-1 cytokines such as IFN-γ induce class switching to IgG2a, TH-2-type cytokines (IL-4) induce class switching to IgG1 and TGF-β induces switching to IgG2b. Since these cytokines also influence Fcγ-receptor expression, the pathogenicity of an autoimmune response will be determined by both cytokine-mediated regulation of class switching and the changes in expression levels of the responsible activating versus inhibitory Fcγ receptors. This dual regulation might allow the development of new therapeutic strategies to treat autoimmune disease. As will be discussed in the next chapter, the infusion of high doses of the pooled fraction of IgG from thousands of donors is used as an anti-inflammatory treatment and has been shown to alter FcγRIIB expression.

Another factor that can influence the interaction of antibodies with cellular FcRs is the sugar moicty attached to the asparagine residue at position 297 in the antibody Fcγ fragment. Deletion of this sugar side chain results in loss of FcR binding. In addition, this core sugar structure contains variable amounts of branching and terminal sugar residues such as N-acetylglucosamine, fucose, galactose and sialic acid. Presence or absence of these terminal or branching sugar residues can significantly influence antibody–FcR interactions. Thus, antibodies without fucose bind approximately 10-fold stronger to mouse-activating FcγRIV and human FcγRIIIa. In contrast, high levels of terminal sialic acid residues impair antibody binding to FcRs. This concept is underlying novel strategies to improve the function of therapeutic antibodies such as rituximab or Herceptin, which will be discussed in the last chapter of the book.

4.10.3 FcγRIII Knockout Mice

In mice, the IgG1 subclass binds selectively to activating FcγRIII (and the inhibitory FcγRIIB). FcγRIII is expressed on virtually all cells of the innate immune system and on dendritic cells. It is absent from cells of the adaptive immune system. FcγRIII-/- mice are completely protected from IgG1-mediated injury, including autoimmune haemolytic anaemia, thrombocytopenia and immune complex injury (Nimmerjahn and Ravetch 2005; Hazenbos et al. 1996; Meyer et al. 1998). It was demonstrated that mice deficient for the ligand-binding α-chain of FcγRIII lack NK cell-mediated antibody-dependent cytotoxicity and phagocytosis of IgG1-coated particles by macrophages. Strikingly, these mice lack IgG-mediated mast cell degranulation, are resistant to IgG-dependent passive cutaneous anaphylaxis and exhibit an impaired Arthus reaction. This is consistent with FcγRIII expression on mast cells in addition to the more well-known high-affinity Fcγ receptor for IgE on this cell type (FcγRI). With respect to the other IgG subclasses, FcγRIII may be responsible for IgG2a and IgG2b activity in combination with FcγRIV.

4.10.4 FcRn Knockout Mice

FcRn-/- mice demonstrated that the neonatal FcR is critical for IgG half-life. In the absence of this receptor, IgG molecules have a half-life of a few hours compared to the normal half-life of 7–12 days. FcRn is broadly expressed on endothelial cells but can also be found on haematopoietic cells. In addition to prolonging the half-life of IgG, FcRn was also demonstrated to extend the half-life of serum albumin (Chaudhury et al. 2003).

4.10.5 FcγRIV Knockout Mice

FcγRIV shows a restricted expression pattern and can only be found on neutrophils, on most tissue macrophages and on a subpopulation of monocytes, called classical monocytes. FcγRIV-/- mice established the role of this receptor for IgG2a and IgG2b activity in vivo. Thus, autoantibody as well as therapeutic IgG2a and IgG2b activity was impaired in FcγRIV knockout mice. Whereas the single deletion of FcγRIV was sufficient to block IgG2b activity, the IgG2a subclass often required other FcγRs for its activity. Thus, it was shown that in models of autoimmune haemolytic anaemia, in which red blood cell-specific antibodies are injected in mice resulting in removal of red blood cells, FcγRIII and FcγRIV are involved in IgG2a activity. In models of therapeutic IgG2a activity, such as B cell or melanoma-specific antibodies of the IgG2a subclass, it was suggested that FcγRI and FcγRIV are critical for therapy. This was shown earlier with blocking antibodies to FcγRIV in Nimmerjahn et al. 2005.

4.11 Role of FcγR in Anaphylaxis or Type I Inflammation

Anaphylaxis is a hyperacute allergic reaction that can have fatal consequences. A frequently used mouse model to study anaphylaxis is to inject mice locally (mostly intradermally) with an antigen-specific IgE or IgG antibody followed by systemic injection of the antigen (this model is also referred to as reverse anaphylaxis as normally the antibody would extravasate from the bloodstream and the antigen would be at a local site). Upon extravasation of the antigen form the bloodstream and encounter of the antibody at the site of injection, an immediate (within a few minutes) inflammatory reaction will follow resulting in an influx of mostly neutrophils causing local skin swelling. Classically, it depends on IgE, FcγRI, mast cells and histamine. Interestingly, FcγRIIB-deficient mice show an enhanced IgE-dependent anaphylactic reaction, suggesting that low-affinity binding of IgE to FcγRIIB is sufficient for this negative regulation. However, anaphylaxis can also be induced by IgG antibodies, and an IgG1-induced passive type of systemic anaphylaxis has been reported to depend on basophils. Moreover, mast cells themselves express FcγRIII and can become activated through FcγRIII cross-linking by immune complexes (Hazenbos et al. 1996). In FcγRI-deficient mice, FcγRIII becomes upregulated on mast cells resulting in enhanced IgG-dependent anaphylaxis (Dombrowicz et al. 1997) Recently, it was discovered that also neutrophils and, to a lesser extent, basophils can play a crucial role in induction of anaphylaxis. Neutrophil depletion inhibited active, and also passive, systemic anaphylaxis. Importantly, mouse and human neutrophils each restored anaphylaxis in anaphylaxis-resistant mice, demonstrating that neutrophils are sufficient to induce anaphylaxis in this model system in mice. This type of anaphylaxis was dependent on FcγRIII and

FcγRIV on neutrophils (Jonsson et al. 2011). Taken together, several innate immune effector cells and most notably mast cells, basophils and neutrophils are critical for anaphylaxis.

4.12 Type II Inflammation: In Vivo Mechanism of Cytotoxic IgG

In type II hypersensitivity (or cytotoxic hypersensitivity), the antibodies produced by the immune response bind to antigens on the patient's own cell surfaces. The antigens recognised in this way may either be intrinsic ('self' antigen, innately part of the patient's cells) or extrinsic (adsorbed onto the cells during exposure to some foreign antigen, possibly as part of infection with a pathogen). These cells are recognised by macrophages or dendritic cells, which act as antigen-presenting cells. This causes a B cell response, wherein antibodies are produced against the foreign antigen. FcγRIIB can be an important negative regulator in this process, by dampening the B cell response. Examples of a type II inflammatory/cytotoxic IgG response are autoantibodies directed against red blood cells (causing autoimmune haemolytic anaemia, AIHA) or against platelets (causing immunothrombocytopenia, ITP). The result of autoantibody binding to platelets or red blood cells is the phagocytosis of the opsonised material via the cells of the mononuclear phagocytic system. Besides resident macrophages in the spleen and liver, more recently monocyte subsets called resident or nonclassical monocytes were shown to be involved in phagocytosis of opsonised material directly from the blood. Despite the fact that phagocytosis may be happening via both, complement and FcγRs expressed on these phagocytic cells, it seems that for most of the IgG subclasses, the FcγR-dependent pathway is critical. For IgG2b a contribution of the complement pathway for red blood cell phagocytosis was noted, although again FcγRs were the dominant factor. In tumour therapy cytotoxic IgG is mostly used in the form of tumour-specific antibodies directed against B cells, breast cancer or colon cancer cells. Upon binding of the therapeutic antibody to the target cell, either cellular (antibody-dependent cellular cytotoxicity, ADCC) or humoral (complement-dependent cytotoxicity) pathways may result in induction of apoptosis in the target cell. Interestingly, the data from mouse models of antitumour antibodies directed against B cells or melanoma cells again indicates that cellular FcγRs are essential for therapeutic IgG activity. Thus, even IgG2a or IgG2b antibodies which have the capacity to activate the classical pathway of complement activation were fully dependent on the presence of FcγRs. In contrast, therapeutic IgG activity was normal in complement C3 knockout mice. Whether the same type of pathway is responsible for IgG activity in humans remains to be established. Thus, B cell-specific therapeutic antibodies such as ofatumumab, which was optimised for activation of the complement pathway, show clinical efficacy in humans. On the other hand, tumour patients with allelic variants of FcγRIIA and FcγRIIIA which confer a higher affinity for the therapeutic antibody seem to respond better to IgG-dependent antitumour therapy. Building on this concept, many companies are in the process of generating next generation therapeutic IgG molecules with enhanced binding to activating FcγRs. Evidence that antitumour antibodies with enhanced binding to activating FcγRs (and therefore less prone to negative regulation via the inhibitory FcγRIIB) was provided by a study using anti-melanoma antibodies with absent fucosylation. In humans and mice afucosylated IgG glycovariants bind with enhanced affinity to human FcγRIIIA and mouse FcγRIV. Correlating with this enhanced affinity for the activating FcγR, these antibodies showed a strong increase in therapeutic activity. Similar results were observed for human antitumour antibodies in immunodeficient mice with human tumour xenografts.

Of note, IgG antibodies can also initiate anti-inflammatory responses. In the form of intravenous immunoglobulin (IVIG), IgGs are routinely administered to treat inflammatory autoimmune diseases, such as idiopathic thrombocytopenia purpura (ITP), chronic inflammatory demyelinating polyneuropathy (CIDP), Kawasaki disease and Guillain–Barre syndrome. Importantly, the N-linked glycans on the IgG Fcγ are absolutely required for initiating these IgG effector functions which will be the focus of the next chapter.

4.13 Type III Inflammation (Immune Complex-Mediated)

Typical examples for this type of inflammation are immune complex diseases such as systemic lupus erythematodes and rheumatoid arthritis in which soluble antigen–antibody (IgG or IgM) complexes, which are normally removed by macrophages in the spleen and liver, are formed in large amounts and overwhelm the phagocytic system of the body. These small complexes lodge in the capillaries, pass between the endothelial cells of blood vessels – especially those in the skin, joints and kidneys – and become trapped on the surrounding basement membrane beneath these cells, resulting in triggering of Fcγ-dependent effector functions, which frequently include (1) massive inflammation, (2) influx of neutrophils, (3) MAC lysis, (4) aggregation of platelets and (5) activation of macrophages.

A lot of data concerning this pathway was obtained in a passive model of rheumatoid arthritis originally developed by Diane Mathis and Christophe Benoist (68). In this model system T cell receptor transgenic mice (so-called KRN mice) are crossed to NOD (nonobese diabetic) mice. The offspring of these animals develop a chronic arthritis characterised by ankle and joint swelling, bone erosion and a massive recruitment of neutrophils, monocytes and macrophages. Transferring the serum of these animals into healthy mice results in a transient arthritis with all the features of the chronic disease, thus enabling the use of different knockout animals to study the molecular details of the inflammatory pathway. As for cytotoxic IgG, mice deficient in all activating FcγRs were protected from disease development. By using mice deficient in individual FcγRs, it was demonstrated that FcγRIII and to a smaller extend FcγRIV were involved in development of disease consistent with the predominant IgG subclasses of the autoantibodies, which are largely of the IgG1 and to a much smaller extend of the IgG2a subclass (Ji et al. 2002; Nimmerjahn and Ravetch 2010). In contrast to cytotoxic IgG, mice deficient in several complement components of the alternative and not the classical pathway were also critical for induction of joint inflammation (Ji et al. 2002). Similar results were observed in models of nephrotoxic nephritis and certain skin blistering diseases. The exact details of how this FcγR and complement codependent pathway works are currently unknown. There is evidence, however, that a local production of complement components C3 and C5 in myeloid cells may be induced by immune complexes in an FcγR-dependent pathway (Syed et al. 2009). Activated C3a and C5a would then further enhance FcγR expression, which may be crucial for full-blown inflammation.

4.14 Differences Between Mouse and Human FcγR Functional Properties

Most human FcγRs bear the same name and CD as mouse FcγRs. They are, however, quite dissimilar in binding abilities and expression pattern. Recent analyses revealed three major differences between human and mouse FcγRs binding abilities: (1) all human-activating FcγRs bind the major human IgG subclass IgG1, whereas only mouse-activating FcγR FcγRIII binds mouse IgG1; (2) human inhibitory hFcγRIIB has a lower affinity for IgG1, IgG2 and IgG3 than all other hFcγRs, which is not the case in mice for IgG1 and IgG2b; and (3) no human FcγR binds human IgE, whereas 3 of the 4 mouse FcγRs (FcγRIIB, FcγRIII, FcγRIV) bind mouse IgE (Nimmerjahn et al. 2005; Takizawa et al. 1992; Mancardi et al. 2008; Bruhns et al. 2009). Interestingly, mouse FcγRs and FcRn efficiently bind human IgG subclasses (Overdijk et al. 2012), whereas human FcγRs and FcRn do not or only bind some mouse IgG subclasses. For example, human FcγRI binds mouse IgG2a and IgG2b with high affinity, but not mouse IgG1 (van der Poel et al. 2010). Human FcγRIIA binds mouse IgG1, 2a and 2b, but not mouse IgG3, and human FcγRIIIB does not bind mouse IgG (Jonsson et al. 2011). Notably, human FcRn has no affinity for mouse IgG1, but its low affinity for mouse IgG2a and IgG2b39 is sufficient to restore a mouse IgG2-dependent autoimmune arthritis in a transgenic mouse model (mice deficient for FcRn and transgenic for hFcRn) (Petkova et al. 2006). Major differences between human FcγR expression patterns and that of their mouse homolog also exist: (1) the expression of FcγRI is easily induced on human PMN, but almost absent on mouse PMN; (2) the expression of human, but not mouse, FcγRIIB is mainly restricted to B cells and basophils; (3) the expression of human FcγRIIIA, but not mouse FcγRIII, is restricted to NK cells and monocytes/ macrophages; (4) FcγRIV exists in mice but not in humans; and (5) FcγRIIA, FcγRIIC and FcγRIIIB exist in humans but not in mice (Ravetch and Kinet 1991). FcγR homologs between humans and mice may thus be defined based on expression and ligand binding, rather than on amino acid similarity. Illustrating this concept, based on binding IgG, mouse FcγRIV was proposed to be a 'functional' homolog of FcγRIIIA (Nimmerjahn et al. 2005).

4.15 Future Perspectives: Humanised Mice

Two approaches have thus far been followed to reach FcR humanised mice: (1) genetic approach, in which the human FcR are introduced as transgenes, or (2) replacement of mouse bone marrow with human stem cells.

Studying human FcRs in vivo constitutes a major challenge that many have attempted by generating human FcR-transgenic mice. Human FcγRI tg mice were reported in 1996 (Heijnen et al. 1996), hFcγRIIA tg mice in 1999 (McKenzie et al. 1999), hFcγRIIB tg in 2011 (Li and Ravetch 2011), hFcγRIIIA tg and hFcγRIIIB tg mice in 1996 (Li et al. 1996) and hFcRn tg in 2003 (Chaudhury et al. 2003). hFcγRIIC tg mice have not yet been generated. hFcγRIIA tg and hFcγRIIIB tg mice closely recapitulated their respective expression patterns in humans. However, hFcγRI is constitutively expressed on neutrophils in hFcγRItg mice, whereas an inducible expression is reported on human neutrophils (Heijnen et al. 1996). hFcγRIIIA and hFcγRIIIB have recently been reported to be both expressed on spleen and circulating DCs, and hFcγRIIIA on eosinophils from hFcγRIIIA tg, hFcγRIIIB tg mice but are not expressed on these cell populations in humans (Smith et al. 2012). hFcγRIIB tg mice have high hFcγRIIB expression on more than 90 % mouse monocytes and granulocytes compared with the poor expression on minor fractions of these cell types reported in humans (Veri et al. 2007).

Crossing these transgenic mice together resulted in a mouse model expressing multiple hFcγRs (i.e. hFc_RI/IIA/IIB/IIIA/ IIIB, except hFcγRIIC and hFcRn) that conserved the original expression patterns of these human transgenes and, unfortunately, also their abnormal expression on several cell populations (Smith et al. 2012). The in vivo properties of these human FcγRs can, nevertheless, be studied in these transgenic mice. However, the results should be carefully evaluated.

Alternatively, cotransplantation of human fetal thymic tissue and $CD34^{+}$ fetal liver cells in nonobese diabetic (NOD)/severe combined immunodeficiency (SCID) or NOD/SCID/$\gamma c^{-/-}$ mice results in the development of multilineage human hematopoietic cells (Manz and Di Santo 2009; Hu and Yang 2012). This can provide a model to study the function of human FcR in mice exemplified by Baerenwaldt (Baerenwaldt et al. 2011).

References

Aalberse RC, van der Gaag R, van Leeuwen J (1983) Serologic aspects of IgG4 antibodies. I. Prolonged immunization results in an IgG4-restricted response. J Immunol 130(2):722–726, PubMed PMID: 6600252

Baerenwaldt A, Lux A, Danzer H, Spriewald BM, Ullrich E, Heidkamp G et al (2011) Fcgamma receptor IIB (FcgammaRIIB) maintains humoral tolerance in the human immune system in vivo. Proc Natl Acad Sci USA 108(46):18772–18777, PubMed PMID: 22065769. Pubmed Central PMCID: 3219118

Barnes N, Gavin AL, Tan PS, Mottram P, Koentgen F, Hogarth PM (2002) FcgammaRI-deficient mice show multiple alterations to inflammatory and immune responses. Immunity 16(3):379–389, PubMed PMID: 11911823

Baudino L, Nimmerjahn F, Azeredo da Silveira S, Martinez-Soria E, Saito T, Carroll M et al (2008) Differential contribution of three activating IgG Fc receptors (FcgammaRI, FcgammaRIII, and FcgammaRIV) to IgG2a- and IgG2b-induced autoimmune hemolytic anemia in mice. J Immunol 180(3):1948–1953, PubMed PMID: 18209093

Bevaart L, Jansen MJ, van Vugt MJ, Verbeek JS, van de Winkel JG, Leusen JH (2006) The high-affinity IgG receptor, FcgammaRI, plays a central role in antibody therapy of experimental melanoma. Cancer Res 66(3):1261–1264, PubMed PMID: 16452176

Bjornson AB, Lobel JS (1987) Direct evidence that decreased serum opsonization of Streptococcus pneumoniae via the alternative complement pathway in sickle cell disease is related to antibody deficiency. J Clin Invest 79(2):388–398, PubMed PMID: 3805275. Pubmed Central PMCID: 424081

Bolland S, Ravetch JV (2000) Spontaneous autoimmune disease in Fc(gamma)RIIB-deficient mice results from strain-specific epistasis. Immunity 13(2):277–285, PubMed PMID: 10981970

Boross P, Arandhara VL, Martin-Ramirez J, Santiago-Raber ML, Carlucci F, Flierman R et al (2011) The inhibiting Fc receptor for IgG, FcgammaRIIB, is a modifier of autoimmune susceptibility. J Immunol 187(3):1304–1313, PubMed PMID: 21724994

Boruchov AM, Heller G, Veri MC, Bonvini E, Ravetch JV, Young JW (2005) Activating and inhibitory IgG Fc receptors on human DCs mediate opposing functions. J Clin Invest 115(10):2914–2923, PubMed PMID: 16167082

Bredius RG, Derkx BH, Fijen CA, de Wit TP, de Haas M, Weening RS et al (1994) Fc gamma receptor IIa (CD32) polymorphism in fulminant meningococcal septic shock in children. J Infect Dis 170(4):848–853

Brownlie RJ, Lawlor KE, Niederer HA, Cutler AJ, Xiang Z, Clatworthy MR et al (2008) Distinct cell-specific control of autoimmunity and infection by FcgammaRIIb. J Exp Med 205(4):883–895, PubMed PMID: 18362174. Pubmed Central PMCID: 2292226

Bruhns P, Iannascoli B, England P, Mancardi DA, Fernandez N, Jorieux S et al (2009) Specificity and affinity of human Fcgamma receptors and their polymorphic variants for human IgG subclasses. Blood 113(16):3716–3725, PubMed PMID: 19018092

Burton DR, Gregory L, Jefferis R (1986) Aspects of the molecular structure of IgG subclasses. Monogr Allergy 19:7–35, PubMed PMID: 2945094

Cartron G, Dacheux L, Salles G, Solal-Celigny P, Bardos P, Colombat P et al (2002) Therapeutic activity of humanized anti-CD20 monoclonal antibody and polymorphism in IgG Fc receptor FcgammaRIIIa gene. Blood 99(3):754–758, PubMed PMID: 11806974

Chaudhury C, Mehnaz S, Robinson JM, Hayton WL, Pearl DK, Roopenian DC et al (2003) The major histocompatibility complex-related Fc receptor for IgG (FcRn) binds albumin and prolongs its lifespan. J Exp Med 197(3):315–322, PubMed PMID: 12566415

Dai X, Jayapal M, Tay HK, Reghunathan R, Lin G, Too CT et al (2009) Differential signal transduction, membrane trafficking, and immune effector functions mediated by FcgammaRI versus FcgammaRIIa. Blood 114(2):318–327, PubMed PMID: 19420354

Dombrowicz D, Flamand V, Miyajima I, Ravetch JV, Galli SJ, Kinet JP (1997) Absence of Fc epsilonRI alpha chain results in upregulation of Fc gammaRIII-dependent mast cell degranulation and anaphylaxis. Evidence of competition between Fc epsilonRI and Fc gammaRIII for limiting amounts of FcR beta and gamma chains. J Clin Invest 99(5):915–925, PubMed PMID: 9062349

Erbe DV, Pfefferkorn ER, Fanger MW (1991) Functions of the various IgG Fc receptors in mediating killing of Toxoplasma gondii. J Immunol 146(9):3145–3151, PubMed PMID: 1826707

Ferrante A, Beard LJ, Feldman RG (1990) IgG subclass distribution of antibodies to bacterial and viral antigens. Pediatr Infect Dis J 9(8 Suppl):S16–S24, PubMed PMID: 2216603

Fijen CA, Bredius RG, Kuijper EJ (1993) Polymorphism of IgG Fc receptors in meningococcal disease. Ann Intern Med 119(7 Pt 1):636, PubMed PMID: 8363182

Flanagan JG, Rabbitts TH (1982) Arrangement of human immunoglobulin heavy chain constant region genes implies evolutionary duplication of a segment containing gamma, epsilon and alpha genes. Nature 300(5894):709–713, PubMed PMID: 6817141

Floto RA, Clatworthy MR, Heilbronn KR, Rosner DR, MacAry PA, Rankin A et al (2005) Loss of function of a lupus-associated FcgammaRIIb polymorphism through exclusion from lipid rafts. Nat Med 11(10):1056–1058, PubMed PMID: 16170323

Hazenbos WL, Gessner JE, Hofhuis FM, Kuipers H, Meyer D, Heijnen IA et al (1996) Impaired IgG-dependent anaphylaxis and Arthus reaction in Fc gamma RIII (CD16) deficient mice. Immunity 5(2):181–188

Heijnen IA, van Vugt MJ, Fanger NA, Graziano RF, de Wit TP, Hofhuis FM et al (1996) Antigen targeting to myeloid-specific human Fc gamma RI/CD64 triggers enhanced antibody responses in transgenic mice. J Clin Invest 97(2):331–338, PubMed PMID: 8567952

Hu Z, Yang YG (2012) Full reconstitution of human platelets in humanized mice after macrophage depletion. Blood 120(8):1713–1716, PubMed PMID: 22773384. Pubmed Central PMCID: 3429310

Ioan-Facsinay A, de Kimpe SJ, Hellwig SM, van Lent PL, Hofhuis FM, van Ojik HH et al (2002) FcgammaRI (CD64) contributes substantially to severity of arthritis, hypersensitivity responses, and protection from bacterial infection. Immunity 16(3):391–402, PubMed PMID: 11911824

Ji H, Ohmura K, Mahmood U, Lee DM, Hofhuis FM, Boackle SA et al (2002) Arthritis critically dependent on innate immune system players. Immunity 16(2):157–168, PubMed PMID: 11869678

Jonsson F, Mancardi DA, Kita Y, Karasuyama H, Iannascoli B, Van Rooijen N et al (2011) Mouse and human neutrophils induce anaphylaxis. J Clin Invest 121(4):1484–1496, PubMed PMID: 21436586. Pubmed Central PMCID: 3069785

Kenneth M, Paul T, Mark W. Janeway`s Immunobiology 7th edition, 2008. Publisher: Garland Science, Taylor and Franscis Group, LLC

Keler T, Guyre PM, Vitale LA, Sundarapandiyan K, van De Winkel JG, Deo YM et al (2000) Targeting weak antigens to CD64 elicits potent humoral responses in human CD64 transgenic mice. J Immunol 165(12):6738–6742

Kono H, Kyogoku C, Suzuki T, Tsuchiya N, Honda H, Yamamoto K et al (2005) FcgammaRIIB Ile232Thr transmembrane polymorphism associated with human systemic lupus erythematosus decreases affinity to lipid rafts and attenuates inhibitory effects on B cell receptor signaling. Hum Mol Genet 14(19):2881–2892, PubMed PMID: 16115811

Kuijpers TW, Weening RS, Out TA (1992) IgG subclass deficiencies and recurrent pyogenic infections, unresponsiveness against bacterial polysaccharide antigens. Allergol Immunopathol 20(1):28–34, PubMed PMID: 1509985

Kumar V, Ali SR, Konrad S, Zwirner J, Verbeek JS, Schmidt RE et al (2006) Cell-derived anaphylatoxins as key mediators of antibody-dependent type II autoimmunity in mice. J Clin Invest 116(2):512–520, PubMed PMID: 16453025

Kyogoku C, Dijstelbloem HM, Tsuchiya N, Hatta Y, Kato H, Yamaguchi A et al (2002) Fcgamma receptor gene polymorphisms in Japanese patients with systemic lupus erythematosus: contribution of FCGR2B to genetic susceptibility. Arthritis Rheum 46(5):1242–1254, PubMed PMID: 12115230

Li F, Ravetch JV (2011) Inhibitory Fcgamma receptor engagement drives adjuvant and anti-tumor activities of agonistic CD40 antibodies. Science 333(6045):1030–1034, PubMed PMID: 21852502. Pubmed Central PMCID: 3164589

Li M, Wirthmueller U, Ravetch JV (1996) Reconstitution of human Fc gamma RIII cell type specificity in transgenic mice. J Exp Med 183(3):1259–1263, PubMed PMID: 8642269. Pubmed Central PMCID: 2192335

Liu C, Goldstein J, Graziano RF, He J, O'Shea JK, Deo Y et al (1996) F(c)gammaRI-targeted fusion proteins result in efficient presentation by human monocytes of antigenic and antagonist T cell epitopes. J Clin Invest 98(9):2001–2007, PubMed PMID: 8903318

Mancardi DA, Iannascoli B, Hoos S, England P, Daeron M, Bruhns P (2008) FcgammaRIV is a mouse IgE receptor that resembles macrophage FcepsilonRI in humans and promotes IgE-induced lung inflammation. J Clin Invest 118(11):3738–3750, PubMed PMID: 18949059. Pubmed Central PMCID: 2571035

Manz MG, Di Santo JP (2009) Renaissance for mouse models of human hematopoiesis and immunobiology. Nat Immunol 10(10):1039–1042, PubMed PMID: 19767720

McGaha TL, Sorrentino B, Ravetch JV (2005) Restoration of tolerance in lupus by targeted inhibitory receptor expression. Science 307(5709):590–593, PubMed PMID: 15681388

McIntosh RS, Shi J, Jennings RM, Chappel JC, de Koning-Ward TF, Smith T et al (2007) The importance of human FcgammaRI in mediating protection to malaria. PLoS Pathog 3(5):e72, PubMed PMID: 17511516. Pubmed Central PMCID: 1868954

McKenzie SE, Taylor SM, Malladi P, Yuhan H, Cassel DL, Chien P et al (1999) The role of the human Fc receptor Fc gamma RIIA in the immune clearance of platelets: a transgenic mouse model. J Immunol 162(7):4311–4318, PubMed PMID: 10201963

Metes D, Ernst LK, Chambers WH, Sulica A, Herberman RB, Morel PA (1998) Expression of functional CD32 molecules on human NK cells is determined by an allelic polymorphism of the FcgammaRIIC gene. Blood 91(7):2369–2380, PubMed PMID: 9516136

Meyer D, Schiller C, Westermann J, Izui S, Hazenbos WL, Verbeek JS et al (1998) FcgammaRIII (CD16)-deficient mice show IgG isotype-dependent protection to experimental autoimmune hemolytic anemia. Blood 92(11):3997–4002, PubMed PMID: 9834201

Minard-Colin V, Xiu Y, Poe JC, Horikawa M, Magro CM, Hamaguchi Y et al (2008) Lymphoma depletion during CD20 immunotherapy in mice is mediated by macrophage FcgammaRI, FcgammaRIII, and FcgammaRIV. Blood 112(4):1205–1213, PubMed PMID: 18495955. Pubmed Central PMCID: 2515149

Morell A, Doran JE, Skvaril F (1990) Ontogeny of the humoral response to group A streptococcal carbohydrate: class and IgG subclass composition of antibodies in children. Eur J Immunol 20(7):1513–1517, PubMed PMID: 2201548

Mosser DM, Edwards JP (2008) Exploring the full spectrum of macrophage activation. Nat Rev Immunol 8(12):958–969, PubMed PMID: 19029990. Pubmed Central PMCID: 2724991

Musolino A, Naldi N, Bortesi B, Pezzuolo D, Capelletti M, Missale G et al (2008) Immunoglobulin G fragment C receptor polymorphisms and clinical efficacy of trastuzumab-based therapy in patients with HER-2/neu-positive metastatic breast cancer. J Clin Oncol 26(11):1789–1796, PubMed PMID: 18347005

Nieto A, Caliz R, Pascual M, Mataran L, Garcia S, Martin J (2000) Involvement of Fcgamma receptor IIIA genotypes in susceptibility to rheumatoid arthritis. Arthritis Rheum 43(4): 735–739, PubMed PMID: 10765917

Nimmerjahn F, Ravetch JV (2005) Divergent immunoglobulin g subclass activity through selective Fc receptor binding. Science 310(5753):1510–1512, PubMed PMID: 16322460

Nimmerjahn F, Ravetch JV (2008) Anti-inflammatory actions of intravenous immunoglobulin. Annu Rev Immunol 26:513–533, PubMed PMID: 18370923

Nimmerjahn F, Ravetch JV (2010) Antibody-mediated modulation of immune responses. Immunol Rev 236:265–275, PubMed PMID: 20636822

Nimmerjahn F, Bruhns P, Horiuchi K, Ravetch JV (2005) FcgammaRIV: a novel FcR with distinct IgG subclass specificity. Immunity 23(1):41–51, PubMed PMID: 16039578

Nimmerjahn F, Lux A, Albert H, Woigk M, Lehmann C, Dudziak D et al (2010) FcgammaRIV deletion reveals its central role for IgG2a and IgG2b activity in vivo. Proc Natl Acad Sci USA 107(45):19396–19401, PubMed PMID: 20974962. Pubmed Central PMCID: 2984189

Overdijk MB, Verploegen S, Ortiz Buijsse A, Vink T, Leusen JH, Bleeker WK et al (2012) Crosstalk between human IgG isotypes and murine effector cells. J Immunol 189(7): 3430–3438, PubMed PMID: 22956577

Petkova SB, Akilesh S, Sproule TJ, Christianson GJ, Al Khabbaz H, Brown AC et al (2006) Enhanced half-life of genetically engineered human IgG1 antibodies in a humanized FcRn mouse model: potential application in humorally mediated autoimmune disease. Int Immunol 18(12):1759–1769, PubMed PMID: 17077181

Pricop L, Redecha P, Teillaud JL, Frey J, Fridman WH, Sautes-Fridman C et al (2001) Differential modulation of stimulatory and inhibitory Fc gamma receptors on human monocytes by Th1 and Th2 cytokines. J Immunol 166(1):531–537, PubMed PMID: 11123333

Ravetch JV, Clynes RA (1998) Divergent roles for Fc receptors and complement in vivo. Annu Rev Immunol 16:421–432, PubMed PMID: 9597136
Ravetch JV, Kinet JP (1991) Fc receptors. Annu Rev Immunol 9:457–492
Roux KH, Strelets L, Michaelsen TE (1997) Flexibility of human IgG subclasses. J Immunol 159(7):3372–3382, PubMed PMID: 9317136
Schmidt RE, Gessner JE (2005) Fc receptors and their interaction with complement in autoimmunity. Immunol Lett 100(1):56–67, PubMed PMID: 16125792
Shushakova N, Skokowa J, Schulman J, Baumann U, Zwirner J, Schmidt RE et al (2002) C5a anaphylatoxin is a major regulator of activating versus inhibitory FcgammaRs in immune complex-induced lung disease. J Clin Invest 110(12):1823–1830, PubMed PMID: 12488432
Siber GR, Santosham M, Reid GR, Thompson C, Almeido-Hill J, Morell A et al (1990) Impaired antibody response to Haemophilus influenzae type b polysaccharide and low IgG2 and IgG4 concentrations in Apache children. N Engl J Med 323(20):1387–1392, PubMed PMID: 2233905
Smith KG, Clatworthy MR (2010) FcgammaRIIB in autoimmunity and infection: evolutionary and therapeutic implications. Nat Rev Immunol 10(5):328–343, PubMed PMID: 20414206
Smith P, DiLillo DJ, Bournazos S, Li F, Ravetch JV (2012) Mouse model recapitulating human Fcgamma receptor structural and functional diversity. Proc Natl Acad Sci USA 109(16): 6181–6186, PubMed PMID: 22474370. Pubmed Central PMCID: 3341029
Syed SN, Konrad S, Wiege K, Nieswandt B, Nimmerjahn F, Schmidt RE et al (2009) Both FcgammaRIV and FcgammaRIII are essential receptors mediating type II and type III autoimmune responses via FcRgamma-LAT-dependent generation of C5a. Eur J Immunol 39(12):3343–3356, PubMed PMID: 19795417
Takai T (1996) Multiple loss of effector cell functions in FcR gamma-deficient mice. Int Rev Immunol 13(4):369–381, PubMed PMID: 8884432
Takai T (2002) Roles of Fc receptors in autoimmunity. Nat Rev Immunol 2(8):580–592, PubMed PMID: 12154377
Takai T, Ono M, Hikida M, Ohmori H, Ravetch JV (1996) Augmented humoral and anaphylactic responses in Fc gamma RII-deficient mice. Nature 379(6563):346–349, PubMed PMID: 8552190
Takizawa F, Adamczewski M, Kinet JP (1992) Identification of the low affinity receptor for immunoglobulin E on mouse mast cells and macrophages as Fc gamma RII and Fc gamma RIII. J Exp Med 176(2):469–475, PubMed PMID: 1386873. Pubmed Central PMCID: 2119311
Tridandapani S, Wardrop R, Baran CP, Wang Y, Opalek JM, Caligiuri MA et al (2003) TGF-beta 1 suppresses [correction of supresses] myeloid Fc gamma receptor function by regulating the expression and function of the common gamma-subunit. J Immunol 170(9):4572–4577, PubMed PMID: 12707335
Ujike A, Ishikawa Y, Ono M, Yuasa T, Yoshino T, Fukumoto M et al (1999) Modulation of immunoglobulin (Ig)E-mediated systemic anaphylaxis by low-affinity Fc receptors for IgG. J Exp Med 189(10):1573–1579, PubMed PMID: 10330436. Pubmed Central PMCID: 2193631
Van de Winkel JGJ, Hogarth MP (1998) Fc receptors in immunity. Kluwer, Great Britain
Van den Herik-Oudijk IE, Capel PJ, van der Bruggen T, Van de Winkel JG (1995) Identification of signaling motifs within human Fc gamma RIIa and Fc gamma RIIb isoforms. Blood 85(8):2202–2211
van der Poel CE, Karssemeijer RA, Boross P, van der Linden JA, Blokland M, van de Winkel JG et al (2010) Cytokine-induced immune complex binding to the high-affinity IgG receptor, FcgammaRI, in the presence of monomeric IgG. Blood 116(24):5327–5333, PubMed PMID: 20805361
van der Poel CE, Spaapen RM, van de Winkel JG, Leusen JH (2011) Functional characteristics of the high affinity IgG receptor, FcgammaRI. J Immunol 186(5):2699–2704, PubMed PMID: 21325219
Van der Pol W, Van de Winkel JGJ (1998) IgG receptor polymorphisms: risk factors for disease. Immunogenetics 48(3):222–232, PubMed PMID: 9683667

Veri MC, Gorlatov S, Li H, Burke S, Johnson S, Stavenhagen J et al (2007) Monoclonal antibodies capable of discriminating the human inhibitory Fcgamma-receptor IIB (CD32B) from the activating Fcgamma-receptor IIA (CD32A): biochemical, biological and functional characterization. Immunology 121(3):392–404, PubMed PMID: 17386079. Pubmed Central PMCID: 2265948

Vidarsson G, van de Winkel JG (1998) Fc receptor and complement receptor-mediated phagocytosis in host defence. Curr Opin Infect Dis 11(3):271–278, PubMed PMID: 17033391

Weinberg GA, Granoff DM, Nahm MH, Shackelford PG (1986) Functional activity of different IgG subclass antibodies against type b capsular polysaccharide of Haemophilus influenzae. J Immunol 136(11):4232–4236, PubMed PMID: 3486228

Weng WK, Levy R (2003) Two immunoglobulin G fragment C receptor polymorphisms independently predict response to rituximab in patients with follicular lymphoma. J Clin Oncol 21(21):3940–3947, PubMed PMID: 12975461

Wu J, Edberg JC, Redecha PB, Bansal V, Guyre PM, Coleman K et al (1997) A novel polymorphism of FcgammaRIIIa (CD16) alters receptor function and predisposes to autoimmune disease. J Clin Invest 100(5):1059–1070, PubMed PMID: 9276722. Pubmed Central PMCID: 508280

Zhang W, Gordon M, Schultheis AM, Yang DY, Nagashima F, Azuma M et al (2007) FCGR2A and FCGR3A polymorphisms associated with clinical outcome of epidermal growth factor receptor expressing metastatic colorectal cancer patients treated with single-agent cetuximab. J Clin Oncol 25(24):3712–3718, PubMed PMID: 17704420

Chapter 5
Molecular and Cellular Pathways Involved in the Anti-inflammatory Activity of IgG

Falk Nimmerjahn

Abstract Despite recent advances in defining and understanding the pathways involved in the anti-inflammatory activity of IgG, much has still to be learned to get a clear picture of how IgG molecules can mediate both pro- and anti-inflammatory effector functions. Complicating the situation, suitable in vivo model systems for all the different diseases where IVIg was suggested to be beneficial for the patient are lacking in many instances. Here, well-defined clinical studies are critical to ultimately prove the beneficial effects for the respective autoimmune disease. In other instances, human studies have already proven that the IgG constant fragment can have an anti-inflammatory activity, and mouse studies have been used to delineate the molecular and cellular pathways of IVIg activity, elucidating an unexpected role of IgG glycosylation-specific Fc-receptors in this immunomodulatory pathway. As this is a very active field of research, this chapter should be viewed as a first insight into this exciting and rapidly developing field, and the interested reader can get a more in-depth insight into the varying mechanisms of IVIg activity in recent more specialized review articles.

Keywords Fc-receptors • ITP • Immunoglobulin G • IVIg • Inflammation • Sialic acid

As we have discussed in the previous chapter, immunoglobulin G (IgG) molecules can be potent inducers of proinflammatory effector functions which are critical for an efficient elimination of invading pathogens and for the elimination of target cells by therapeutic antibodies used in the therapy of human autoimmune and malignant diseases (Nimmerjahn and Ravetch 2011). However, the intravenous infusion of

F. Nimmerjahn (✉)
Institute of Genetics, Department of Biology, University of Erlangen-Nuremberg, Erwin-Rommel-Str. 3, 91058 Erlangen, Germany
e-mail: fnimmerj@biologie.uni-erlangen.de

F. Nimmerjahn (ed.), *Molecular and Cellular Mechanisms of Antibody Activity*,
DOI 10.1007/978-1-4614-7107-3_5,

Licensed	Off-label (selection)
Immunothrombocytopenia (ITP)	Rheumatoid arthritis
Chronic inflammatory demyelinating polyneuropathy (CIDP)	Multiple sclerosis
Guillan-Barré syndrome	Systemic vasculitis
Kawasaki disease	Systemic lupus erythematosus
	Polymyesitis
	Dermatomyesitis
	Graft versus Host disease
	Sepsis
	Myesthenia gravis
	Antiphospholipid syndrome

Table 5.1 Clinical use of IgG as an anti-inflammatory treatment

high amounts of IgG (intravenous IgG or IVIg therapy) is used since nearly 30 years to suppress autoantibody-induced inflammation in a variety of autoimmune diseases indicating that IgG molecules can also have an active anti-inflammatory activity (Negi et al. 2007; Nimmerjahn and Ravetch 2008). Pooled IgG preparations from several thousands of donors are used since the early 1950s to protect immunodeficient patients (e.g., due to genetic predisposition or post hematopoetic stem cell transplantation) from infections. Interestingly, many immunodeficient patients having low circulating IgG levels also have a higher incidence of autoantibodies (Grammatikos and Tsokos 2012). Thus, it was found more or less by chance by Imbach and colleagues that upon treating an immunodeficient child with simultaneous immunothrombocytopenia (ITP) with IVIg, platelet counts increased immediately after polyclonal IgG infusion (Imbach et al. 1981). Since then, IVIg has been used to treat an enormous amount of inflammatory and autoimmune diseases (Table 5.1). Most of these treatments occur off-label as IVIg is only licensed for the treatment of ITP, Guillain-Barre syndrome, Kawasaki disease, and chronic inflammatory demyelinating polyneuropathy (CIDP). Given the variety of autoimmune diseases that have been treated by IVIg infusion, it seems likely that depending on the specific autoimmune disease, different molecular and cellular pathways may be involved in the anti-inflammatory activity (Table 5.2). Most importantly, for many of these diseases, suitable animal models are lacking, making it hard to study the cellular and molecular mechanisms involved in IVIg activity in vivo. In these cases,

F(ab)2-dependent	Fc-dependent
Modulation of antibody production	Blocking of activating Fcγ-receptors
Modulation of signaling pathways	Upregulation of inhibitory FcγRIIB expression
Autoantibody neutralization	Blocking neonatal Fc-receptor function
Modulation of cytokine production	Modulation of cytokine production
Modulation of dendritic cell activation	Modulation of dendritic cell activation
Modulation of cell survival	Immunomodulation via glycosylation specific IgG receptors (SIGNR1)
Inhibition of complement dependent cell activation	Inhibition of complement dependent cell activation

Table 5.2 Proposed mechanisms of IVIg activity

our current models of IVIg activity have to rely on in vitro data until better in vivo model systems become available. Thus, understanding the anti-inflammatory activity of IgG is a challenge and a very active field of research. The aim of this chapter is to provide a snapshot of our current understanding of which pathways may be involved in the immunoregulatory activity of IVIg and will focus largely on experimental models where suitable in vivo systems are available. We will start with a brief description of how therapeutic IgG preparations are generated, followed by an overview of the potential mechanisms involved in the anti-inflammatory activity of IgG and the relevance of these observations for the human immune system.

5.1 How Therapeutic IVIg Preparations Are Generated

Despite the availability of several highly specific novel IgG purification procedures during the last 30 years, the protocol to enrich the IgG fraction from human serum for the IVIg preparation has remained nearly the same since its introduction (Buchacher and Iberer 2006). Initially, IVIg was given as subcutaneous or intramuscular injections, followed by a shift towards intravenous infusions in later years. For this, adjustments in salt content and osmolarity were made to prevent adverse reactions upon intravenous infusions. Indeed, most of the mild side effects that can occur upon IVIg infusion, including a headache, fever, fatigue, and a sore throat, are thought to be the result of the additives present in IVIg to stabilize the preparation

(Hamrock 2006). Instead of using Protein A- or Protein G-based affinity purification protocols, an ethanol- or caprylate-based precipitation step is used to predominantly enrich serum IgG. In addition to this IgG enrichment, several measures have to be taken to inactivate potential viral contaminations, such as HIV and hepatitis (Schleis 2005; Carbone 2007). The last major outbreak of a hepatitis C infection as a result of inadequate viral inactivation protocols happened nearly 20 years ago (Bjoro et al. 1994). As different IVIg providers use varying protocols for the IVIg preparation, it is not surprising that apart from the main IgG fraction, the presence of other molecules can greatly vary. IgA, for example, can either be virtually absent or be present in relatively high concentrations (1–2 mg/ml) in IVIg preparations. The critical steps in the IVIg quality control process (apart from preventing microbial contamination) are to avoid high levels of IgG aggregates and the contamination of blood coagulation factors, which may result in a cytokine storm or in the induction of cardiovascular complications, respectively. Although these complications are quite rare, they can be fatal, which has triggered the interest to understand the molecular and cellular mechanisms of the anti-inflammatory activity of IVIg to be able to replace this primary blood product by recombinantly generated preparations in the future. In addition to the safety issue, this would allow to prevent shortages in supply that have occurred more frequently in recent years resulting in ever rising costs for IVIg products.

5.2 F(ab)2-Dependent Mechanisms of IVIg Activity

As IVIg is generated from the serum of thousands of healthy immunocompetent donors, the repertoire of IgG specificities contained in this preparation will be quite diverse. It is estimated that more than 10,000,000 antibody specificities may be present in an IVIg preparation. Given that the blood donors have received vaccinations or have been infected with a variety of microbial pathogens, the majority of antibody specificities may be expected to be directed against environmental antigens, explaining its potency in protecting immunodeficient patients from reoccurring bacterial and viral infections. In addition, the so-called natural IgG repertoire of the different blood donors is present in IVIg preparations. The natural IgG repertoire is generated during the steady state in the absence of infections and can contain a variety of self-reactive antibody specificities. Moreover, anti-idiotypic antibodies, that is, antibodies directed against the mutated variable region of other antibodies generated during immune responses, and other molecules including cytokines and soluble receptors can be found consistently (albeit in varying amounts) in IVIg preparations (Negi et al. 2007; Rossi and Kazatchkine 1989; Le Pottier et al. 2007). Examples for this are antibodies directed against the T cell receptor, Siglec-9, integrins, FAS, and CD5 (Rossi and Kazatchkine 1989; Marchalonis et al. 1992; Prasad et al. 1998; Vassilev et al. 1993; von Gunten et al. 2006; von Gunten and Simon 2008; von Gunten et al. 2007). Moreover, a recent study demonstrated that during the process of IVIg fractionation, certain cryptic natural antibody specificities can become activated (Djoumerska-Alexieva et al. 2010). This includes cytokine- and April (anti-proliferation-inducing ligand)-specific antibodies which have been found to be produced in autoimmune and malignant diseases (Table 5.2).Depending on the disease, it is quite likely that such specificities in the IVIg preparation may be responsible for some of the therapeutic effects. In myasthenia gravis, for example, autoantibodies directed against the acetylcholine receptor block signal transduction via acetylcholine between the neuron and the muscle, resulting in a severe impairment of muscle innervations and function. If antibodies within the IVIg preparation would neutralize these autoantibodies, this would clearly impact the disease phenotype (Gilhus 2011). Another example is toxic epidermal necrolysis, where the interaction of CD95 with the CD95 receptor (Fas and Fas ligand) is critical for the death of keratinocytes (French et al. 2006). It was shown that CD95-specific blocking antibodies can be present in IVIg preparations which can block induction of keratinocyte apoptosis (Viard et al. 1998). Interestingly, several groups have found that incubating a variety of human cell types including monocytes and B cells in vitro with IVIg preparations can also induce apoptosis and demonstrated the presence of CD95-specific agonistic antibodies in the IVIg preparation (Table 5.2) (Altznauer et al. 2003). The major problem with the potential role of different antibody specificities is that they are present usually only in minute amounts and may vary between IVIg preparations. Thus, it remains to be shown whether these are general mechanisms of IVIg activity. Nonetheless, these studies clearly open new avenues for therapeutic strategies. Taken together, there is convincing evidence for the presence of a wide variety of self-reactive antibodies within the IVIg preparation, and depending on the disease, this may be one explanation for the therapeutic activity. In addition to the role of the antigen specificity, it was also shown that the IgG F(ab)2 domain can bind to activated complement components C3a or C5a (Fig. 5.1). As these activated complement factors are produced during active inflammation and might be responsible for IgG-dependent tissue inflammation, it was suggested that IVIg might neutralize activated complement components. Indeed, IVIg-mediated neutralization of activated complement components was able to interfere with asthma in a mouse model of allergic airway inflammation (Basta et al. 1989). Of note, however, IVIg failed to show improvement of human asthma in two out of three clinical trials, and allergy is not the major indication for IVIg therapy. Besides allergy, activated complement C3a and C5a have been demonstrated to be critical for the development of autoimmune diseases in mice, including rheumatoid arthritis and bullous pemphigoid (an autoantibody-mediated skin blistering disease). While this provides an attractive model for an anti-inflammatory mechanism, it should be kept in mind that in several autoimmune diseases including rheumatoid arthritis, the IVIg Fc-fragment, and not the F(ab)2-fragment, was shown to be sufficient for the anti-inflammatory activity (Bruhns et al. 2003; Debre et al. 1993; Kaneko et al. 2006a, b; Samuelsson et al. 2001).

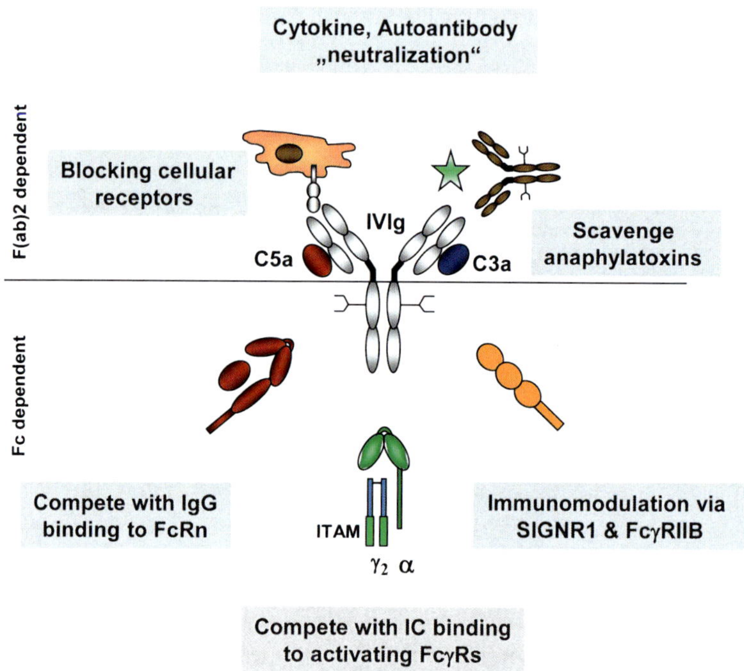

Fig. 5.1 *Potential pathways for the direct interference of IVIg with autoantibody-mediated proinflammatory responses.* Depicted are a selection of F(ab)2- and Fc-dependent mechanisms which have been proposed to be involved in the anti-inflammatory activity of IVIg. F(ab)2-dependent mechanisms include the blocking of cellular receptors involved in autoimmune pathology, autoantibody "neutralization" (anti-idiotype antibodies), and scavenging of proinflammatory mediators such as the activated complement proteins C3a and C5a. Fc-dependent mechanisms include competition with immune complexes for binding to activating FcγRs and to the neonatal FcRn. Moreover, upregulation of FcγRIIB via a SIGNR1- and IgG glycosylation-dependent pathway was demonstrated to be responsible for the anti-inflammatory activity of IVIg in ITP and rheumatoid arthritis. See text for further details

5.3 Fc-Dependent Mechanisms of IVIg Activity

There is clear experimental evidence that the isolated Fc-fragment of the IVIg preparation can be sufficient to achieve therapeutic effects in mice and humans in vivo. As a matter of fact, this was first demonstrated in humans in a clinical trial with ITP patients in 1993 where the infusion of IVIg Fc-fragments was sufficient to induce a rise in platelet counts (Debre et al. 1993). Later, these results were confirmed in a variety of animal studies including mouse models of ITP, rheumatoid arthritis, and nephrotoxic nephritis (Bruhns et al. 2003; Kaneko et al. 2006a, b; Samuelsson et al. 2001). The major question, addressed by many groups over the last years, was how the isolated IgG Fc-fragment may achieve such therapeutic effects. In summary, three different types of potential mechanism have been proposed, based on blocking one of the three pathways critical for IgG function as we have discussed before (Fig. 5.1). These include:

1. Involvement of activating Fc-receptors in the anti-inflammatory activity of IVIg
2. Lowering the half-life of autoantibodies due to saturating the interaction with the neonatal FcRn
3. IVIg-mediated immunomodulation via the inhibitory FcγRIIB

In the next paragraphs, we will discuss arguments, speaking either for or against an involvement in the anti-inflammatory activity of IVIg.

1. *Involvement of activating Fcγ-receptors in the anti-inflammatory activity of IVIg*

In the previous chapter, we discussed in detail that binding of IgG immune complexes to activating FcγRs is essential for their activity. Thus, eliminating this interaction would be the most straightforward way to interfere with the clearance of IgG-coated red blood cells in autoimmune hemolytic anemia (AIHA) or of IgG-coated platelets in immunothrombocytopenia (ITP). Indeed, using a blocking antibody for FcγRIIIA in human ITP patients was able to interfere with autoantibody-mediated platelet depletion (Clarkson et al. 1986). These results were confirmed in transgenic mice expressing human FcγRIIIA (Samuelsson et al. 2001). Of note, however, injecting humans with FcγRIIIA antibodies had severe side effects, including neutropenia and a severe systemic cytokine release. If IVIg would work via the same mechanism, one would expect similar side effects, which are usually not seen upon intravenous infusion. Moreover, IVIg consists of predominantly monomeric IgG which does not have the capacity to bind to low affinity FcγRs unless present in the form of an immune complex. However, we have discussed before that all kinds of self-reactive antibody specificities including anti-idiotypic antibodies are present in the IVIg preparation. Accordingly, it was shown that antibodies specific for soluble proteins or for cell surface proteins on red blood cells, for example, can exhibit IVIg-like activity in ITP (Bussel et al. 1991). This finding is consistent with earlier data from human ITP patients treated with anti-D IgG hyperimmune serum, which is a pooled polyclonal IgG fraction collected from donors immunized to the rhesus D antigen [anti-Rh0(D) positive]. Anti-D, which, compared with IVIg, is used at a more than 40,000-fold lower dose, can prevent platelet consumption in human ITP (Salama et al. 1984; Oksenhendler et al. 1988; Godeau et al. 1996). However, the use of a monoclonal anti-D antibody of the IgG1 subclass that can efficiently interact with human FcRs was not able to prevent platelet depletion in seven Rh0(D)-positive patients, arguing against a simple IC-formation mechanism (Godeau et al. 1996). Clearly, the use of anti-D or other hyperimmune sera in the presence of the respective antigen will create a certain amount of immune complexes that may compete with the autoantibody-antigen complexes and inhibit their phagocytosis by a mechanism similar to directly blocking activating FcγRs with an FcγR-specific antibody. Whether the amount of IC formation is high enough to achieve a significant level of binding to activating FcγRs is unclear (Crow and Lazarus 2008). Arguing against this is the fact that the Fc-fragment is usually sufficient for the anti-inflammatory activity and that formation of large amounts of immune complexes would result in more severe side effects than are typically seen upon IVIg infusion.

Apart from a simple blocking mechanism via the formation of immune complexes, it was also suggested that IVIg binding to FcγRIII on dendritic cells may be involved in the anti-inflammatory activity of IVIg (Crow et al. 2009). In this study, ITP induction via autoantibody injection was prevented if dendritic cells were incubated in vitro with IVIg, washed to remove free IVIg, and transferred back into mice (Siragam et al. 2006). This protective effect was lost if dendritic cells were generated from FcγRIII-deficient mice, suggesting that FcγRIII-dependent signaling events may be important for the anti-inflammatory activity of IVIg. While these experiments are very convincing, several studies performed exclusively in vivo without any cell transfer step could not verify a contribution of FcγRIII to the anti-inflammatory activity of IVIg (Kaneko et al. 2006a; Schwab et al. 2012). Thus, IVIg was still able to protect FcγRIII-deficient mice from autoantibody-induced kidney damage in a model of nephrotoxic nephritis and against platelet depletion in a passive model of ITP.

Taken together, it is unclear at present if activating FcγRs is involved in the anti-inflammatory pathway of intravenous immunoglobulins at least in the models discussed in this paragraph. Given the importance of activating FcγRs for autoantibody-induced tissue inflammation, however, strategies aimed at blocking the activation of innate immune effector cells via activating FcγRs are a very promising anti-inflammatory therapy, evidenced by the use of small inhibitors interfering with the signal transduction pathway of activating FcγRs (Podolanczuk et al. 2009).

2. *Lowering the half-life of autoantibodies due to saturating the interaction with the neonatal FcRn*

Another pathway that has been suggested to be modulated by IVIg infusion is the neonatal FcR-dependent regulation of IgG half-life (Fig. 5.1). As will be described in more detail in Chap 6 by Baker and colleagues, the neonatal FcRn is critical for the long serum half-life of IgG molecules (Roopenian and Akilesh 2007). Mice or rats deficient in FcRn or blocked for FcR function have a dramatically reduced IgG half-life of a few hours compared to the 8–12 days observed in normal mice (Getman and Balthasar 2005). Thus, if the high amounts of IgG infused into patients would compete for FcRn binding with autoantibodies, this may result in a faster clearance of the self-reactive antibodies resulting in a lower level of tissue damage. Indeed, a reduction in autoantibody half-life from 80 to approximately 50 h was observed in ITP model systems (Hansen and Balthasar 2002). While this may be relevant during the chronic phase of an autoimmune disease, it cannot explain the immediate IVIg-mediated suppression of passive induction of ITP by autoantibody injection. Here, IVIg is injected 1–2 h before application of the autoantibody and the maximum autoantibody-dependent platelet depletion is observed within 2–4 h after antiplatelet IgG injection, making it unlikely that the reduction in IgG half-life would have a significant impact on this acute induction of disease. Moreover, a critical issue for studies using FcRn knockout mice is that due to the short half-life of IgG in these animals, much higher doses of the autoimmune disease-inducing antibody has to be injected to achieve a similar level of disease. Similarly, the half-life of IVIg will be much shorter in this mouse strain, making it difficult to compare the data to wild-type mice. This may explain that some studies found a role of the neonatal FcRn in mediating IVIg activity, while others could not confirm these findings (Crow et al. 2011).

Regardless if this is the actual mechanism of IVIg activity in some models, inhibiting FcRn function with specific blocking antibodies is a promising new avenue to lower (auto)antibody half-life and interfere with autoimmune pathology.

3. *IVIg-mediated immunomodulation*

Every immune response including IgG-dependent effector mechanisms has to be tightly regulated to prevent overt tissue inflammation. As discussed in the previous chapter dealing with the mechanism of IgG activity, the inhibitory FcγRIIB is critical for balancing the activity of signals transduced by activating FcγRs. Thus, one mechanism of IVIg activity could be in resetting this threshold to a higher level thereby decreasing tissue inflammation (Fig. 5.1). Indeed, it was observed in several animal models and more recently in a study with patients being treated with IVIg to suppress chronic inflammatory demyelinating polyneurophathy (CIDP), in which autoantibodies cause inflammation of peripheral nervous tissue resulting in progressing paralysis, that FcγRIIB becomes upregulated upon IVIg infusion (Tackenberg et al. 2009). This increase in FcγRIIB expression was seen on myeloid cells and B cells, suggesting that IVIg affects innate as well as adaptive immune responses. Considering the important role of the inhibitory FcγRIIB in regulating immune responses in both arms of the immune system and the clear association of decreased FcγRIIB expression in autoimmune-prone mouse strains and human SLE and CIDP patients with the production of autoantibodies and enhanced tissue inflammation, increasing FcγRIIB expression to normal levels may have the capacity to interfere with autoimmune disease (Tackenberg et al. 2009; Floto et al. 2005; Kono et al. 2005; Pritchard et al. 2000; Mackay et al. 2006; Takai et al. 1996). Indeed, several studies could show that correcting FcγRIIB expression levels on B cells of autoimmune-prone mouse strains could restore humoral tolerance and interfere with autoantibody production (Brownlie et al. 2008; McGaha et al. 2005). Furthermore, FcγRIIB-deficient mice were no longer protected by IVIg from the development of autoantibody-induced immunothrombocytopenia, rheumatoid arthirits, and nephrotoxic nephritis (Bruhns et al. 2003; Kaneko et al. 2006a; Samuelsson et al. 2001). The major question is how the change in FcγRIIB expression is achieved by IVIg. A direct binding to FcγRIIB is very unlikely due to the low affinity of monomeric IgG towards the inhibitory FcγR. A possible solution to this conundrum was provided by Kaneko and colleagues in 2006, showing that the sugar moiety attached to the asparagine residue 297 in the IgG constant fragment is essential not only for the pro- but also for the anti-inflammatory activity of IgG (Kaneko et al. 2006b). To understand this potential pathway of IVIg activity and the impact of differential IgG glycosylation in general, we have to make a little detour into IgG glycosylation first, before we come back to the importance of these phenomenon for IgG-dependent immunomodulation.

5.4 Variability in Immunoglobulin Glycosylation

Immunoglobulins of all isotypes are glycoproteins, although the actual level of sugar moieties attached to the individual isotype can differ dramatically. Whereas IgA, IgD, IgM, and IgE molecules have several sugar moieties attached to their fragment crystallizable (Fc) domain, the members of the IgG isotype are characterized by only one sugar moiety attached to the asparagine 297 (N297) residue in each CH2 domain of the two different heavy chains in the IgG molecule (Arnold et al. 2007). This biantennary sugar moiety consists of a heptameric core of three mannose (Man) and four N-acetylglucosamine (GlcNAc) residues with variable additions of terminal galactose and sialic acid residues or branching GlcNAc and fucose residues. The two branches of this structure are designated 1,3- or 1,6-arm according to the linkage of the branching mannose residue of the respective arm to the mannose residue in the linear core (Fig. 5.2). A similar sugar structure can be found on many proteins including fetuin, plasminogen, transferrin, and on other immunoglobulin isotypes including IgA, IgM, and IgE (Kobata 2008; Rudd et al. 2001). The sugar moiety on these other serum proteins is highly processed, containing terminal galactose and sialic acid residues in the majority of cases (60–95 %) (Raju 2008). Serum IgG, however, shows a great variability with respect to the N297-attached sugar moiety (Kobata 2008; Raju 2008). Thus, more than 30 different IgG glycosylation variants can be identified for any of the four different human IgG subclasses generating a great heterogeneity. Creating even more complexity, there is evidence that the sugar moieties of the two different Fc-fragments may differ, theoretically enabling more than 500 glycoforms (Holland et al. 2006; Masuda et al. 2000; Saphire et al. 2002; Sutton and Phillips 1983). Of note, this sugar moiety is not exposed on the surface of the molecule but rather buried in the hydrophobic pocket formed by the two IgG heavy chain constant domains two and three (CH2 and CH3).

Based on the presence or absence of terminal galactose residues, three IgG glycosylation subsets have been defined. Whereas the IgG-G0 form (~20–25 % of IgG pool) has no terminal galactose residues, the IgG-G1 form (35–45 % of IgG pool) has one and the IgG-G2 form (~16–27 % of IgG pool) has two galactose residues (Fig. 5.2a) (Arnold et al. 2007; Wormald et al. 1997). This distribution can change during states of active inflammation and lead to an overrepresentation of IgG-G0 glycoforms (> 55 % of the glycan pool) as will be discussed in the next paragraph in detail. With respect to terminal sialic acid, about 10–12 % of IgG glycans contain one terminal sialic acid residue, whereas desialylated structures can only be identified in trace amounts. There seems to be a predominant sialylation on the 1,3-arm of the biantennary structure (Wormald et al. 1997) and monosialylation on both sugar moieties occurs in about 1 % of the glycan pool (Fig. 5.2) (Stadlmann et al. 2009). In addition to the sugar moiety of the Fc-domain, up to15–20 % of serum IgG molecules may become glycosylated in the heavy or light chain of the Fab-fragment due to the de novo generation of glycosylation sites during somatic hypermutation and diversification of the IgG repertoire (Arnold et al. 2007; Stadlmann

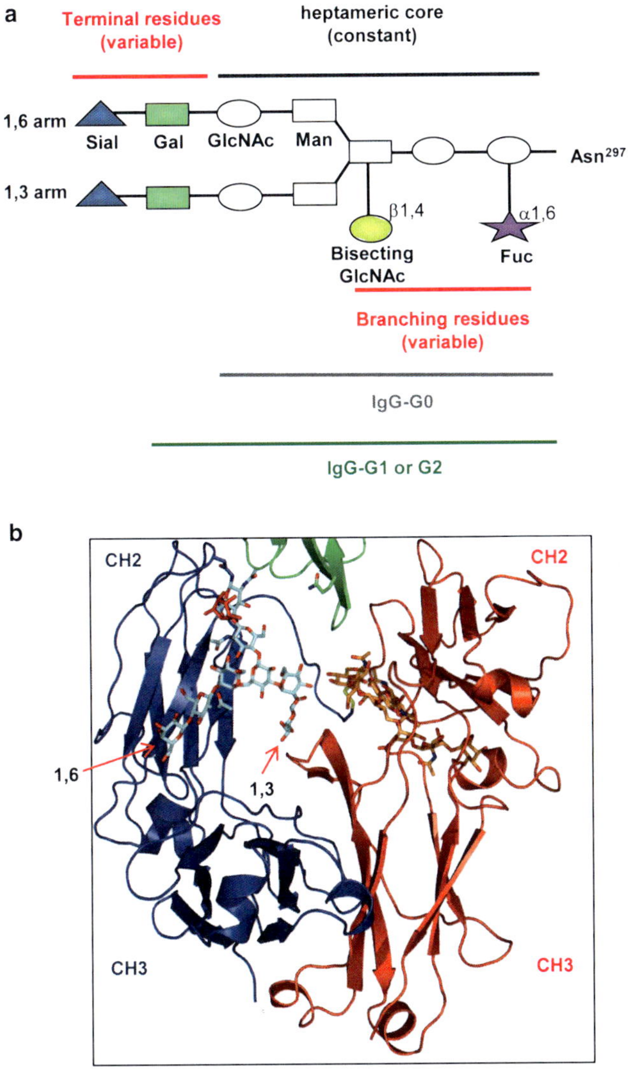

Fig. 5.2 *The sugar moiety of the IgG Fc-fragment*. (**a**) Shown are all the sugar residues that may be present in the different IgG glycoforms that can be found in the serum. The variable branching and terminal residues attached to the core structure are colored. Depending on the presence or absence of terminal galactose residues on the 1,3 or 1,6 arm, IgG-G1/2 or IgG-G0 glycoforms can be distinguished. (**b**) The three-dimensional structure of the IgG Fc-fragment constant domains 2 and 3 (CH2 and CH3) including the structure of the asparagine 297-attached sugar moiety is shown. Arrows indicate the 1,3 and 1,6 arm of the branched sugar structure. See text for further details

et al. 2009; Jefferis 2007). In contrast to the N297-attached sugar moiety, more than 46 % of the Fab-associated sugar structures may contain terminal sialic acid (predominantly on the 1,6-arm) and galactose residues and have a higher level of branching GlcNAc residues (Arnold et al. 2007; Wormald et al. 1997; Stadlmann et al. 2009). In contrast to the high flexibility of the Fab-associated sugar moiety which can hardly be resolved in crystal structure analysis, the Fc-linked sugar moiety is more rigid due to the confined space within the hydrophobic pocket formed by the two Fc-fragments. Whereas the 1,6-arm comes in close contact with the amino acid backbone of the CH2 domain, the 1,3-arms of the two sugar moieties are oriented towards the center of the hydrophobic pocket (Fig. 5.2b) (Deisenhofer 1981; Radaev et al. 2001; Sondermann et al. 2000).

5.5 Changes in IgG Galactosylation and Sialylation During Inflammation

The pool of IgG glycoforms is rather stable during the steady state between different healthy individuals. Dramatic changes in the abundance of certain variants, however, can be observed during proinflammatory responses such as active autoimmune disease. Thus, a reduction of IgG glycoforms containing terminal galactose and sialic acid residues has been observed in young and adult patients with rheumatoid arthritis and primary osteoarthritis in the serum and synovial fluid (Parekh et al. 1985, 1988a; Sumar et al. 1991). The increased appearance of these IgG-G0 glycoforms was also found to be relevant for early diagnosis of disease, especially in combination with other markers such as rheumatoid factor and anti-citrullinated protein antibodies (ACPA) (Lu et al. 2007; Young et al. 1991). More recently, glycosylation of ACPA-specific antibodies was directly assessed after affinity purification. Although there was only a trend towards increased IgG-G0 forms compared to total serum IgG, there was a clear reduction of monosialylated sugar moieties on ACPA antibodies compared to total serum IgG (Scherer et al. 2010). Autoantibodies isolated from synovial fluid demonstrated a highly increased level of IgG-G0 glycoforms lacking terminal sialic acid residues, providing evidence that changes in IgG glycosylation may be more dramatic at the site of inflammation. Although very speculative, one might envisage that B cells at the site of chronic inflammation (in RA: the inflamed joint) might be the major source of differentially glycosylated (auto)antibodies. Lymphoid-like structures containing B cells, T cells, and dendritic cells were described in several autoimmune diseases (including RA) in inflamed organs. Similar observations were made in patients with Crohn's disease, systemic lupus erythematosus, small-vessel vasculitides, and spondyloarthropathy, where a heightened level of IgG-G0 forms correlated with disease activity and also had prognostic value (Dube et al. 1990; Holland et al. 2002; Keusch et al. 1996; Leirisalo-Repo et al. 1999; Tomana et al. 1988). Interestingly, this increased level of IgG-G0 disappeared in women and female mice with RA during pregnancy, paralleled by a reduction in autoimmune disease activity (van de Geijn et al. 2009;

Rook et al. 1991). Similar observations were made after treatment of acute RA with methotrexate and infusion of TNFα-blocking agents, suggesting that proinflammatory cytokines such as TNFα may directly influence IgG glycosylation (Croce et al. 2007; Pasek et al. 2006). A recent in vitro study with human B cells cultured in the presence or absence of different TH1 or TH2 cytokines supports this notion by showing that IgG galactosylation and sialylation may be enhanced in the presence of IL21 and decreased in the presence of all-trans retinoic acid (Wang et al. 2011). Of note, no direct effect of TNFα was observed in this in vitro system. In accordance with the human data, autoimmune-prone mouse strains such as the MRL/lpr strain or KBxN mice which spontaneously develop an RA-like autoimmune disease show an altered serum IgG glycosylation pattern (Kaneko et al. 2006b; Bond et al. 1990). Starting with the onset of disease, an increase of IgG-G0 forms can be detected. Interestingly, this increase in the IgG-G0 glycoforms also occurred in the absence of disease after removal of T-helper cells which are crucial for arthritis development, suggesting that the genetic background of the animals contributed to the rise of the IgG-G0 glycoform independently of disease activity (Kuroda et al. 2001a, b). Data from mouse models in which autoimmune disease is induced by active immunization, however, clearly show that induction of inflammation by pristane oil injection or immunization with complete Freund's adjuvant results in an increase in IgG-G0 glycoforms, suggesting that proinflammatory stimuli can directly influence IgG glycosylation and reduce the level of terminal galactosylation and sialylation (Kaneko et al. 2006b; Rook et al. 1991). Consistently, a similar increase in IgG-G0 was observed in individuals infected with mycobacterium tuberculosis- and in hepatitis C virus-infected patients with fibrosis and cirrhosis (Tomana et al. 1988; Mehta et al. 2008; Parekh et al. 1989). In the latter case and in mouse immunization studies, this decrease in terminal galactose and sialic acid residues could be demonstrated on antigen-specific IgG (Kaneko et al. 2006b; Mehta et al. 2008). In patients with systemic vasculitis, it was demonstrated that the changes in the composition of the sugar moiety were specific for the Fc-linked sugar domain and Fab-linked sugar moieties within the same IgG molecule still contained high levels of terminal galactose and sialic acid residues, suggesting that no general B cell intrinsic defect of antibody glycosylation but rather structural changes in the Fc-fragment may be responsible for the observed glycosylation pattern (Holland et al. 2006). An important question is whether the inflammatory milieu changes IgG glycosylation in all antibody-producing plasma cells or if it affects only select antigen-specific plasma cell subsets. Arguing for the latter scenario, there is evidence that the glycosylation of antigen-specific IgG can differ from serum IgG glycosylation, suggesting that different B cell subsets produce differentially glycosylated IgG (Wuhrer et al. 2009). This finding is corroborated by data from patients with monoclonal immunoglobulin deposition disorders showing that the aberrantly produced IgG has a different glycosylation pattern than total serum IgG (Omtvedt et al. 2006). Moreover, there is data indicating that altered galactosyltransferase activity in patients with RA may be responsible for the enhanced IgG-G0 level (Axford 1988). More studies will be necessary to obtain a better picture of the mechanism underlying altered IgG glycosylation.

5.6 Changes in IgG Glycosylation During Aging

The likelihood to develop an autoimmune disease such as rheumatoid arthritis increases with age. Interestingly, age-related differences in serum IgG glycosylation have been observed by several studies (Parekh et al. 1988b; Shikata et al. 1998; Yamada et al. 1997). It was demonstrated that galactosylation increases from early childhood to 25 years where a maximum of 80 % serum IgG is in either the IgG-G1 or IgG-G2 glycoform, followed by a decrease of these glycoforms later in life. The age-related increase of IgG lacking terminal sialic acid and galactose residues seemed to be strongly influenced by gender, as only women showed a significant increase of IgG-G0 variants, which has to be taken into consideration when interpreting results from studies which included male and female individuals. Thus, the IgG-G0 glycoform was present in 20 % of the serum IgG pool of women in their twenties and it increased to about 40 % in women in their 60s. In contrast, the IgG-G0 level remained at a rather stable level of 25 % in men. Of note, these changes happened independently of the development of overt autoimmune disease comparable to the increase of the IgG-G0 form MRL/lpr mice in the absence of CD4 T cells. However, the age-related increase in IgG-G0 never reached the level that was observed during acute inflammation in patients with rheumatoid arthritis, where IgG-G0 can reach levels of more than 50 %. A more recent study investigating age-related changes in IgG1 and IgG2 subclass glycosylation was able to confirm these findings and furthermore suggested that decreased levels of bisecting GlcNAc residues on IgG-G0 glycoforms below the age of 60 may be associated with human longevity (Ruhaak et al. 2010). One possible interpretation of these results is that age-related physiologic changes and sex hormones may impact IgG glycosylation. Whether this provides the basis or lowers the threshold for development of autoimmune diseases is unknown. However, these observations have fuelled the interest in understanding whether terminal sialic acid and galactose residues might influence IgG activity.

5.7 The Role of Galactose for the Proinflammatory Activity of IgG

As the lack of terminal galactose (and sialic acid) residues coincided with high autoimmune disease activity in rheumatoid arthritis, it was suggested that the lack of galactose might increase the proinflammatory activity of IgG. Indeed, a serum transfer study with arthritogenic serum either left untreated or digested with galactosidase to remove terminal galactose residues induced a more severe arthritis in recipient mice (Rademacher et al. 1994). A subsequent study demonstrated that the IgG-G0 glycoform due to its exposure of the mannose-rich core sugar structure acquires the capacity to bind to mannose-binding lectin (MBL), providing a plausible explanation for this enhanced proinflammatory activity in vivo (Malhotra et al. 1995).

Similar to C1q, MBL is a multimeric protein consisting of two subunits (MBL-A and MBL-B) that can bind to mannose-rich structures on the surface of bacteria resulting in the activation of the complement pathway. Complicating the situation, some studies reported the opposite phenotype showing that IgG glycovariants containing high levels of galactose residues may have a higher proinflammatory or lytic activity, although others could not confirm this enhanced activity in independent model systems (Boyd et al. 1995; Hadley et al. 1995; Kumpel et al. 1994, 1995). At the time of these experiments, MBL knockout mice were not available to test this prediction in vivo. With the generation of these mice, it was possible to revisit the role of the lectin pathway for the proinflammatory activity of IgG-G0. Interestingly, MBL-deficient mice did not show an impaired activity of IgG-G0 or IgG-G1/2 variants in several different model systems in vivo (Ji et al. 2002; Nimmerjahn et al. 2007). Similar results were obtained in complement C3-deficient mice showing no impaired IgG-G0 activity. The complement protein C3 is an essential component of the classical (antibody dependent) and the alternative pathways including the MBL-dependent pathway of complement activation (Carroll 1998). In contrast, IgG-G0 activity was abrogated in the absence of activating FcγRs consistent with results obtained with IgG-G1/2 glycoforms (Nimmerjahn et al. 2007). Moreover, the absence of galactose residues resulted in a slight decrease in affinity for the inhibitory FcγRIIB which may account for an increased activity at least in some model systems (Nimmerjahn et al. 2007). A recent study investigated the presence of different MBL alleles resulting in either normal, intermediate, or low/absent MBL serum levels in women with arthritis during and right after pregnancy (van de Geijn et al. 2009). In accordance with the in vivo studies in MBL knockout mice, no correlation between MBL levels and rheumatoid arthritis disease activity could be observed. Data from cross-sectional studies with rheumatoid arthritis patients even suggested that the absence of MBL may accelerate development of disease and joint destruction, indicating that MBL might rather have a protective role (Garred et al. 2000; Graudal et al. 2000). Taken together, at present, no clear picture for an important role of the MBL pathway as a mediator of the proinflammatory activity of IgG glycoforms lacking terminal galactose residues is emerging. Similarly, the effect of terminal galactose residues on IgG activity in vivo is at best considered an open question. There is no doubt, however, that these glycoforms show a strong correlation with disease activity, necessitating an alternative explanation for an active function of this glycosylation variant.

5.8 The Role of Sialic Acid for the Anti-inflammatory Activity of IgG

Based on the data presented in the previous paragraphs, it was assumed a long time that the IgG-G0 glycoform may acquire a more proinflammatory activity due to the association of increased levels of this glycovariant with exacerbated autoimmune disease activity. As IgG infused intravenously at high doses (1–3 g/kg) can also have an active anti-inflammatory activity, this allows an alternative hypothesis, bringing us back to a potential mechanism of IVIg activity. Thus, not the increased level of IgG-G0 might have an enhanced proinflammatory activity but the loss of IgG-G1/G2 variants with or without terminal sialic acid residues might have an active anti-inflammatory activity that gets lost during autoimmune disease (Nimmerjahn and Ravetch 2008; Durandy et al. 2009). The first evidence that the N297-attached sugar moiety is an important component for the anti-inflammatory activity was provided by studies that compared the ability of normal or aglycosylated IVIg to suppress joint inflammation in a murine model of rheumatoid arthritis (Kaneko et al. 2006b). Whereas the fully glycosylated IVIg preparation was functional, the aglycosylated variant lost its anti-inflammatory activity. A more detailed analysis demonstrated that terminal sialic acid residues might be the major factor determining the anti-inflammatory activity as IVIg treated with neuraminidase also lost its capacity to suppress rheumatoid arthritis. Conversely, enriching the IVIg preparation for the IgG glycovariants rich in sialic acid via lectins such as sambucus nigra agglutinin (SNA) enhanced the proinflammatory activity (Kaneko et al. 2006b). Consistent with previous studies in mice and humans, showing that the IgG Fc-fragment, but not the Fab-fragment, is essential for the anti-inflammatory activity, it was demonstrated that the Fc- and not the Fab-linked sugar moiety is crucial for the anti-inflammatory activity. In a similar manner, other serum proteins such as transferrin and fetuin, containing a sugar moiety with a similar structure and high levels of terminal sialic acid residues, were not able to recapitulate IVIg activity, indicating that not simply the presence of sialic acid on these branched sugar moieties but more likely the structural changes imposed by high levels of sialic acid on the Fc-fragment might be responsible for the anti-inflammatory activity (Kaneko et al. 2006b). Due to the higher level of sialic acid in the fraction of serum IgG glycosylated in the Fab-fragment (15–20 %), it may not be possible to directly enrich for high levels of sialic acid in the Fc-fragment as the lectin will predominantly enrich for sialic acid in the Fab-fragment. Indeed, recent studies showed that a simple SNA enrichment step did not result in a significant enrichment of the sialic acid-rich glycovariants of the Fc-fragment and expectedly not to enhanced anti-inflammatory activity in a murine immunothrombocytopenia model system (Stadlmann et al. 2009; Guhr et al. 2011). This simple method of enrichment may be only possible for IVIg preparations with proven low/absent Fab-glycosylation or with monoclonal antibody preparations known to lack Fab-associated sugar moieties (Kaneko et al. 2006b; Anthony et al. 2008a). That it is indeed possible to generate a recombinant IgG with anti-inflammatory activity was shown by a study using a human IgG1 antibody produced in tissue culture as the starting material (Anthony et al. 2008a). As the majority of IgG produced in tissue culture does not contain terminal sialic acid residues, a high degree of sialylation was achieved by consecutive treatment with galactosyl- and sialyltransferases in vitro. Of note, only the addition of 2,6-linked sialic acid residues, but not 2,3-linked sialic acid residues, was able to recapitulate IVIg activity and suppress rheumatoid arthritis. An interesting open question is whether one or two sialic acid residues are necessary on one or both sugar moieties of the two IgG heavy chains to achieve anti-inflammatory activity. A recent study showed that affinity purification of IgG Fc-fragments via the lectin SNA selectively enriches for sugar moieties containing one sialic acid residue on both sugar moieties (Stadlmann et al. 2009). In contrast, monosialylation on one arm was not sufficient to allow for optimal binding to SNA. As desialylated IgG was only present in trace amounts in the IVIg preparation, it seems likely that the IgG fraction carrying monosialylated sugar structures on both heavy chains may be crucial (Stadlmann et al. 2009). The data from the in vitro sialylated Fc-fragments supports this notion, although here, a level of sialylation (two sialic acid residues on both sugar moieties in the Fc-fragment) was obtained that is usually not present in regular IVIg preparations (Anthony et al. 2008a). Thus, IVIg therapy might restore the level of the immunomodulatory sialic acid-rich IgG glycoforms which are present at decreased levels in patients with autoimmune disease.

It should be noted that independent of this active anti-inflammatory activity, high levels of sialic acid in the Fc-fragment of mouse and human IgG reduce the binding to cellular Fcγ-receptors. This decreased level of binding occurs for both 2,3 and 2,6-linked sialic acid residues and results in decreased IgG-dependent effector functions (Kaneko et al. 2006b; Scallon et al. 2007). Importantly, this decreased activity is not to be interpreted as an active anti-inflammatory activity which requires much higher levels of sialic acid-rich IgG. Moreover, the decreased affinity for cellular FcγRs argues for an anti-inflammatory pathway that is independent of a simple block of activating FcγRs by IVIg or sialic acid-rich IVIg binding.

5.9 The Sialic Acid Dependent Anti-inflammatory Pathway

Although the identification of the essential role of sialic acid-rich IgG glycovariants for the therapeutic activity of IVIg was considered a major breakthrough, the molecular and cellular pathways triggered by this molecule remained unclear. As we have discussed before, the inhibitory Fcγ-receptor IIB (FcγRIIB) was shown to be an important component of the anti-inflammatory pathway triggered by IVIg infusion. Similarly, the anti-inflammatory activity of the sialic acid-rich IVIg fraction and of the recombinantly generated sialylated Fc-fragment was shown to be FcγRIIB dependent in the rheumatoid arthritis model (Kaneko et al. 2006b; Anthony et al. 2011). A critical question is how IVIg infusion triggers this upregulation of FcγRIIB on innate immune effector cells. In principle, IVIg could act directly on the effector cell population or indirectly via other immunomodulatory pathways. At present, the latter possibility seems to be the more likely pathway. Thus, IVIg activity was abrogated in osteopetrotic op/op mice lacking colony-stimulating factor 1 (CSF1)-dependent monocytes and macrophages including the splenic marginal zone macrophage population expressing the cell surface receptors MARCO and SIGNR1 (Bruhns et al. 2003). An important role of the spleen and an intact splenic structure was suggested by the inability of IVIg to protect splenectomized mice and RAG mice lacking B and T cells and therefore a normal splenic architecture (Anthony et al. 2008b). By using a variety of blocking antibodies and knockout mice, the authors could show that the SIGNR1 receptor is crucial for IVIg activity in mice. Moreover, sialic acid-rich IgG, but not asialylated IgG, could bind to SIGNR1 and its human orthologue DC-SIGN. In contrast to the restricted expression pattern of mouse SIGNR1, human DC-SIGN can be detected on most myeloid-derived cells including certain monocyte subpopulations, dendritic cells, and macrophages (Granelli-Piperno et al. 2005; Soilleux et al. 2002). To study the role of human DC-SIGN for IVIg activity in vivo, a transgenic mouse expressing human DC-SIGN on a mouse SIGN-R1 background was generated recently (Anthony et al. 2011). Indeed, the anti-inflammatory activity of sialic acid-rich Fc-fragments (sFC) which was absent in SIGN-R1 knockout mice was restored in human DC-SIGN transgenic animals. Adoptive transfer experiments with sFC pretreated bone marrow-derived macrophages and dendritic cells obtained from human DC-SIGN transgenic mice indicate that these cell types could be involved in the anti-inflammatory pathway, consistent with previous reports (although here, no sialic acid-enriched IVIg was used) (Siragam et al. 2006; Anthony et al. 2011). Moreover, IVIg injection resulted in the production of IL33, which can initiate the production of T-helper cell type 2 (TH2) cytokines including IL4 production which is known to upregulate FcγRIIB and to downmodulate activating FcγRs on innate immune effector cells (Fig. 5.3) (Nimmerjahn and Ravetch 2006; Pricop et al. 2001). Although IL4 production could not be shown directly in vivo, knockout mice for IL4, the IL4Ra chain, and Stat6 which is essential for IL4-dependent signal transduction were no longer protected by IVIg from development of rheumatoid arthritis (Anthony et al. 2011). Basophils were identified as a potential cell type responsible for IL33-dependent production of IL4. Thus, depletion of basophils blocked the anti-inflammatory activity of the sialic acid-rich Fc-fragment and adoptive transfer of IL33 pretreated basophils (primed to produce IL4) was able to suppress development of arthritis.

Besides the clear role of the spleen and the TH2 cytokine pathway in a mouse model of rheumatoid arthritis, it should be kept in mind, however, that an IL4-independent pathway of IVIg protection was described in mouse models of ITP (Fig. 5.3) (Schwab et al. 2012; Crow et al. 2007). By using mice deficient in the common cytokine gamma chain, essential for the activity of IL2, 4, 7, 9, 15, and 21 and knockout mice for IL10, IL12β, TNFα, MIP1α, and the IFNγR, the authors demonstrated that IVIg activity was not impaired in the absence of these cytokines (Crow et al. 2007). In contrast to the essential role of the spleen in the rheumatoid arthritis model, splenectomized human ITP patients and mice induced to develop ITP can still respond to IVIg therapy, suggesting that the spleen and splenic resident cells may not play a critical role in all autoimmune diseases (Schwab et al. 2012). Supporting this notion, DC-SIGN-positive dendritic cells and macrophages are not restricted to the spleen and can be found in peripheral lymph nodes and the bone marrow. Importantly, despite being independent of the spleen, sialic acid and SIGNR1 were still critical for the spleen independent pathway of IVIg-mediated protection from ITP development (Schwab et al. 2012).

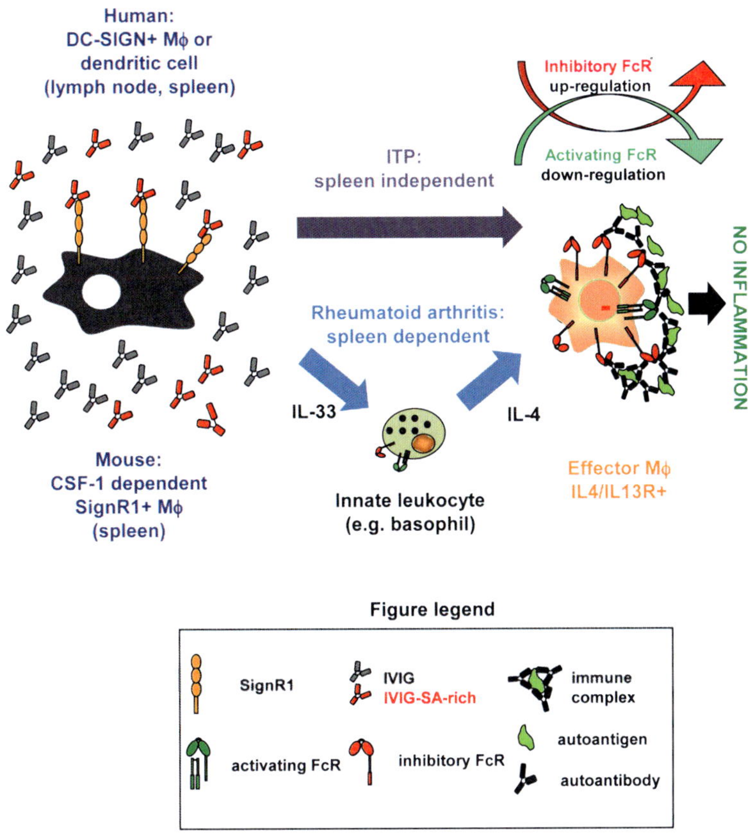

Fig. 5.3 *Model for the anti-inflammatory pathway triggered by sialic acid-rich IgG.* Two sialic acid-rich and SIGNR1-dependent pathways have been proposed to be involved in the anti-inflammatory activity of IVIg. In immunothrombocytopenia, IVIg suppresses autoantibody-mediated platelet depletion independent of the spleen and TH2-cytokines, whereas in rheumatoid arthritis, IVIg-mediated suppression of joint inflammation is dependent on splenic macrophages and an IL33- and IL4-dependent pathway. The result of both pathways is the upregulation of the inhibitory FcγRIIB, thus changing the ratio of activating and inhibitory FcγR expression on effector macrophages resulting in a higher threshold for cell activation by immune complexes

5.10 Conclusions

Despite recent advances in defining and understanding the pathways involved in the anti-inflammatory activity of IgG, much has still to be learned to get a clear picture of how IgG molecules can mediate both pro- and anti-inflammatory effector functions. Complicating the situation, suitable in vivo model systems for all the different diseases where IVIg was suggested to be beneficial for the patient are lacking in many instances. Here, well-defined clinical studies are critical to ultimately prove the beneficial effects for the respective autoimmune disease. In other instances, human studies have already proven that the IgG constant fragment can have an anti-inflammatory activity, and mouse studies have been used to delineate the molecular and cellular pathways of IVIg activity, elucidating an unexpected role of IgG glycosylation-specific Fc-receptors in this immunomodulatory pathway. As this is a very active field of research, this chapter should be viewed as a first insight into this exciting and rapidly developing field, and the interested reader can get a more in-depth insight into the varying mechanisms of IVIg activity in recent more specialized review articles.

Acknowledgements This work was supported by grants from the German Research Foundation (SFB 643, FOR 832, SPP 1468, GK 1660), the Bavarian Genome Research Network, and the Paul-Ehrlich and Ludwig-Darmstädter Foundation.

References

Altznauer F, von Gunten S, Spath P, Simon HU (2003) Concurrent presence of agonistic and antagonistic anti-CD95 autoantibodies in intravenous Ig preparations. J Allergy Clin Immunol 112(6):1185–1190

Anthony RM, Nimmerjahn F, Ashline DJ, Reinhold VN, Paulson JC, Ravetch JV (2008a) Recapitulation of IVIG anti-inflammatory activity with a recombinant IgG Fc. Science 320(5874):373–376

Anthony RM, Wermeling F, Karlsson MC, Ravetch JV (2008b) Identification of a receptor required for the anti-inflammatory activity of IVIG. Proc Natl Acad Sci USA 105(50):19571–19578

Anthony RM, Kobayashi T, Wermeling F, Ravetch JV (2011) Intravenous gammaglobulin suppresses inflammation through a novel T(H)2 pathway. Nature 475(7354):110–113

Arnold JN, Wormald MR, Sim RB, Rudd PM, Dwek RA (2007) The impact of glycosylation on the biological function and structure of human immunoglobulins. Annu Rev Immunol 25:21–50

Axford JS (1988) Decreased B-cell galactosyltransferase activity in rheumatoid arthritis. Br J Rheumatol 27(Suppl 2):170

Basta M, Kirshbom P, Frank MM, Fries LF (1989) Mechanism of therapeutic effect of high-dose intravenous immunoglobulin. Attenuation of acute, complement-dependent immune damage in a guinea pig model. J Clin Invest 84(6):1974–1981

Bjoro K, Froland SS, Yun Z, Samdal HH, Haaland T (1994) Hepatitis C infection in patients with primary hypogammaglobulinemia after treatment with contaminated immune globulin. N Engl J Med 331(24):1607–1611

Bond A, Cooke A, Hay FC (1990) Glycosylation of IgG, immune complexes and IgG subclasses in the MRL-lpr/lpr mouse model of rheumatoid arthritis. Eur J Immunol 20(10):2229–2233

Boyd PN, Lines AC, Patel AK (1995) The effect of the removal of sialic acid, galactose and total carbohydrate on the functional activity of Campath-1H. Mol Immunol 32(17–18):1311–1318

Brownlie RJ, Lawlor KE, Niederer HA, Cutler AJ, Xiang Z, Clatworthy MR et al (2008) Distinct cell-specific control of autoimmunity and infection by FcgammaRIIb. J Exp Med 205(4):883–895

Bruhns P, Samuelsson A, Pollard JW, Ravetch JV (2003) Colony-stimulating factor-1-dependent macrophages are responsible for IVIG protection in antibody-induced autoimmune disease. Immunity 18(4):573–581

Buchacher A, Iberer G (2006) Purification of intravenous immunoglobulin G from human plasma–aspects of yield and virus safety. Biotechnol J 1(2):148–163

Bussel JB, Graziano JN, Kimberly RP, Pahwa S, Aledort LM (1991) Intravenous anti-D treatment of immune thrombocytopenic purpura: analysis of efficacy, toxicity, and mechanism of effect. Blood 77(9):1884–1893

Carbone J (2007) Adverse reactions and pathogen safety of intravenous immunoglobulin. Curr Drug Saf 2(1):9–18

Carroll MC (1998) The role of complement and complement receptors in induction and regulation of immunity. Annu Rev Immunol 16:545–568

Clarkson SB, Bussel JB, Kimberly RP, Valinsky JE, Nachman RL, Unkeless JC (1986) Treatment of refractory immune thrombocytopenic purpura with an anti-Fc gamma-receptor antibody. N Engl J Med 314(19):1236–1239

Croce A, Firuzi O, Altieri F, Eufemi M, Agostino R, Priori R et al (2007) Effect of infliximab on the glycosylation of IgG of patients with rheumatoid arthritis. J Clin Lab Anal 21(5):303–314

Crow AR, Lazarus AH (2008) The mechanisms of action of intravenous immunoglobulin and polyclonal anti-d immunoglobulin in the amelioration of immune thrombocytopenic purpura: what do we really know? Transfus Med Rev 22(2):103–116

Crow AR, Song S, Semple JW, Freedman J, Lazarus AH (2007) A role for IL-1 receptor antagonist or other cytokines in the acute therapeutic effects of IVIg? Blood 109(1):155–158

Crow AR, Brinc D, Lazarus AH (2009) New insight into the mechanism of action of IVIg: the role of dendritic cells. J Thromb Haemost 7(Suppl 1):245–248

Crow AR, Suppa SJ, Chen X, Mott PJ, Lazarus AH (2011) The neonatal Fc receptor (FcRn) is not required for IVIg or anti-CD44 monoclonal antibody-mediated amelioration of murine immune thrombocytopenia. Blood 118(24):6403–6406

Debre M, Bonnet MC, Fridman WH, Carosella E, Philippe N, Reinert P et al (1993) Infusion of Fc gamma fragments for treatment of children with acute immune thrombocytopenic purpura. Lancet 342(8877):945–949

Deisenhofer J (1981) Crystallographic refinement and atomic models of a human Fc fragment and its complex with fragment B of protein A from Staphylococcus aureus at 2.9- and 2.8-A resolution. Biochemistry 20(9):2361–2370

Djoumerska-Alexieva IK, Dimitrov JD, Voynova EN, Lacroix-Desmazes S, Kaveri SV, Vassilev TL (2010) Exposure of IgG to an acidic environment results in molecular modifications and in enhanced protective activity in sepsis. FEBS J 277(14):3039–3050

Dube R, Rook GA, Steele J, Brealey R, Dwek R, Rademacher T et al (1990) Agalactosyl IgG in inflammatory bowel disease: correlation with C-reactive protein. Gut 31(4):431–434

Durandy A, Kaveri SV, Kuijpers TW, Basta M, Miescher S, Ravetch JV et al (2009) Intravenous immunoglobulins–understanding properties and mechanisms. Clin Exp Immunol 158(Suppl 1):2–13

Floto RA, Clatworthy MR, Heilbronn KR, Rosner DR, MacAry PA, Rankin A et al (2005) Loss of function of a lupus-associated FcgammaRIIb polymorphism through exclusion from lipid rafts. Nat Med 11(10):1056–1058

French LE, Trent JT, Kerdel FA (2006) Use of intravenous immunoglobulin in toxic epidermal necrolysis and Stevens-Johnson syndrome: our current understanding. Int Immunopharmacol 6(4):543–549

Garred P, Madsen HO, Marquart H, Hansen TM, Sorensen SF, Petersen J et al (2000) Two edged role of mannose binding lectin in rheumatoid arthritis: a cross sectional study. J Rheumatol 27(1):26–34

Getman KE, Balthasar JP (2005) Pharmacokinetic effects of 4C9, an anti-FcRn antibody, in rats: implications for the use of FcRn inhibitors for the treatment of humoral autoimmune and alloimmune conditions. J Pharm Sci 94(4):718–729

Gilhus NE (2011) Neuromuscular disease: acute treatment for myasthenia gravis. Nat Rev Neurol 7(3):132–134

Godeau B, Oksenhendler E, Brossard Y, Bartholeyns J, Leaute JB, Duedari N et al (1996) Treatment of chronic autoimmune thrombocytopenic purpura with monoclonal anti-D. Transfusion 36(4):328–330

Grammatikos AP, Tsokos GC (2012) Immunodeficiency and autoimmunity: lessons from systemic lupus erythematosus. Trends Mol Med 18(2):101–108

Granelli-Piperno A, Pritsker A, Pack M, Shimeliovich I, Arrighi JF, Park CG et al (2005) Dendritic cell-specific intercellular adhesion molecule 3-grabbing nonintegrin/CD209 is abundant on macrophages in the normal human lymph node and is not required for dendritic cell stimulation of the mixed leukocyte reaction. J Immunol 175(7):4265–4273

Graudal NA, Madsen HO, Tarp U, Svejgaard A, Jurik G, Graudal HK et al (2000) The association of variant mannose-binding lectin genotypes with radiographic outcome in rheumatoid arthritis. Arthritis Rheum 43(3):515–521

Guhr T, Bloem J, Derksen NI, Wuhrer M, Koenderman AH, Aalberse RC et al (2011) Enrichment of sialylated IgG by lectin fractionation does not enhance the efficacy of immunoglobulin G in a murine model of immune thrombocytopenia. PLoS One 6(6):e21246

Hadley AG, Zupanska B, Kumpel BM, Pilkington C, Griffiths HL, Leader KA et al (1995) The glycosylation of red cell autoantibodies affects their functional activity in vitro. Br J Haematol 91(3):587–594

Hamrock DJ (2006) Adverse events associated with intravenous immunoglobulin therapy. Int Immunopharmacol 6(4):535–542

Hansen RJ, Balthasar JP (2002) Intravenous immunoglobulin mediates an increase in anti-platelet antibody clearance via the FcRn receptor. Thromb Haemost 88(6):898–899

Holland M, Takada K, Okumoto T, Takahashi N, Kato K, Adu D et al (2002) Hypogalactosylation of serum IgG in patients with ANCA-associated systemic vasculitis. Clin Exp Immunol 129(1):183–190

Holland M, Yagi H, Takahashi N, Kato K, Savage CO, Goodall DM et al (2006) Differential glycosylation of polyclonal IgG, IgG-Fc and IgG-Fab isolated from the sera of patients with ANCA-associated systemic vasculitis. Biochim Biophys Acta 1760(4):669–677

Imbach P, Barandun S, d'Apuzzo V, Baumgartner C, Hirt A, Morell A et al (1981) High-dose intravenous gammaglobulin for idiopathic thrombocytopenic purpura in childhood. Lancet 1(8232):1228–1231

Jefferis R (2007) Antibody therapeutics: isotype and glycoform selection. Expert Opin Biol Ther 7(9):1401–1413

Ji H, Ohmura K, Mahmood U, Lee DM, Hofhuis FM, Boackle SA et al (2002) Arthritis critically dependent on innate immune system players. Immunity 16(2):157–168

Kaneko Y, Nimmerjahn F, Madaio MP, Ravetch JV (2006a) Pathology and protection in nephrotoxic nephritis is determined by selective engagement of specific Fc receptors. J Exp Med 203(3):789–797, Epub 2006 Mar 6

Kaneko Y, Nimmerjahn F, Ravetch JV (2006b) Anti-inflammatory activity of immunoglobulin G resulting from Fc sialylation. Science 313(5787):670–673

Keusch J, Levy Y, Shoenfeld Y, Youinou P (1996) Analysis of different glycosylation states in IgG subclasses. Clin Chim Acta 252(2):147–158

Kobata A (2008) The N-linked sugar chains of human immunoglobulin G: their unique pattern, and their functional roles. Biochim Biophys Acta 1780(3):472–478

Kono H, Kyogoku C, Suzuki T, Tsuchiya N, Honda H, Yamamoto K et al (2005) FcgammaRIIB Ile232Thr transmembrane polymorphism associated with human systemic lupus erythematosus decreases affinity to lipid rafts and attenuates inhibitory effects on B cell receptor signaling. Hum Mol Genet 14(19):2881–2892, Epub 2005 Aug 22

Kumpel BM, Rademacher TW, Rook GA, Williams PJ, Wilson IB (1994) Galactosylation of human IgG monoclonal anti-D produced by EBV-transformed B-lymphoblastoid cell lines is dependent on culture method and affects Fc receptor-mediated functional activity. Hum Antibodies Hybridomas 5(3–4):143–151

Kumpel BM, Wang Y, Griffiths HL, Hadley AG, Rook GA (1995) The biological activity of human monoclonal IgG anti-D is reduced by beta-galactosidase treatment. Hum Antibodies Hybridomas 6(3):82–88

Kuroda Y, Nakata M, Hirose S, Shirai T, Iwamoto M, Izui S et al (2001a) Abnormal IgG galactosylation in MRL-lpr/lpr mice: pathogenic role in the development of arthritis. Pathol Int 51(12):909–915

Kuroda Y, Nakata M, Nose M, Kojima N, Mizuochi T (2001b) Abnormal IgG galactosylation and arthritis in MRL-Fas(lpr) or MRL-FasL(gld) mice are under the control of the MRL genetic background. FEBS Lett 507(2):210–214

Le Pottier L, Sapir T, Bendaoud B, Youinou P, Shoenfeld Y, Pers JO (2007) Intravenous immunoglobulin and cytokines: focus on tumor necrosis factor family members BAFF and APRIL. Ann N Y Acad Sci 1110:426–432

Leirisalo-Repo M, Hernandez-Munoz HE, Rook GA (1999) Agalactosyl IgG is elevated in patients with active spondyloarthropathy. Rheumatol Int 18(5–6):171–176

Lu MC, Hsieh SC, Lai NS, Li KJ, Wu CH, Yu CL (2007) Comparison of anti-agalactosyl IgG antibodies, rheumatoid factors, and anti-cyclic citrullinated peptide antibodies in the differential diagnosis of rheumatoid arthritis and its mimics. Clin Exp Rheumatol 25(5):716–721

Mackay M, Stanevsky A, Wang T, Aranow C, Li M, Koenig S et al (2006) Selective dysregulation of the FcgammaIIB receptor on memory B cells in SLE. J Exp Med 203(9):2157–2164

Malhotra R, Wormald MR, Rudd PM, Fischer PB, Dwek RA, Sim RB (1995) Glycosylation changes of IgG associated with rheumatoid arthritis can activate complement via the mannose-binding protein. Nat Med 1(3):237–243

Marchalonis JJ, Kaymaz H, Dedeoglu F, Schluter SF, Yocum DE, Edmundson AB (1992) Human autoantibodies reactive with synthetic autoantigens from T-cell receptor beta chain. Proc Natl Acad Sci USA 89(8):3325–3329

Masuda K, Yamaguchi Y, Kato K, Takahashi N, Shimada I, Arata Y (2000) Pairing of oligosaccharides in the Fc region of immunoglobulin G. FEBS Lett 473(3):349–357

McGaha TL, Sorrentino B, Ravetch JV (2005) Restoration of tolerance in lupus by targeted inhibitory receptor expression. Science 307(5709):590–593

Mehta AS, Long RE, Comunale MA, Wang M, Rodemich L, Krakover J et al (2008) Increased levels of galactose-deficient anti-Gal immunoglobulin G in the sera of hepatitis C virus-infected individuals with fibrosis and cirrhosis. J Virol 82(3):1259–1270

Negi VS, Elluru S, Siberil S, Graff-Dubois S, Mouthon L, Kazatchkine MD et al (2007) Intravenous immunoglobulin: an update on the clinical use and mechanisms of action. J Clin Immunol 27(3):233–245

Nimmerjahn F, Ravetch JV (2006) Fcgamma receptors: old friends and new family members. Immunity 24(1):19–28

Nimmerjahn F, Ravetch JV (2008) Anti-inflammatory actions of intravenous immunoglobulin. Annu Rev Immunol 26:513–533

Nimmerjahn F, Ravetch JV (2011) FcgammaRs in health and disease. Curr Top Microbiol Immunol 350:105–125

Nimmerjahn F, Anthony RM, Ravetch JV (2007) Agalactosylated IgG antibodies depend on cellular Fc receptors for in vivo activity. Proc Natl Acad Sci USA 104(20):8433–8437

Oksenhendler E, Bierling P, Brossard Y, Schenmetzler C, Girard PM, Seligmann M et al (1988) Anti-RH immunoglobulin therapy for human immunodeficiency virus-related immune thrombocytopenic purpura. Blood 71(5):1499–1502

Omtvedt LA, Royle L, Husby G, Sletten K, Radcliffe CM, Harvey DJ et al (2006) Glycan analysis of monoclonal antibodies secreted in deposition disorders indicates that subsets of plasma cells differentially process IgG glycans. Arthritis Rheum 54(11):3433–3440

Parekh RB, Dwek RA, Sutton BJ, Fernandes DL, Leung A, Stanworth D et al (1985) Association of rheumatoid arthritis and primary osteoarthritis with changes in the glycosylation pattern of total serum IgG. Nature 316(6027):452–457

Parekh RB, Roitt IM, Isenberg DA, Dwek RA, Ansell BM, Rademacher TW (1988a) Galactosylation of IgG associated oligosaccharides: reduction in patients with adult and juvenile onset rheumatoid arthritis and relation to disease activity. Lancet 1(8592):966–969

Parekh R, Roitt I, Isenberg D, Dwek R, Rademacher T (1988b) Age-related galactosylation of the N-linked oligosaccharides of human serum IgG. J Exp Med 167(5):1731–1736
Parekh R, Isenberg D, Rook G, Roitt I, Dwek R, Rademacher T (1989) A comparative analysis of disease-associated changes in the galactosylation of serum IgG. J Autoimmun 2(2):101–114
Pasek M, Duk M, Podbielska M, Sokolik R, Szechinski J, Lisowska E et al (2006) Galactosylation of IgG from rheumatoid arthritis (RA) patients–changes during therapy. Glycoconj J 23(7–8):463–471
Podolanczuk A, Lazarus AH, Crow AR, Grossbard E, Bussel JB (2009) Of mice and men: an open-label pilot study for treatment of immune thrombocytopenic purpura by an inhibitor of Syk. Blood 113(14):3154–3160
Prasad NK, Papoff G, Zeuner A, Bonnin E, Kazatchkine MD, Ruberti G et al (1998) Therapeutic preparations of normal polyspecific IgG (IVIg) induce apoptosis in human lymphocytes and monocytes: a novel mechanism of action of IVIg involving the Fas apoptotic pathway. J Immunol 161(7):3781–3790
Pricop L, Redecha P, Teillaud JL, Frey J, Fridman WH, Sautes-Fridman C et al (2001) Differential modulation of stimulatory and inhibitory Fc gamma receptors on human monocytes by Th1 and Th2 cytokines. J Immunol 166(1):531–537
Pritchard NR, Cutler AJ, Uribe S, Chadban SJ, Morley BJ, Smith KG (2000) Autoimmune-prone mice share a promoter haplotype associated with reduced expression and function of the Fc receptor FcgammaRII. Curr Biol 10(4):227–230
Radaev S, Rostro B, Brooks AG, Colonna M, Sun PD (2001) Conformational plasticity revealed by the cocrystal structure of NKG2D and its class I MHC-like ligand ULBP3. Immunity 15(6):1039–1049
Rademacher TW, Williams P, Dwek RA (1994) Agalactosyl glycoforms of IgG autoantibodies are pathogenic. Proc Natl Acad Sci USA 91(13):6123–6127
Raju TS (2008) Terminal sugars of Fc glycans influence antibody effector functions of IgGs. Curr Opin Immunol 20(4):471–478
Rook GA, Steele J, Brealey R, Whyte A, Isenberg D, Sumar N et al (1991) Changes in IgG glycoform levels are associated with remission of arthritis during pregnancy. J Autoimmun 4(5):779–794
Roopenian DC, Akilesh S (2007) FcRn: the neonatal Fc receptor comes of age. Nat Rev Immunol 7(9):715–725
Rossi F, Kazatchkine MD (1989) Antiidiotypes against autoantibodies in pooled normal human polyspecific Ig. J Immunol 143(12):4104–4109
Rudd PM, Elliott T, Cresswell P, Wilson IA, Dwek RA (2001) Glycosylation and the immune system. Science 291(5512):2370–2376
Ruhaak LR, Uh HW, Beekman M, Koeleman CA, Hokke CH, Westendorp RG et al (2010) Decreased levels of bisecting GlcNAc glycoforms of IgG are associated with human longevity. PLoS One 5(9):e12566
Salama A, Kiefel V, Amberg R, Mueller-Eckhardt C (1984) Treatment of autoimmune thrombocytopenic purpura with rhesus antibodies (anti-Rh0(D). Blut 49(1):29–35
Samuelsson A, Towers TL, Ravetch JV (2001) Anti-inflammatory activity of IVIG mediated through the inhibitory Fc receptor. Science 291(5503):484–486
Saphire EO, Stanfield RL, Crispin MD, Parren PW, Rudd PM, Dwek RA et al (2002) Contrasting IgG structures reveal extreme asymmetry and flexibility. J Mol Biol 319(1):9–18
Scallon BJ, Tam SH, McCarthy SG, Cai AN, Raju TS (2007) Higher levels of sialylated Fc glycans in immunoglobulin G molecules can adversely impact functionality. Mol Immunol 44(7):1524–1534
Scherer HU, van der Woude D, Ioan-Facsinay A, el Bannoudi H, Trouw LA, Wang J et al (2010) Glycan profiling of anti-citrullinated protein antibodies isolated from human serum and synovial fluid. Arthritis Rheum 62(6):1620–1629
Schleis TG (2005) The process: new methods of purification and viral safety. Pharmacotherapy 25(11 Pt 2):73S–77S

Schwab I, Biburger M, Kronke G, Schett G, Nimmerjahn F (2012) IVIg-mediated amelioration of ITP in mice is dependent on sialic acid and SIGNR1. Eur J Immunol 42(4):826–830

Shikata K, Yasuda T, Takeuchi F, Konishi T, Nakata M, Mizuochi T (1998) Structural changes in the oligosaccharide moiety of human IgG with aging. Glycoconj J 15(7):683–689

Siragam V, Crow AR, Brinc D, Song S, Freedman J, Lazarus AH (2006) Intravenous immunoglobulin ameliorates ITP via activating Fc gamma receptors on dendritic cells. Nat Med 12(6):688–692, Epub 2006 May 21

Soilleux EJ, Morris LS, Leslie G, Chehimi J, Luo Q, Levroney E et al (2002) Constitutive and induced expression of DC-SIGN on dendritic cell and macrophage subpopulations in situ and in vitro. J Leukoc Biol 71(3):445–457

Sondermann P, Huber R, Oosthuizen V, Jacob U (2000) The 3.2-A crystal structure of the human IgG1 Fc fragment-Fc gammaRIII complex. Nature 406(6793):267–273

Stadlmann J, Weber A, Pabst M, Anderle H, Kunert R, Ehrlich HJ et al (2009) A close look at human IgG sialylation and subclass distribution after lectin fractionation. Proteomics 9(17):4143–4153

Sumar N, Isenberg DA, Bodman KB, Soltys A, Young A, Leak AM et al (1991) Reduction in IgG galactose in juvenile and adult onset rheumatoid arthritis measured by a lectin binding method and its relation to rheumatoid factor. Ann Rheum Dis 50(9):607–610

Sutton BJ, Phillips DC (1983) The three-dimensional structure of the carbohydrate within the Fc fragment of immunoglobulin G. Biochem Soc Trans 11(Pt 2):130–132

Tackenberg B, Jelcic I, Baerenwaldt A, Oertel WH, Sommer N, Nimmerjahn F et al (2009) Impaired inhibitory Fc gamma receptor IIB expression on B cells in chronic inflammatory demyelinating polyneuropathy. Proc Natl Acad Sci USA 106(12):4788–4792

Takai T, Ono M, Hikida M, Ohmori H, Ravetch JV (1996) Augmented humoral and anaphylactic responses in Fc gamma RII-deficient mice. Nature 379(6563):346–349

Tomana M, Schrohenloher RE, Koopman WJ, Alarcon GS, Paul WA (1988) Abnormal glycosylation of serum IgG from patients with chronic inflammatory diseases. Arthritis Rheum 31(3):333–338

van de Geijn FE, Wuhrer M, Selman MH, Willemsen SP, de Man YA, Deelder AM et al (2009) Immunoglobulin G galactosylation and sialylation are associated with pregnancy-induced improvement of rheumatoid arthritis and the postpartum flare: results from a large prospective cohort study. Arthritis Res Ther 11(6):R193

Vassilev T, Gelin C, Kaveri SV, Zilber MT, Boumsell L, Kazatchkine MD (1993) Antibodies to the CD5 molecule in normal human immunoglobulins for therapeutic use (intravenous immunoglobulins, IVIg). Clin Exp Immunol 92(3):369–372

Viard I, Wehrli P, Bullani R, Schneider P, Holler N, Salomon D et al (1998) Inhibition of toxic epidermal necrolysis by blockade of CD95 with human intravenous immunoglobulin. Science 282(5388):490–493

von Gunten S, Simon HU (2008) Natural anti-Siglec autoantibodies mediate potential immunoregulatory mechanisms: implications for the clinical use of intravenous immunoglobulins (IVIg). Autoimmun Rev 7(6):453–456

von Gunten S, Schaub A, Vogel M, Stadler BM, Miescher S, Simon HU (2006) Immunologic and functional evidence for anti-Siglec-9 autoantibodies in intravenous immunoglobulin preparations. Blood 108(13):4255–4259

von Gunten S, Vogel M, Schaub A, Stadler BM, Miescher S, Crocker PR et al (2007) Intravenous immunoglobulin preparations contain anti-Siglec-8 autoantibodies. J Allergy Clin Immunol 119(4):1005–1011

Wang J, Balog CI, Stavenhagen K, Koeleman CA, Scherer HU, Selman MH et al (2011) Fc-glycosylation of IgG1 is modulated by B-cell stimuli. Mol Cell Proteomics 10(5):M110 004655

Wormald MR, Rudd PM, Harvey DJ, Chang SC, Scragg IG, Dwek RA (1997) Variations in oligosaccharide-protein interactions in immunoglobulin G determine the site-specific glycosylation profiles and modulate the dynamic motion of the Fc oligosaccharides. Biochemistry 36(6):1370–1380

Wuhrer M, Porcelijn L, Kapur R, Koeleman CA, Deelder A, de Haas M et al (2009) Regulated glycosylation patterns of IgG during alloimmune responses against human platelet antigens. J Proteome Res 8(2):450–456
Yamada E, Tsukamoto Y, Sasaki R, Yagyu K, Takahashi N (1997) Structural changes of immunoglobulin G oligosaccharides with age in healthy human serum. Glycoconj J 14(3):401–405
Young A, Sumar N, Bodman K, Goyal S, Sinclair H, Roitt I et al (1991) Agalactosyl IgG: an aid to differential diagnosis in early synovitis. Arthritis Rheum 34(11):1425–1429

Chapter 6
Example of the Pathogenic Potential of Two Sets of Autoantibodies: Anti-RBC and IgG3 RF Cryoglobulins

Shozo Izui

Abstract The pathogenic potential of IgG autoantibodies is determined by effector functions associated with the Fc regions of the different IgG subclasses, as documented through the analysis of two different sets of monoclonal autoantibodies. The IgG2a and IgG2b subclasses of autoantibodies are most pathogenic because of the efficient activation of stimulatory IgG Fc receptors and complement. In addition, because of the unique property of IgG3 to form self-associating complexes and generate cryoglobulins, this subclass of autoantibodies could become highly pathogenic. Since the switch to IgG2a and IgG3 is promoted by Th1 cytokine IFN-γ, Th1 autoimmune responses could be critically involved in the generation of more pathogenic IgG antibodies. This underlines the importance of the regulation of IgG subclass responses in antibody-mediated inflammatory disorder.

Keywords Autoantibody • Cryoglobulin • IgG Fc receptor • Complement • Hemolytic anemia • Rheumatoid factor

6.1 Introduction

The pathogenesis of autoantibody-mediated cellular and tissue lesions in autoimmune diseases is most straightforwardly attributed to the combined action of self-antigen binding properties and effector functions of the different immunoglobulin (Ig) isotypes. On the one hand, the structure of the Fab region defines the specificity and affinity with which autoantibodies bind to self-antigen. On the other hand, effector functions associated with the Fc regions of the different Ig isotypes play a key role in determining autoantibody pathogenicity by activating IgG Fc receptor

S. Izui, M.D (✉)
Department of Pathology and Immunology, University of Geneva,
1, rue Michel-Servet, 1211 Geneva 4, Switzerland
e-mail: Shozo.Izui@unige.ch

F. Nimmerjahn (ed.), *Molecular and Cellular Mechanisms of Antibody Activity*,
DOI 10.1007/978-1-4614-7107-3_6, © Springer Science+Business Media New York 2013

(FcγR)-bearing effector cells, triggering the complement cascade, and inducing multivalency-dependent agglutination of target cells.

It has been demonstrated in vitro that individual IgG subclasses display considerable differences in their capacity to interact with different classes of FcγR and activate complement. However, until about the year 2000, the relative contributions of different classes of FcγR and complement to the development of autoantibody-mediated tissue and cellular injuries have not been well defined because of a lack of suitable experimental tools and models. This has changed during the last decade, thanks to the availability of various mutant mice lacking different classes of FcγRs or complement components. Moreover, extensive analyses of different experimental models of autoimmune diseases have revealed the importance of the qualitative aspects of autoantibodies in the development of autoimmune cellular and tissue lesions. This chapter exemplifies the pathogenic role of a subset of autoantibodies in autoimmune diseases by describing the molecular and cellular basis for the pathogenic activity of two different sets of monoclonal antibodies (mAb) derived from autoimmune mice – anti-red blood cell (RBC) autoantibodies and anti-IgG2a rheumatoid factors (RF) – in relation to the effector functions associated with the Fc region of the different Ig isotypes.

6.2 Pathogenicity of Anti-RBC mAb

The experimental model of autoimmune hemolytic anemia induced by a single injection of Coombs' anti-RBC autoantibodies derived from autoimmune-prone NZB mice presents remarkable advantages for the assessment of the pathogenic activity of autoantibodies. First, the binding of anti-RBC autoantibodies and the activation of complement in vivo after injection of anti-RBC autoantibodies into mice can be monitored by flow cytometric analysis of opsonized RBC (Fig. 6.1a). Second, the pathogenic effect of anti-RBC autoantibodies (i.e., the development of anemia) can be followed by the simple measurement of hematocrit values (Fig. 6.1b). Finally, the pathogenic mechanisms of anemia can be determined through histological analysis of spleen and liver: erythrophagocytosis by hepatic Kupffer cells which is associated with iron deposits revealed by Perls iron staining (Fig. 6.1c) or sequestration of agglutinated RBC in spleen and liver (Fig. 6.1d).

6.2.1 Differential Contributions of Different Classes of FcγRs to IgG Subclass-Dependent Pathogenicity of Anti-RBC Autoantibodies

Murine immune effector cells express three different classes of stimulatory FcγR: the high-affinity FcγRI, the intermediate-affinity FcγRIV, and the low-affinity

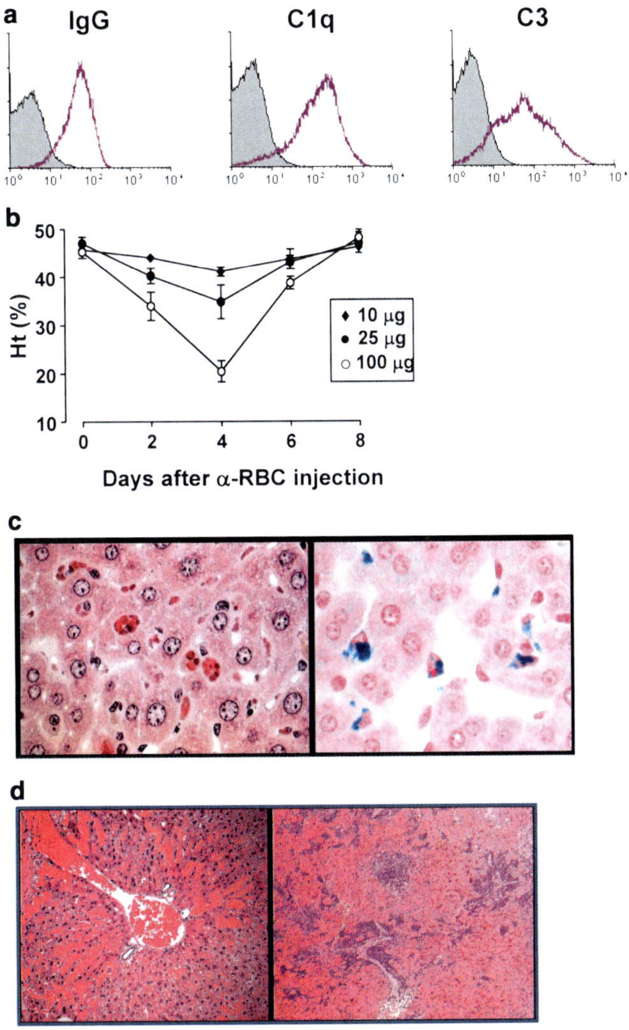

Fig. 6.1 *Induction of autoimmune hemolytic anemia in mice after a single injection of anti-RBC mAb*. (**a**) Flow cytometric analysis of opsonized RBC in mice 24 h after injection of 100 µg of 34-3C IgG2a anti-RBC mAb. Mouse RBC were stained with biotinylated anti-mouse κ-chain mAb, anti-C1q antibodies, or anti-C3 antibodies, followed by phycoerythrin-conjugated streptavidin. Shaded areas indicate the background staining with phycoerythrin-conjugated streptavidin. (**b**) Development of anemia, documented by measuring hematocrit (Ht) values, after injection of different doses of by 34-3C IgG2a anti-RBC mAb. Mean Ht values (± SD) of three mice are shown. (**c**) *Left panel*: representative histological appearance of liver 4 days after injection of 100 µg of 34-3C IgG2a anti-RBC mAb. Note the presence of marked erythrophagocytosis by Kupffer cells (HE staining). *Right panel*: representative histological appearance of iron deposits in Kupffer cells as a result of RBC destruction 8 days after injection of 34-3C IgG2a mAb (Perls iron staining). (**d**) Representative histological appearance of liver and spleen 2 days after injection of 100 µg of 34-3C IgM anti-RBC mAb. Note an enormous accumulation of agglutinated RBC in sinusoids of liver (*left panel*) and red pulp of spleen (*right panel*; HE staining)

Table 6.1 Effector functions of 34-3C anti-RBC Ig class-switch variants

Ig	FcγRI	FcγRIII	FcγRIV	FcγRIIB	C′
IgG1	–	+	–	++	–
IgG2a	+	++	+	+	++
IgG2b	–	++	+	++	++
IgG3	–	–	–	–	+
IgM[a]	–	–	–	–	++
IgA[b]	–	–	–	–	–

[a]In contrast to polymeric IgM, monomeric IgM fails to activate complement

[b]Neither polymeric nor monomeric IgA activate complement

FcγRIII (Nimmerjahn and Ravetch 2006). These three stimulatory FcγRs display a distinct pattern of interaction with four different IgG subclasses in mice, as FcγRI interacts only with IgG2a; FcγRIII with IgG1, IgG2a, and IgG2b; and FcγRIV with IgG2a and IgG2b (Table 6.1). In contrast, the IgG3 subclass is unable to interact with any stimulatory FcγRs. The stimulatory FcγRs are hetero-oligomeric complexes, in which the respective ligand-binding α-chains are associated with the common γ-chain (FcRγ) which contains an immunoreceptor tyrosine-based activation motif. FcRγ is critical for the cell surface expression of these three stimulatory receptors and for the triggering of their various effector functions, including phagocytosis by macrophages (Takai et al. 1994). FcγR-mediated erythrophagocytosis by Kupffer cells results in intracellular destruction of opsonized RBC, i.e., extravascular hemolysis. On the other hand, IgG2a, IgG2b, and IgG3, but not IgG1, efficiently activate complement (Table 6.1). The activation of complement could then trigger erythrophagocytosis by phagocytes through complement receptors (CR) which recognize C3 fragments deposited on RBC, and also induce intravascular hemolysis as a result of the generation of membrane attack complexes.

In view of the remarkable heterogeneities in Fc-associated effector functions between the four different IgG subclasses, one could expect considerable differences in the pathogenic potency between the IgG subclasses in mice. Indeed, the analysis of IgG subclass-switch variants of a 34-3C IgG2a anti-RBC mAb derived from NZB mice has revealed striking differences in pathogenic activity across the four different IgG subclasses, in correlation with the capacity of the individual IgG subclasses to activate the three different classes of stimulatory FcγRs and/or complement (Tables 6.1 and 6.2). IgG2a and IgG2b, which both efficiently activate FcγRIII, FcγRIV, and complement, display the highest pathogenic activity (Azeredo da Silveira et al. 2002; Baudino et al. 2008). In contrast, the contribution of the high-affinity FcγRI to IgG2a-induced anemia is limited because of the competition with excess amounts of circulating monomeric IgG2a having a high-affinity interaction with this receptor. However, increased avidity due to extensive RBC opsonization can result in more efficient binding to FcγRI and thus contribute to the development of severe forms of anemia provoked by injection of higher doses of 34-3C IgG2a (Baudino et al. 2008). The IgG3 subclass, which is unable to interact with any stimulatory FcγRs, induces anemia through the activation of complement (Azeredo da Silveira et al. 2002). Consequently, the pathogenic potency of IgG3 is

Table 6.2 Pathogenic activities of 34-3C anti-RBC Ig class-switch variants and respective contributions of stimulatory FcγRs and complement to the development of autoimmune hemolytic anemia

Ig	Anemic dose[a]	Effector mechanisms	Pathogenesis
IgG1[b]	500 μg	FcγRIII	Erythrophagocytosis
IgG2a[c]	25 μg	FcγRIII, FcγRIV, FcγRI, C′	Erythrophagocytosis
IgG2b	25 μg	FcγRIII, FcγRIV, C′	Erythrophagocytosis
IgG3	100 μg	C′	Erythrophagocytosis
IgM[d]	25 μg	Agglutination	Hemagglutination
IgA[d]	25 μg	Agglutination	Hemagglutination

[a]The quantity of mAb required for inducing hematocrit values <40 %

[b]FcγRIIB efficiently counteracts FcγRIII-triggered erythrophagocytosis induced by IgG1 anti-RBC mAb

[c]FcγRI contributes to the development of severe anemia induced by a high dose (200 μg) of 34-3C IgG2a

[d]Only polymeric forms of IgM and IgA are pathogenic

far less than that of IgG2a and IgG2b but higher than that of the IgG1 subclass activating only FcγRIII but not complement (Azeredo da Silveira et al. 2002; Meyer et al. 1998). Overall, IgG2a and IgG2b are a 20-fold more potent than IgG1.

The weak pathogenic potential of IgG1 anti-RBC autoantibodies could also be due to the high-binding capacity of this subclass to the low-affinity inhibitory FcγRIIB (Table 6.1) (Nimmerjahn and Ravetch 2005), which is additionally expressed on phagocytes. FcγRIIB is a single α-chain receptor carrying an immunoreceptor tyrosine-based inhibitory motif (Takai et al. 1996). Upon coligation of the inhibitory FcγRIIB with stimulatory FcγRs, cellular activation is inhibited by the recruitment of an inositol polyphosphate phosphatase, SHIP, which mediates the inhibition of calcium fluxes. The importance of FcγRIIB for the negative regulation of FcγRIII- and FcγRIV-dependent effector functions in vivo has been demonstrated by markedly enhanced anaphylactic and inflammatory immune responses in FcγRIIB-deficient mice (Nimmerjahn and Ravetch 2005; Takai et al. 1996; Ujike et al. 1999; Clynes et al. 1999; Schiller et al. 1999). Indeed, studies with 34-3C IgG switch variants have demonstrated that injection of IgG1, but not IgG2a and IgG2b, anti-RBC antibodies provokes more severe anemia in FcγRIIB-deficient mice than in FcγRIIB-sufficient mice (Yamada et al. Submitted for publication). The failure of the IgG2a and IgG2b subclasses to induce more severe anemia in FcγRIIB-deficient mice is likely due to the engagement of at least three different phagocytic receptors (FcγRIII, FcγRIV, and CR), which contrasts with the engagement of only FcγRIII by IgG1. Notably, similar results have been obtained in an experimental model of autoimmune thrombocytopenia using IgG subclass-switch variants of a 6A6 anti-platelet mAb (Nimmerjahn and Ravetch 2005). Collectively, these data indicate that FcγRIIB more efficiently inhibits the pathogenic effect of IgG1, but not IgG2a and IgG2b, autoantibodies, thereby further contributing to the weak pathogenic potential of the IgG1 subclass.

Given the importance of competitive engagement of stimulatory and inhibitory FcγRs and of relative expression levels of different classes of FcγR on immune effector cells, it is worth noting that lupus-prone NZB, BXSB, and MRL strains share the NZB-type defective allele of the *Fcgr2b* gene encoding FcγRIIB (Jiang et al. 2000; Pritchard et al. 2000). Because of deletion polymorphism in the promoter region (Jiang et al. 2000; Pritchard et al. 2000; Xiu et al. 2002), expression levels of FcγRIIB are markedly downregulated on their Gr-1^{lo} monocyte subset and peritoneal macrophages as compared with mice carrying the wild-type *Fcgr2b* allele (Kikuchi et al. 2006; Santiago-Raber et al. 2009). Gr-1^{lo} monocytes are one of the two major subsets of monocytes and considered to be a source of tissue-resident macrophages including Kupffer cells (Geissmann et al. 2003). Significantly, they display a functionally more mature phenotype than Gr-1^{hi} monocytes since all three sets of stimulatory FcγRs are expressed on Gr-1^{lo} monocytes, while Gr-1^{hi} monocytes fail to express FcγRIV (Santiago-Raber et al. 2009; Biburger et al. 2011). In addition, lupus-prone mice develop an age-dependent monocytosis characterized by an accumulation of Gr-1^{lo} monocytes (Amano et al. 2005), the development of which is dependent on the expression of stimulatory FcγRs (Santiago-Raber et al. 2009). Further studies have demonstrated that the activation of TLR7 and TLR9 induces the maturation of Gr-1^{hi} monocytes towards Gr-1^{lo} monocytes, enhances the expression of FcγRIV on Gr-1^{lo} monocytes, and also promotes the generation of monocytes per se (Santiago-Raber et al. 2011). Thus, it is tempting to assume that IgG immune complexes (IC) containing endogenous nuclear antigens interact with stimulatory FcγRs and then TLR7/9, thereby promoting the development of monocytosis associated with an expansion of the Gr-1^{lo} monocyte subset in lupus-prone mice. In addition, because of low expression of FcγRIIB in lupus-prone mice, Gr-1^{lo} monocytes could be excessively activated in the presence of circulating IgG IC and hence actively participate in the development and progression of IC-mediated tissue injury in systemic lupus erythematosus (SLE). This idea is consistent with the demonstration that Gr-1^{lo} monocytes play a critical role in IgG-dependent effector functions in vivo (Biburger et al. 2011).

6.2.2 Contribution of CR-Mediated Phagocytosis but Not Intravascular Hemolysis to the Development of Autoimmune Hemolytic Anemia

Early studies reported that the development of anemia with a relatively low dose of 34-3C IgG2a or polyclonal rabbit IgG anti-mouse RBC antibodies was not affected in C3-depleted or C3-deficient mice (Shibata et al. 1990; Clynes and Ravetch 1995; Sylvestre et al. 1996). However, the subsequent analysis of different doses of IgG switch variants of the 34-3C mAb has provided new insights into this issue (Table 6.2): Complement activation is indeed involved in the development of anemia through CR-mediated erythrophagocytosis in a dose-dependent and IgG subclass-specific manner (Azeredo da Silveira et al. 2002). For example, C3 deficiency only minimally affects the development of mild anemia provoked by a low dose (50 μg) of 34-3C IgG2a, while CR-dependent erythrophagocytosis significantly contributes to the development of severe anemia induced by a higher dose (200 μg). This indicates that CR-mediated erythrophagocytosis requires an extensive opsonization of RBC by C3 fragments, while a limited opsonization with IgG2a antibodies is sufficient to trigger FcγR-dependent erythrophagocytosis. It is also worth stressing that the IgG3 subclass, which fails to interact with FcγRs, is able to induce anemia as a result of CR-mediated erythrophagocytosis.

It is somehow unexpected that none of the anti-RBC mAbs, even of the IgM isotype, are able to induce complement-mediated intravascular hemolysis since the development of anemia is not affected in C5-deficient mice which are unable to generate the membrane attack complexes required for complement-mediated intravascular hemolysis (Shibata et al. 1990). The absence of complement-mediated intravascular hemolysis in mice injected with IgG2a/2b/3 and IgM anti-RBC, all of which efficiently activate complement (Table 6.1), is likely due to the presence of a number of complement inhibitory proteins on RBC membranes. This idea is supported by the observations that the injection of 34-3C IgG2a induces a more rapid elimination of RBC deficient in membrane-bound complement regulatory proteins, such as decay-accelerating factor (DAF) and CR1-related gene/protein y (Crry), than wild-type RBC (Molina et al. 2002). Although this study did not directly demonstrate that deficiency in DAF or Crry results in intravascular hemolysis of 34-3C IgG2a-opsonized RBC, it has been shown that DAF-deficient RBC are highly sensitive to antibody-induced complement-dependent hemolysis in vitro (Miwa et al. 2002). In addition, the lack of induction of complement-mediated intravascular hemolysis by anti-RBC mAb could be related to the nature of the target RBC antigen. Since ABO-incompatible transfusion results in typical acute intravascular hemolysis, it is likely that certain antigenic sites on RBC are more vulnerable to complement-mediated immune hemolysis.

It should be stressed that in mice injected with IgM anti-RBC mAb, no sign of CR-mediated erythrophagocytosis by Kupffer cells is detectable despite strong activation of complement, as documented by a marked deposition of C3 fragments on RBC (Baudino et al. 2007). In fact, the pathogenic mechanism responsible for the induction of anemia by IgM is a mechanical sequestration of RBC in spleen and occasionally in liver due to a massive hemagglutination promoted by its pentameric form (Fig. 6.1d and Table 6.2). Thus, the formation of large agglutinated RBC likely interferes with the process of recognition and subsequent phagocytosis by CR. Significantly, the monomeric form of 34-3C IgM variant completely fails to induce anemia, despite its high-affinity binding to RBC in vivo (Baudino et al. 2007). The lack of pathogenicity of the monomeric IgM anti-RBC mAb can be explained by its inability to induce hemagglutination, its poor activation of complement, and the absence of phagocytic receptors specific for IgM. Notably, the analysis of IgA 34-3C anti-RBC variants, which are unable to activate complement (Table 6.1), has also revealed that only polymeric, but not monomeric, IgA is able to induce anemia through sequestration of agglutinated RBC in spleen, as in the case of IgM (Table 6.2) (Baudino et al. 2007). In agreement with studies in murine models of autoimmune hemolytic anemia, the presence of agglutinating IgM or IgA autoantibodies is indeed associated with severe anemia due to hemagglutination in spleen and liver rather than intravascular hemolysis in human patients (Waterbury et al. 1986; Freedman et al. 1987; Shirey et al. 1987; Clark et al. 1984; Chadebech et al. 2010).

6.3 Pathogenicity of IgG3 mAb with Cryoglobulin Activity

Cryoglobulins are Ig which precipitate at temperatures lower than 37 °C, with resolution upon warming (Grey and Kohler 1973; Brouet et al. 1974). Cryoglobulins can be classified into three types according to the precipitation of monoclonal Ig (type I), both monoclonal and polyclonal Ig (type II), or polyclonal Ig with other serum proteins or exogenous antigens (type III). Monoclonal type I cryoglobulins are mostly associated with various lymphoproliferative disorders. Mixed type II or type III cryoglobulinemia is often found in patients with autoimmune diseases such as SLE and rheumatoid arthritis or with a variety of chronic infectious diseases.

MRL-*Fas*^{lpr} mice deficient in the Fas apoptosis receptor (Watanabe-Fukunaga et al. 1992) spontaneously develop an autoimmune syndrome characterized by immunopathological manifestations such as lupus-like glomerulonephritis, arthritic-like lesions, and necrotizing vascular lesions (Andrews et al. 1978). They also produce the highest amounts of cryoglobulins among several lupus-prone mice. Cryoglobulins from MRL-*Fas*^{lpr} mice are composed almost exclusively of IgG of polyclonal origin, with a markedly enriched IgG3 content in comparison with other IgG subclasses (Abdelmoula et al. 1989). This preferential precipitation of IgG3 in polyclonal cryoglobulins is also found in non-autoimmune mice after polyclonal stimulation of B cells with bacterial lipopolysaccharides or infection with malaria parasites (Abdelmoula et al. 1989). Indeed, antibodies of the IgG3 subclass in mice and humans have a unique physicochemical property, which allows them to self-associate via Fc-Fc interactions and display cryoglobulin activity, independently of their specificities (Abdelmoula et al. 1989; Grey et al. 1968; Capra and Kunkel 1970; Saluk and Clem 1975; Nishimura and Nakamura 1984; Spertini et al. 1989). Thus, the self-association and subsequent cryoprecipitation of IgG3 is the principal mechanism responsible for the generation of cryoglobulins in mice.

6.3.1 Contribution of Charged Amino Acid Residues and Sialylation to the IgG3 Cryoglobulin Activity

The self-associating property of the IgG3 subclass is necessary but not sufficient by itself to confer cryoglobulin activity since not all monoclonal IgG3 proteins form cryoglobulins. The molecular mechanism responsible for the generation of monoclonal IgG3 cryoglobulins has been explored using murine IgG3 mAbs.

Nucleotide sequence analysis of the variable regions of a panel of cryogenic and non-cryogenic IgG3 mAbs, in combination with the assessment of mutant antibodies, has revealed that cryoglobulin activity of IgG3 is associated with the presence of more positively charged amino acid residues at positions 6 and 23 of the heavy-chain variable (VH) domain (Panka et al. 1995; Kuroki et al. 2002). The implication of positive charges in IgG3 cryoprecipitation is supported by the finding that cryoglobulin activity of IgG3 anti-dinitrophenyl (DNP) mAbs is inhibited or enhanced following the binding of negatively or positively charged DNP-labeled amino acid conjugates, respectively (Spertini et al. 1989).

The CH2 domain of each heavy chain of IgG bears an asparagine-linked, biantennary complex-type oligosaccharide chain, which is essential for IgG Fc-dependent effector functions (Nose and Wigzell 1983; Jefferis et al. 1998). Each of these oligosaccharide chains, most of which are non-sialylated and neutral, ends by two *N*-acetylglucosamine residues, one *N*-acetylglucosamine and one galactose, or two galactoses (Parekh et al. 1985; Mizuochi et al. 1990). However, a significant, though minor, fraction of galactosylated oligosaccharide chains bear one or two terminal sialic acids and hence become more negatively charged. In view of the established implication of more positively charged amino acid residues of the VH region (Panka et al. 1995; Kuroki et al. 2002) in IgG3 cryoglobulin activity, the possible inhibitory effect that terminal sialic acids on oligosaccharide chains may have on the cryogenic activity of IgG3 mAbs has been explored. Indeed, comparative oligosaccharide structural analysis of different cryogenic and non-cryogenic IgG3 mAbs has shown an inverse correlation between the extent of sialylation and cryogenic activity (Kuroda et al. 2005). The inhibitory role of sialylation has further been confirmed by the demonstration of enrichment of less and more sialylated IgG3 in cryoprecipitated and non-cryoprecipitated fractions from cryogenic IgG3 mAbs, respectively. Moreover, IgG3 cryoglobulin activity is abolished after introduction of an F243A mutation (phenylalanine to alanine at position 243) in the CH2 domain which leads to a marked increase in sialylation of oligosaccharide chains (Otani et al. 2012). All these results suggest that the content of negatively charged sialic acids in oligosaccharide side chains is an additional factor determining IgG3 cryoglobulin activity. This is in agreement with reports of a markedly reduced content of sialic acids in cryoglobulins in several cases of human cryoglobulinemia (Brouet et al. 1974; Zinneman et al. 1968; McIntosh et al. 1971; Zlotnick et al. 1972; Hansson and Lindstrom 1973). In this regard, it is worth noting that cryoglobulinemia is frequently observed in patients with liver injuries (Levo et al. 1977a, b). This might be due to a lack of efficient elimination of asialylated IgG with cryogenic potential in these patients, since the liver is known to be the major site for removal of serum asialoglycoproteins because of the expression of the asialoglycoprotein receptor on hepatocytes (Van Lenten and Ashwell 1972; Marshall et al. 1978). Thus, the formation of self-associating IgG3 complexes via Fc-Fc interaction is likely to be the first step required for cryoglobulin formation, although the molecular basis for IgG3 self-association has not yet been defined. The second step could then be the electrostatic interaction of these IgG3 complexes which would lead to the generation of larger complexes precipitating at lower temperature. Thus, the presence

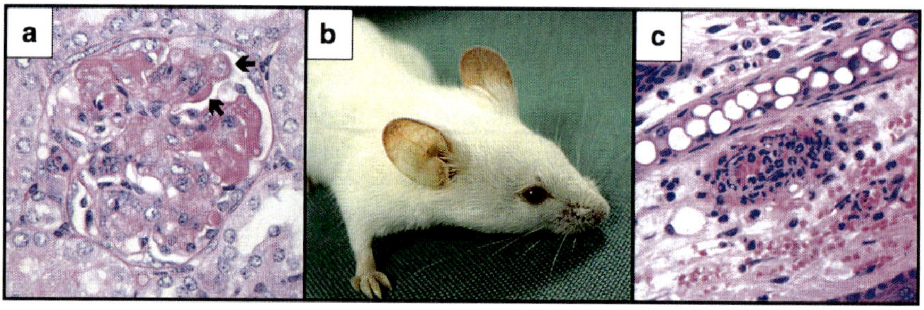

Fig. 6.2 *Induction of glomerulonephritis and cutaneous vasculitis in mice after intraperitoneal implantation of 6–19 IgG3 anti-IgG2a RF hybridoma cells*. (**a**) Representative histological appearance of glomerular lesions 8 days after injection of 6–19 hybridoma cells. Note the presence of PAS-positive deposits along the glomerular capillary walls (*arrows*), resembling "wire-loop" lesions (PAS staining). (**b**) Development of vascular purpura in the skin of ears 5 days after injection of 6–19 hybridoma cells. (**c**) Representative histological appearance of skin lesions induced by 6–19 IgG3 RF mAb. Leukocytoclastic vasculitis is characterized by the infiltration of PMN and the extravasation of RBC (HE staining)

of more positively charged amino acid residues in the VH region and less sialic acid residues in the oligosaccharide side chains of the Fc region could both promote the generation of IgG3 cryoglobulins.

6.3.2 Induction of Glomerular and Vascular Lesions by IgG3 Anti-IgG2a RF Cryoglobulins

Cryoglobulins have long been suggested to be a potential contributor to tissue injuries in SLE, rheumatoid arthritis, and related autoimmune diseases (Brouet et al. 1974). Indeed, a fraction of cryoprecipitating IgG3 anti-IgG2a RF and anti-DNA mAbs derived from lupus-prone mice are highly pathogenic, generating glomerular and vascular lesions (Gyotoku et al. 1987; Berney et al. 1992; Lemoine et al. 1992; Itoh et al. 1993; Takahashi et al. 1993). For example, the implantation of hybridoma cells secreting IgG3 anti-IgG2a RF derived from MRL-*Fas*^{lpr} mice into non-autoimmune mice rapidly induces the development of "wire-loop"-like glomerular lesions characterized by subendothelial immune deposits in glomeruli (Fig. 6.2a) and of cutaneous leukocytoclastic vasculitis (Fig. 6.2b, c) which are also observed in patients with SLE and mixed cryoglobulinemia. The pathogenic potential of IgG3 autoantibodies in association with cryoglobulin activity has been supported by the analysis of a panel of anti-IgG2a RF mAbs including Ig class-switch variants of 6–19 anti-IgG2a RF mAb (Berney et al. 1992; Fulpius et al. 1993; Moll et al. 1995). Moreover, the essential role of cryoglobulin activity for the pathogenic potential of IgG3 autoantibodies has been confirmed by the finding that the highly sialylated 6–19 IgG3 F243A mutant lacking cryoglobulin activity induces neither glomerular nor skin vascular lesions (Otani et al. 2012).

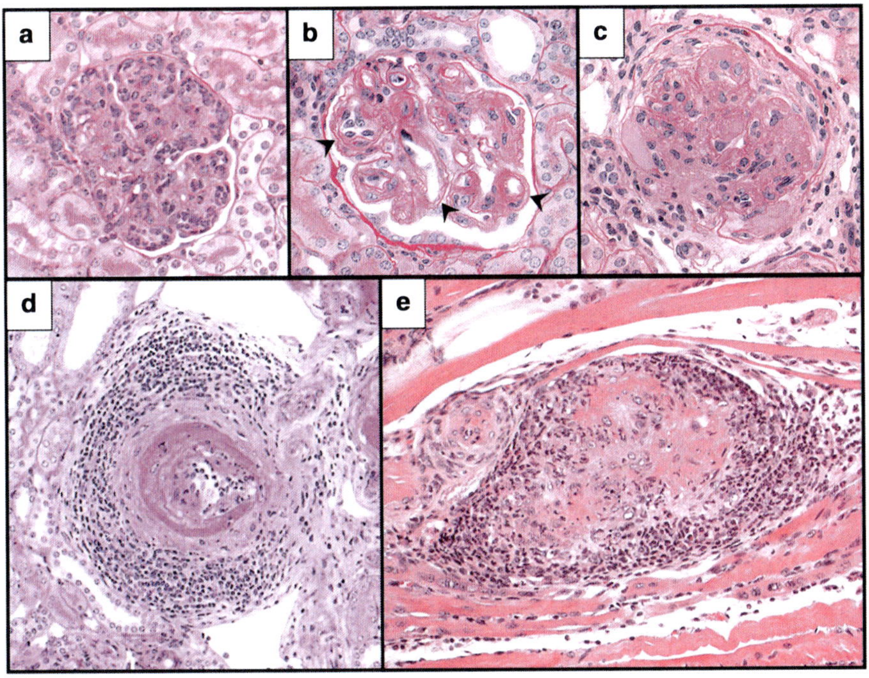

Fig. 6.3 *Histological appearance of glomerular and vascular lesions in 6–19 IgG3 RF transgenic mice*. (**a**) Glomerular lesions characterized by an increase of glomerular cellularity (PAS staining). (**b**) Glomerular lesions characterized by membranoproliferative changes associated with capillary wall thickening due to characteristic reduplication of glomerular basement membrane (*arrowheads*) (PAS staining). (**c**) Glomerular lesions characterized by crescent formation as a result of proliferation of epithelial cells of Bowman's capsule (PAS staining). (**d**) and (**e**) Necrotizing arteritis with fibrinoid necrosis and PMN infiltration associated with perivascular mononuclear cell infiltration in kidney (**d**) and skeletal muscle (**e**) (D, PAS staining; E, HE staining)

To better define the immunopathological consequences associated with the chronic presence of pathogenic 6–19 IgG3 autoantibodies in sera, as opposed to the acute and rapid elevation following experimental implantation of 6–19 hybridoma, a transgenic mouse line expressing IgG3 6–19 RF mAb has been established (Kikuchi et al. 2002). These 6–19 transgenic mice develop severe chronic glomerulonephritis with highly heterogeneous lesions, including diffuse proliferative or membranoproliferative changes, crescent formation, and glomerular sclerosis, within 8–10 months of age (Fig. 6.3a, b and c). In addition, they present necrotizing arteritis in small- and medium-sized arteries of the kidney and skeletal muscle which are characterized by fibrinoid necrosis and polymorphonuclear neutrophil (PMN) infiltration in association with perivascular mononuclear cell infiltration (Fig. 6.3d, e). These changes markedly contrast with the renal and vascular lesions resulting from the rapid induction of high serum levels of 6–19 RF mAb within 8–10 days after implantation of 6–19 hybridoma. Overall, these results highlight that a single autoantibody (such as IgG3 6–19 RF) can induce remarkably different types of glomerular and vascular lesions, depending on its production levels and kinetics as well as on the duration of its presence in the circulating blood.

6.3.3 Differential Pathogenicity for Glomerular and Vascular Lesions Induced by IgG3 Anti-IgG2a RF Cryoglobulins

Significantly, the implantation of 6–19 IgG3 anti-IgG2a RF hybridoma induces the development of glomerular, but not cutaneous vascular, lesions in Ig-deficient mice lacking the corresponding IgG2a autoantigens (Reininger et al. 1990; Izui et al. 1998). This indicates that the RF autoantibody activity, i.e., the formation of IgG3-IgG2a IC, in association with cryoglobulin formation critically contributes to the development of skin vasculitis, while the cryoglobulin activity is solely responsible for renal pathogenicity. Thus, different pathogenetic mechanisms govern the development of skin leukocytoclastic vasculitis and glomerulonephritis induced by IgG3 RF cryoglobulins.

PMN are the principal inflammatory cells in skin vascular lesions caused by 6–19 IgG3 RF mAb since no skin purpuric lesions develop in mice depleted of PMN (Fulpius et al. 1996). In contrast, typical cutaneous vascular lesions are still induced in mice deficient in C3 (Izui et al. 1998). Thus, complement activation does not play a significant role in the pathogenesis of PMN-dependent skin vascular lesions induced by 6–19 IgG3 RF cryoglobulins, thereby indicating the involvement of other types of PMN chemoattractants. Like their PMN-depleted counterparts, mice deficient in FcγRIII are unable to develop skin vasculitis following the implantation of 6–19 RF hybridoma cells but are capable of doing so after the transfer of wild-type FcγRIII-expressing mast cells (Watanabe et al. 1999). Thus, the activation of mast cells through interaction of FcγRIII with IgG2a present in IgG3 RF cryoglobulin IC likely results in the release of inflammatory mediators (such as TNF) which recruit and activate PMN, thereby generating leukocytoclastic vascular lesions. This mechanism would also explain why RF monoclonal cryoglobulins are the only IgG3 mAb capable of inducing skin vascular lesions, as other non-RF IgG3 monoclonal cryoglobulins, including IgG3 anti-DNA mAb, cannot do so, the latter observation being in line with the fact that IgG3 per se is unable to activate FcγRs.

In contrast to cutaneous vascular lesions, the development of glomerular lesions induced by 6–19 IgG3 RF is not dependent on the activation of stimulatory FcγR-expressing cells and of complement (Izui et al. 1998; Watanabe et al. 1999). This is consistent with the fact that the development of glomerular lesions is independent of the formation of 6–19 IgG3-IgG2a IC (Reininger et al. 1990; Izui et al. 1998). It has been shown that FcγR-mediated inflammatory responses play a critical role in the pathogenesis of lupus-like nephritis developing in (NZB x NZW)F1 and BXSB, but not in MRL-*Fas*^{lpr}, mice (Clynes et al. 1998; Matsumoto et al. 2003; Lin et al. 2006). Notably, MRL-*Fas*^{lpr} mice spontaneously generate extremely large amounts of IgG3 cryoglobulins (Abdelmoula et al. 1989), the level of which is well correlated with the progression of lupus nephritis in this strain (Takahashi et al. 1991, 1996; Fossati et al. 1993). These data suggest that the development of lupus nephritis in MRL-*Fas*[lpr] mice might be more dependent on the production of IgG3 autoantibodies (Greenspan et al. 2012), which are unable to activate stimulatory FcγRs (Azeredo da Silveira et al. 2002), and hence less dependent on the activation of FcγR-bearing effector cells.

It should, however, be stressed that additional studies have shown that 6–19 heavy (H)-chain single transgenic mice completely fail to develop any glomerular lesions, despite the presence of substantial amounts of cryoglobulins, and that the nephritogenicity of the 6–19 RF mAb is determined by a combination of 6–19 H and light (L) chains (Kikuchi et al. 2002). This indicates that the majority of IgG3 cryoglobulins resulting from a heterogeneous combination of the 6–19 H chains with endogenous L chains in the 6–19 H-chain transgenic mice are poorly nephritogenic and that 6–19 H chains acquire their nephritogenic activity only when combined with appropriate L chains. This notion is consistent with the fact that the nephritogenicity of 6–19 mAb is conserved in the case of a 6–19H/J558L hybrid IgG3 (made of 6–19 H chains and J558 anti-α1,3-dextran L chains) but not in that of a 6–19H/9A6L hybrid (made of 6–19 H chains and 9A6 anti-DNP L chains) (Mizuochi et al. 2001; Kikuchi et al. 2002) and that only a fraction of IgG3 monoclonal cryoglobulins is able to provoke significant glomerular lesions (Gyotoku et al. 1987; Lemoine et al. 1992; Itoh et al. 1993; Takahashi et al. 1993; Fulpius et al. 1994; Rengers et al. 2000). Thus, the cryoglobulin activity is necessary but not sufficient by itself for the expression of the pathogenic potential of IgG3 autoantibodies. This indicates that qualitative features of cryoglobulins are critical to determine their pathogenic activity.

6–19 IgG3 RF mAb is poorly galactosylated and consequently hardly sialylated (Mizuochi et al. 2001; Kuroki et al. 2002), as the terminal sialylation is dependent on the presence of galactose residues in oligosaccharide chains. Thus, it has been speculated that conformational change(s) in the Fc region resulting from aberrant glycosylation of IgG3 could promote their glomerular deposition as well as subsequent inflammation. Indeed, it has previously been claimed that galactose-less IgG can be more pathogenic based on the following findings. First, it can more efficiently activate complement through the lectin pathway in vitro, in which mannose-binding lectin (MBL) can interact with *N*-acetylglucosamines exposed as a result of absence of galactosylation (Malhotra et al. 1995). Second, in a murine model of collagen-induced arthritis, in vitro treatment of anti-collagen antibodies with β-galactosidase enhances their arthritogenic activity (Rademacher et al. 1994). However, results obtained in MBL-deficient mice argued against the contribution of the MBL pathway to the pathogenic activity of agalactosylated IgG (Nimmerjahn et al. 2007). Moreover, studies on different 6–19 IgG3 variants differing in the content of galactose or sialic acid residues have shown that the nephritogenic activity of 6–19 IgG3 RF mAb is dependent on the extent of terminal sialylation, which negatively regulates the IgG3 cryoglobulin activity, but independent of the extent of galactosylation per se (Otani et al. 2012). In this regard, it should be emphasized that the development of glomerular lesions induced by 6–19 IgG3 cryoglobulins is independent of the activation of FcγR-bearing immune effector cells and of complement (Fulpius et al. 1996; Izui et al. 1998; Watanabe et al. 1999). Thus, the lack of nephritogenic activity of the highly sialylated, non-cryogenic 6–19 variant is unrelated to the reported anti-inflammatory property of highly sialylated IgG antibodies, which is owing to their reduced capacity to interact with stimulatory FcγRs (Kaneko et al. 2006; Scallon et al. 2007). Nevertheless, all these results are consistent with the idea that the level of IgG sialylation could be an important factor in determining the pathogenic potential of autoantibodies. This would also be in line with the increased levels of agalactosylated and thus asialylated IgG in patients with rheumatoid arthritis and in MRL-*Fas*^{lpr} mice, in parallel with the progression of the disease (Parekh et al. 1985, 1989; Mizuochi et al. 1990).

6.4 Concluding Remarks

The pathogenesis of autoantibody-mediated cellular and tissue lesions in autoimmune diseases is most straightforwardly attributed to the combined action of self-antigen binding properties and effector functions associated with the Fc region of the different Ig isotypes and IgG subclasses. The analysis of two different sets of monoclonal autoantibodies derived from lupus-prone mice (34-3C anti-RBC and 6–19 anti-IgG2a RF) has revealed remarkable differences in the pathogenic potential of different IgG subclasses. The IgG2a and IgG2b subclasses of anti-RBC autoantibodies are most pathogenic because of the efficient activation of two classes of stimulatory FcγRs (FcγRIII and FcγRIV) and complement, while the IgG1 subclass is the least pathogenic as it interacts only with FcγRIII and its activation is more strongly blocked by the inhibitory FcγRIIB. In addition, because of the unique property of IgG3 to form self-associating complexes and generate cryoglobulins, RF of this subclass become highly pathogenic and induce lupus-like nephritis and vasculitis.

It has been established that the production of IgG2a and IgG3 subclasses is promoted by the Th1 cytokine IFN-γ, while IL-4, the Th2 cytokine, promotes the switching to IgG1. Thus, results obtained through the analysis of the two aforementioned sets of autoantibodies strongly suggest that Th1-type autoimmune responses may be critically involved in the development and progression of autoantibody-mediated autoimmune diseases, such as SLE, by promoting the switch towards more pathogenic IgG subclasses of autoantibodies. Indeed, this notion has been supported by several studies demonstrating that the progression of murine lupus nephritis is correlated with the relative dominance of Th1 autoimmune responses in association with increases in serum levels of IgG2a and IgG3 vs. IgG1 autoantibodies (Jacob et al. 1987; Takahashi et al. 1996; Santiago et al. 1997; Peng et al. 2002). However, this does not necessarily exclude an active role of Th2 responses in autoimmune diseases since Th2 cytokines such as IL-5, IL-6, and IL-10 have also been suggested to be involved in the induction of autoimmune manifestations in SLE (Finck et al. 1994; Ishida et al. 1994; Nisitani et al. 1995). These effects may be related to the ability of Th2 cytokines to augment overall production of pathogenic autoantibodies by promoting the terminal differentiation of activated B cells into plasma cells.

References

Abdelmoula M, Spertini F, Shibata T, Gyotoku Y, Luzuy S, Lambert PH, Izui S (1989) IgG3 is the major source of cryoglobulins in mice. J Immunol 143:526–532

Amano H, Amano E, Santiago-Raber ML, Moll T, Martinez-Soria E, Fossati-Jimack L, Iwamoto M, Rozzo SJ, Kotzin BL, Izui S (2005) Selective expansion of a monocyte subset expressing the CD11c dendritic cell marker in the *Yaa* model of systemic lupus erythematosus. Arthritis Rheum 52:2790–2798

Andrews BS, Eisenberg RA, Theofilopoulos AN, Izui S, Wilson CB, McConahey PJ, Murphy ED, Roths JB, Dixon FJ (1978) Spontaneous murine lupus-like syndromes. Clinical and immunopathological manifestations in several strains. J Exp Med 148:1198–1215

Azeredo da Silveira S, Kikuchi S, Fossati-Jimack L, Moll T, Saito T, Verbeek JS, Botto M, Walport MJ, Carroll M, Izui S (2002) Complement activation selectively potentiates the pathogenicity of the IgG2b and IgG3 isotypes of a high affinity anti-erythrocyte autoantibody. J Exp Med 195:665–672

Baudino L, Fossati-Jimack L, Chevalley C, Martinez-Soria E, Shulman MJ, Izui S (2007) IgM and IgA anti-erythrocyte autoantibodies induce anemia in a mouse model through multivalency-dependent hemagglutination but not through complement activation. Blood 109:5355–5362

Baudino L, Nimmerjahn F, Azeredo da Silveira S, Martinez-Soria E, Saito T, Carroll M, Ravetch JV, Verbeek JS, Izui S (2008) Differential contribution of three activating IgG Fc receptors (FcγRI, FcγRIII, and FcγRIV) to IgG2a- and IgG2b-induced autoimmune hemolytic anemia in mice. J Immunol 180:1948–1953

Berney T, Fulpius T, Shibata T, Reininger L, Van Snick J, Shan H, Weigert M, Marshak-Rothstein A, Izui S (1992) Selective pathogenicity of murine rheumatoid factors of the cryoprecipitable IgG3 subclass. Int Immunol 4:93–99

Biburger M, Aschermann S, Schwab I, Lux A, Albert H, Danzer H, Woigk M, Dudziak D, Nimmerjahn F (2011) Monocyte subsets responsible for immunoglobulin G-dependent effector functions in vivo. Immunity 35:932–944

Brouet JC, Clauvel JP, Danon F, Klein M, Seligmann M (1974) Biological and clinical significance of cryoglobulins. A report of 86 cases. Am J Med 57:775–788

Capra JD, Kunkel HG (1970) Aggregation of γG3 proteins: relevance to the hyperviscosity syndrome. J Clin Invest 49:610–621

Chadebech P, Michel M, Janvier D, Yamada K, Copie-Bergman C, Bodivit G, Bensussa A, Fournié JJ, Godeau B, Bierling P, Izui S, Noizat-Pirenne F (2010) IgA-mediated human autoimmune hemolytic anemia as a result of hemagglutination in the spleen, but independently of complement activation and FcaRI. Blood 116:4141–4147

Clark DA, Dessypris EN, Jenkins DE Jr, Krantz SB (1984) Acquired immune hemolytic anemia associated with IgA erythrocyte coating: investigation of hemolytic mechanisms. Blood 64:1000–1005

Clynes R, Ravetch JV (1995) Cytotoxic antibodies trigger inflammation through Fc receptors. Immunity 3:21–26

Clynes R, Dumitru C, Ravetch JV (1998) Uncoupling of immune complex formation and kidney damage in autoimmune glomerulonephritis. Science 279:1052–1054

Clynes R, Maizes JS, Guinamard R, Ono M, Takai T, Ravetch JV (1999) Modulation of immune complex-induced inflammation in vivo by the coordinate expression of activation and inhibitory Fc receptors. J Exp Med 189:179–185

Finck BK, Chan B, Wofsy D (1994) Interleukin 6 promotes murine lupus in NZB/NZW F1 mice. J Clin Invest 94:585–591

Fossati L, Takahashi S, Merino R, Iwamoto M, Aubry J-P, Nose M, Spach C, Motta R, Izui S (1993) An MRL/MpJ-*lpr/lpr* substrain with a limited expansion of *lpr* double-negative T cells and a reduced autoimmune syndrome. Int Immunol 5:525–532

Freedman J, Wright J, Lim FC, Garvey MB (1987) Hemolytic warm IgM autoagglutinins in autoimmune hemolytic anemia. Transfusion 27:464–467

Fulpius T, Spertini F, Reininger L, Izui S (1993) Immunoglobulin heavy chain constant region determines the pathogenicity and the antigen-binding activity of rheumatoid factor. Proc Natl Acad Sci USA 90:2345–2349

Fulpius T, Berney T, Lemoine R, Pastore Y, Reininger L, Brighouse G, Izui S (1994) Glomerulopathy induced by IgG3 anti-trinitrophenyl monoclonal cryoglobulins derived from non-autoimmune mice. Kidney Int 45:962–971

Fulpius T, Lemoine R, Berney T, Pastore Y, Moll S, Izui S (1996) Polymorphonuclear leukocytes play a key role in the generation of "wire-loop" lesions induced by a murine IgG3 rheumatoid factor. Kidney Int 49:647–655

Geissmann F, Jung S, Littman DR (2003) Blood monocytes consist of two principal subsets with distinct migratory properties. Immunity 19:71–82

Greenspan NS, Lu MA, Shipley JW, Ding X, Li Q, Sultana D, Kollaros M, Schreiber JR, Fu P, Putterman C, Emancipator SN (2012) IgG3 deficiency extends lifespan and attenuates progression of glomerulonephritis in MRL/lpr mice. Biol Direct 7:3

Grey HM, Kohler PF (1973) Cryoimmunoglobulins. Semin Hematol 10:87–112

Grey HM, Kohler PF, Terry WD, Franklin EC (1968) Human monoclonal γG-cryoglobulins with anti-γ-globulin activity. J Clin Invest 47:1875–1884

Gyotoku Y, Abdelmoula M, Spertini F, Izui S, Lambert PH (1987) Cryoglobulinemia induced by monoclonal immunoglobulin G rheumatoid factors derived from autoimmune MRL/MpJ-*lpr/lpr* mice. J Immunol 138:3785–3792

Hansson UB, Lindstrom FD (1973) Some factors affecting precipitation and complex formation of an IgG cryoglobulin. Clin Exp Immunol 14:427–435

Ishida H, Muchamuel T, Sakaguchi S, Andrade S, Menon S, Howard M (1994) Continuous administration of anti-interleukin 10 antibodies delays onset of autoimmunity in NZB/W F1 mice. J Exp Med 179:305–310

Itoh J, Nose M, Takahashi S, Ono M, Terasaki S, Kondoh E, Kyogoku M (1993) Induction of different types of glomerulonephritis by monoclonal antibodies derived from an MRL/*lpr* lupus mouse. Am J Pathol 143:1436–1443

Izui S, Fulpius T, Reininger L, Pastore Y, Kobayakawa T (1998) Role of neutrophils in murine cryoglobulinemia. Inflamm Res 47:S145–S150

Jacob CO, van der Meide PH, McDevitt HO (1987) In vivo treatment of (NZB x NZW)F1 lupus-like nephritis with monoclonal antibody to g interferon. J Exp Med 166:798–803

Jefferis R, Lund J, Pound JD (1998) IgG-Fc-mediated effector functions: molecular definition of interaction sites for effector ligands and the role of glycosylation. Immunol Rev 163:59–76

Jiang Y, Hirose S, Abe M, Sanokawa-Akakura R, Ohtsuji M, Mi X, Li N, Xiu Y, Zhang D, Shirai J, Hamano Y, Fujii H, Shirai T (2000) Polymorphisms in IgG Fc receptor IIB regulatory regions associated with autoimmune susceptibility. Immunogenetics 51:429–435

Kaneko Y, Nimmerjahn F, Ravetch JV (2006) Anti-inflammatory activity of immunoglobulin G resulting from Fc sialylation. Science 313:670–673

Kikuchi S, Pastore Y, Fossati-Jimack L, Kuroki A, Yoshida H, Fulpius T, Araki K, Takahashi S, Lemoine R, Reininger L, Izui S (2002) A transgenic mouse model of autoimmune glomerulonephritis and necrotizing arteritis associated with cryoglobulinemia. J Immunol 169:4644–4650

Kikuchi S, Santiago-Raber ML, Amano H, Amano E, Fossati-Jimack L, Moll T, Kotzin BL, Izui S (2006) Contribution of *Nba2* to *Yaa*-induced monocytosis in association with murine systemic lupus. J Immunol 176:3240–3247

Kuroda Y, Kuroki A, Kikuchi S, Funase T, Nakata M, Izui S (2005) A critical role for sialylation in cryoglobulin activity of murine IgG3 monoclonal antibodies. J Immunol 175:1056–1061

Kuroki A, Kuroda Y, Kikuchi S, Lajaunias F, Fulpius T, Pastore Y, Fossati-Jimack L, Reininger L, Toda T, Nakata M, Kojima N, Mizuochi T, Izui S (2002) Level of galactosylation determines cryoglobulin activity of murine IgG3 monoclonal rheumatoid factor. Blood 99:2922–2928

Lemoine R, Berney T, Shibata T, Fulpius T, Gyotoku Y, Shimada H, Sawada S, Izui S (1992) Induction of "wire-loop" lesions by murine monoclonal IgG3 cryoglobulins. Kidney Int 41:65–72

Levo Y, Gorevic PD, Kassab HJ, Zucker-Franklin D, Franklin EC (1977a) Association between hepatitis B virus and essential mixed cryoglobulinemia. N Engl J Med 296:1501–1504

Levo Y, Gorevic PD, Kassab HJ, Tobias H, Franklin EC (1977b) Liver involvement in the syndrome of mixed cryoglobulinemia. Ann Intern Med 87:287–292

Lin Q, Xiu Y, Jiang Y, Tsurui H, Nakamura K, Kodera S, Ohtsuji M, Ohtsuji N, Shiroiwa W, Tsukamoto K, Amano H, Amano E, Kinoshita K, Sudo K, Nishimura H, Izui S, Shirai T, Hirose S (2006) Genetic dissection of the effects of stimulatory and inhibitory IgG Fc receptors on murine lupus. J Immunol 177:1646–1654

Malhotra R, Wormald MR, Rudd PM, Fischer PB, Dwek RA, Sim RB (1995) Glycosylation changes of IgG associated with rheumatoid arthritis can activate complement via the mannose-binding protein. Nat Med 1:237–243

Marshall JS, Williams S, Jones P, Hepner GW (1978) Serum desialylated glycoproteins in patients with hepatobiliary dysfunction. J Lab Clin Med 92:30–37

Matsumoto K, Watanabe N, Akikusa B, Kurasawa K, Matsumura R, Saito Y, Iwamoto I, Saito T (2003) Fc receptor-independent development of autoimmune glomerulonephritis in lupus-prone MRL/*lpr* mice. Arthritis Rheum 48:486–494

McIntosh RM, Kulvinskas C, Kaufman DB (1971) Cryoglobulins. II. The biological and chemical properties of cryoproteins in acute post-streptococcal glomerulonephritis. Int Arch Allergy Appl Immunol 41:700–715

Meyer D, Schiller C, Westermann J, Izui S, Hazenbos WLW, Verbeek JS, Schmidt RE, Gessner JE (1998) FcγRIII (CD16)-deficient mice show IgG isotope-dependent protection to experimental autoimmune hemolytic anemia. Blood 92:3997–4002

Miwa T, Zhou L, Hilliard B, Molina H, Song WC (2002) Crry, but not CD59 and DAF, is indispensable for murine erythrocyte protection in vivo from spontaneous complement attack. Blood 99:3707–3716

Mizuochi T, Hamako J, Nose M, Titani K (1990) Structural changes in the oligosaccharide chains of IgG in autoimmune MRL/Mp-*lpr/lpr* mice. J Immunol 145:1794–1798

Mizuochi T, Pastore Y, Shikata K, Kuroki A, Kikuchi S, Fulpius T, Nakata M, Fossati-Jimack L, Reininger L, Matsushita M, Fujita T, Izui S (2001) Role of galactosylation in the renal pathogenicity of murine IgG3 monoclonal cryoglobulins. Blood 97:3537–3543

Molina H, Miwa T, Zhou L, Hilliard B, Mastellos D, Maldonado MA, Lambris JD, Song WC (2002) Complement-mediated clearance of erythrocytes: mechanism and delineation of the regulatory roles of Crry and DAF. Blood 100:4544–4549

Moll S, Menoud PA, Fulpius T, Pastore Y, Takahashi S, Fossati L, Vassalli JD, Sappino AP, Schifferli JA, Izui S (1995) Induction of plasminogen activator inhibitor type 1 in murine lupus-like glomerulonephritis. Kidney Int 48:1459–1468

Nimmerjahn F, Ravetch JV (2005) Divergent immunoglobulin-G subclass activity through selective Fc receptor binding. Science 310:1510–1512

Nimmerjahn F, Ravetch JV (2006) Fcγ receptors: old friends and new family members. Immunity 24:19–28

Nimmerjahn F, Anthony RM, Ravetch JV (2007) Agalactosylated IgG antibodies depend on cellular Fc receptors for in vivo activity. Proc Natl Acad Sci USA 104:8433–8437

Nishimura Y, Nakamura H (1984) Human monoclonal cryoimmunoglobulins. I. Molecular properties of IgG3*k* (Jir protein) and the cryo-coprecipitability of its molecular fragments by papain. J Biochem (Tokyo) 95:255–265

Nisitani S, Tsubata T, Murakami M, Honjo T (1995) Administration of interleukin-5 or −10 activates B-1 cells and induces autoimmune hemolytic anemia in anti-erythrocyte autoantibody-transgenic mice. Eur J Immunol 25:3047–3052

Nose M, Wigzell H (1983) Biological significance of carbohydrate chains on monoclonal antibodies. Proc Natl Acad Sci USA 80:6632–6636

Otani M, Kuroki A, Kikuchi S, Kihara M, Nakata J, Ito K, Furukawa J-I, Shinohara Y, Izui S (2012) Sialylation determines the nephritogenicity of IgG3 cryoglobulins. J Am Soc Nephrol 23:1869–1878

Panka DJ, Salant DJ, Jacobson BA, Minto AW, Marshak-Rothstein A (1995) The effect of VH residues 6 and 23 on IgG3 cryoprecipitation and glomerular deposition. Eur J Immunol 25:279–284

Parekh RB, Dwek RA, Sutton BJ, Fernandes DL, Leung A, Stanworth D, Rademacher TW, Mizuochi T, Taniguchi T, Matsuta K, Takeuchi F, Nagano Y, Miyamoto T, Kobata A (1985) Association of rheumatoid arthritis and primary osteoarthritis with changes in the glycosylation pattern of total serum IgG. Nature 316:452–457

Parekh R, Isenberg D, Rook G, Roitt I, Dwek R, Rademacher T (1989) A comparative analysis of disease-associated changes in the glycosylation of serum IgG. J Autoimmun 2:101–114

Peng SL, Szabo SJ, Glimcher LH (2002) T-bet regulates IgG class switching and pathogenic autoantibody production. Proc Natl Acad Sci USA 99:5545–5550

Pritchard NR, Cutler AJ, Uribe S, Chadban SJ, Morley BJ, Smith KG (2000) Autoimmune-prone mice share a promoter haplotype associated with reduced expression and function of the Fc receptor FcγRII. Curr Biol 10:227–230

Rademacher TW, Williams P, Dwek RA (1994) Agalactosyl glycoforms of IgG autoantibodies are pathogenic. Proc Natl Acad Sci USA 91:6123–6127

Reininger L, Berney T, Shibata T, Spertini F, Merino R, Izui S (1990) Cryoglobulinemia induced by a murine IgG3 rheumatoid factor: skin vasculitis and glomerulonephritis arise from distinct pathogenic mechanisms. Proc Natl Acad Sci USA 87:10038–10042

Rengers JU, Touchard G, Decourt C, Deret S, Michel H, Cogné M (2000) Heavy and light chain primary structures control IgG3 nephritogenicity in an experimental model for cryocrystalglobulinemia. Blood 95:3467–3472

Saluk PH, Clem W (1975) Studies on the cryoprecipitation of a human IgG3 cryoglobulin: the effects of temperature-induced conformational changes on the primary interaction. Immunochemistry 12:29–37

Santiago ML, Fossati L, Jacquet C, Müller W, Izui S, Reininger L (1997) Interleukin-4 protects against a genetically linked lupus-like autoimmune syndrome. J Exp Med 185:65–70

Santiago-Raber ML, Amano H, Amano E, Baudino L, Otani M, Lin Q, Nimmerjahn F, Verbeek JS, Ravetch JV, Takasaki Y, Hirose S, Izui S (2009) FcγR-dependent expansion of a hyperactive monocyte subset in lupus-prone mice. Arthritis Rheum 60:2408–2417

Santiago-Raber ML, Baudino L, Alvarez M, van Rooijen N, Nimmerjahn F, Izui S (2011) TLR7/9-mediated monocytosis and maturation of Gr-1^{hi} inflammatory monocytes towards Gr-1^{lo} resting monocytes implicated in murine lupus. J Autoimmun 37:171–179

Scallon BJ, Tam SH, McCarthy SG, Cai AN, Raju TS (2007) Higher levels of sialylated Fc glycans in immunoglobulin G molecules can adversely impact functionality. Mol Immunol 44:1524–1534

Schiller C, Janssen-Graalfs I, Baumann U, Schwerter-Strumpf K, Izui S, Takai T, Schmidt RE, Gessner JE (1999) Mouse FcγRII is a negative regulator of FcγRIII in IgG immune complex triggered inflammation but not in autoantibody induced hemolysis. Eur J Immunol 30:481–490

Shibata T, Berney T, Reininger L, Chicheportiche Y, Ozaki S, Shirai T, Izui S (1990) Monoclonal anti-erythrocyte autoantibodies derived from NZB mice cause autoimmune hemolytic anemia by two distinct pathogenic mechanisms. Int Immunol 2:1133–1141

Shirey RS, Kickler TS, Bell W, Little B, Smith B, Ness PM (1987) Fatal immune hemolytic anemia and hepatic failure associated with a warm-reacting IgM autoantibody. Vox Sang 52:219–222
Spertini F, Coulie PG, Van Snick J, Davidson E, Lambert PH, Izui S (1989) Inhibition of cryoprecipitation of murine IgG3 anti-dinitrophenyl (DNP) monoclonal antibodies by anionic DNP-amino acid conjugates. Eur J Immunol 19:273–278
Sylvestre DL, Clynes R, Ma M, Warren H, Carroll MC, Ravetch JV (1996) Immunoglobulin G-mediated inflammatory responses develop normally in complement deficient mice. J Exp Med 184:2385–2392
Takahashi S, Nose M, Sasaki J, Yamamoto T, Kyogoku M (1991) IgG3 production in MRL/lpr mice is responsible for development of lupus nephritis. J Immunol 147:515–519
Takahashi S, Itoh J, Nose M, Ono M, Yamamoto T, Kyogoku M (1993) Cloning and cDNA sequence analysis of nephritogenic monoclonal antibodies derived from an MRL/*lpr* lupus mouse. Mol Immunol 30:177–182
Takahashi S, Fossati L, Iwamoto M, Merino R, Motta R, Kobayakawa T, Izui S (1996) Imbalance towards Th1 predominance is associated with acceleration of lupus-like autoimmune syndrome in MRL mice. J Clin Invest 97:1597–1604
Takai T, Li M, Sylvestre D, Clynes R, Ravetch JV (1994) FcR γ chain deletion results in pleiotrophic effector cell defects. Cell 76:519–529
Takai T, Ono M, Hikida M, Ohmori H, Ravetch JV (1996) Augmented humoral and anaphylactic responses in FcγRII-deficient mice. Nature 379:346–349
Ujike A, Ishikawa Y, Ono M, Yuasa T, Yoshino T, Fukumoto M, Ravetch JV, Takai T (1999) Modulation of IgE-mediated systemic anaphylaxis by low-affinity Fc receptors for IgG. J Exp Med 189:1573–1579
Van Lenten L, Ashwell G (1972) The binding of desialylated glycoproteins by plasma membranes of rat liver. Development of a quantitative inhibition assay. J Biol Chem 247:4633–4640
Watanabe N, Akikusa B, Park SY, Ohno H, Fossati L, Vecchietti G, Gessner JE, Schmidt RE, Verbeek JS, Ryffel B, Iwamoto I, Izui S, Saito T (1999) Mast cells induce autoantibody-mediated vasculitis syndrome through tumor necrosis factor production upon triggering Fcγ receptors. Blood 94:3855–3863
Watanabe-Fukunaga R, Brannan CI, Copeland NG, Jenkins NA, Nagata S (1992) Lymphoproliferation disorder in mice explained by defects in Fas antigen that mediates apoptosis. Nature 356:314–317
Waterbury L, Parnes H, Katz RS (1986) Fatal warm antibody autoimmune hemolysis with marked erythrocyte autoagglutination and liver necrosis. South Med J 79:646–647
Xiu Y, Nakamura K, Abe M, Li N, Wen XS, Jiang Y, Zhang D, Tsurui H, Matsuoka S, Hamano Y, Fujii H, Ono M, Takai T, Shimokawa T, Ra C, Shirai T, Hirose S (2002) Transcriptional regulation of *Fcgr2b* gene by polymorphic promoter region and its contribution to humoral immune responses. J Immunol 169:4340–4346
Yamada K, Ito K, Furukawa J-I, Nimmerjahn F, Verbeek JS, Shinohara Y, Izui S IgG1-restricted FcγRIIB-mediated inhibition on the development of autoimmune hemolytic anemia. Submitted for publication
Zinneman HH, Levi D, Seal US (1968) On the nature of cryoglobulins. J Immunol 100:594–603
Zlotnick A, Slavin S, Eliakim M (1972) Mixed cryoglobulinemia with a monoclonal IgM component, associated with chronic liver disease. Isr J Med Sci 8:1968–1979

Chapter 7
Cross-Talk Between Antibodies, IgG Fc Receptors, and the Complement System

Christian M. Karsten and Jörg Köhl

Abstract Complement is one of the oldest parts of the immune system. It has evolved during the past 1,000 million years to form a network of more than 30 serum and cell surface proteins with pleiotropic functions. Primarily, complement has been considered as an important danger sensing system, recognizing and flagging pathogens and eventually eliminating the intruders. Further, complement senses and eliminates endogenous danger such as apoptotic cells or cell debris. In addition to these well-appreciated properties, complement plays an integral role in the regulation of innate and adaptive immune responses and is involved in tissue regeneration, angiogenesis, lipid metabolism, solubilization and clearance of immune complexes (IC), mobilization of hematopoietic progenitor cells, and neurologic synapse maturation among other functions.

Keywords Complement system • Factor B • Factor C3 • Antibody-mediated activation • Immunoglobulins • Immune adherence

C.M. Karsten
Institute for Systemic Inflammation Research, University of Lübeck, Lübeck, Germany

J. Köhl, M.D
Institute for Systemic Inflammation Research, University of Lübeck, Lübeck 23538, Germany

Division of Cellular and Molecular Immunology, Cincinnati Children's Hospital and University of Cincinnati, College of Medicine, Cincinnati, OH, USA
e-mail: joerg.koehl@uksh.de

F. Nimmerjahn (ed.), *Molecular and Cellular Mechanisms of Antibody Activity*, DOI 10.1007/978-1-4614-7107-3_7, © Springer Science+Business Media New York 2013

7.1 Introduction

Complement is one of the oldest parts of the immune system. It has evolved during the past 1,000 million years (Nonaka and Kimura 2006) to form a network of more than 30 serum and cell surface proteins with pleiotropic functions. Primarily, complement has been considered as an important danger sensing system, recognizing and flagging pathogens and eventually eliminating the intruders. Further, complement senses and eliminates endogenous danger such as apoptotic cells or cell debris. In addition to these well-appreciated properties, complement plays an integral role in the regulation of innate and adaptive immune responses and is involved in tissue regeneration, angiogenesis, lipid metabolism, solubilization and clearance of immune complexes (IC), mobilization of hematopoietic progenitor cells, and neurologic synapse maturation among other functions (Köhl 2006; Ricklin et al. 2010).

During evolution, the complement system has been built around Factor B and C3 (Janssen et al. 2005) stressing the critical importance of C3 and complement in general for the survival of complex organisms. The central C3 molecule can be activated by different mechanisms. The evolutionary oldest pathway is the alternative pathway engaging C3, factors B and D, and properdin. The initial step of alternative complement pathway activation is the thioester activation within C3 by polysaccharide hydroxyl groups (Tack et al. 1980; Sahu et al. 1994), nucleating the formation of the alternative pathway C3 convertase, a protease complex comprising C3bBb. As we will outline below, immunoglobulin G (IgG) IC can also activate the alternative pathway, which is critical for the solubilization of precipitated ICs. The lectin pathway and the classical pathway have evolved before the establishment of adaptive immunity and the appearance of immunoglobulins. The classical complement pathway is activated at the level of C1q when natural and induced IgG and IgM antibodies form ICs with their cognate antigens. Mannan-binding lectin (MBL), a C-type lectin, recognizes conserved carbohydrate patterns on pathogens and binds to MBL-associated serine proteases MASP-1 and MASP-2 (Fujita et al. 2004). Once the immune sensors of the complement system have recognized their targets, a cascade of proteolytic events is triggered in which complement components and factors in the fluid phase are cleaved to release smaller fragments which in turn nucleate the assembly of new protease complexes required for the cascade to proceed to the next step. Proteases of the lectin and the classical pathway promote the cleavage of C4 and C3 resulting in the formation of a C3 convertase composed of C4b2a (Fig. 7.1). Cleavage of C3 by either pathway drives the generation of the two fragments C3b and C3a. Depending on the environment, C3b can bind to different partner molecules which may either result in the amplification of the complement response, formation of the C5 convertase and activation of the terminal pathway leading to the generation of C5a and the membrane attack complex (MAC), or the irreversible degradation of C3b (reviewed in (Gros et al. 2008)). The MAC has an important function in destroying bacterial invaders. This is of particular importance with regard to infections with *Neisseria meningitidis*, as evidenced by a high frequency

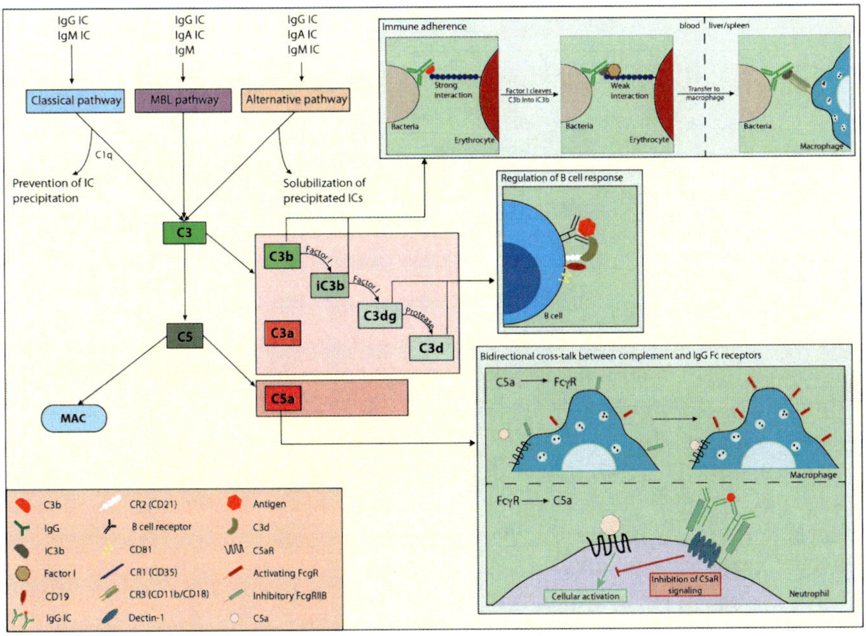

Fig. 7.1 *Network of the interactions between complement and immunoglobulins*. Immune complexes comprising IgM, IgG, or IgA and antigen activate the different pathways of complement. Complement activation is critical for appropriate handling of ICs, i.e., it prevents the formation of immune precipitates, can solubilize such precipitates, and promotes IC clearance through activation of either the classical or the alternative pathway. Further tagging of antigens with C3 fragments and activation of complement receptors on B cells and follicular dendritic cells is an important mechanism of B cell regulation. More downstream in the complement cascade, C5a cross-talks with IgG Fc receptors on neutrophils and macrophages to lower the threshold for FcγR-mediated activation of such cells. Vice versa, an anti-inflammatory feedback loop exists, by which IC inhibit proinflammatory properties of C5a through engagement of the inhibitory FcγRIIB

of early life meningococcal infections in people with deficiencies in one of the components of the terminal pathway (Skattum et al. 2011). On the other hand, MAC-mediated cell destruction can result in life-threatening disease in case of paroxysmal nocturnal hemoglobinuria (PNH), which is due to an acquired defect of glycosylphosphatidylinositol-linked membrane proteins, resulting in the absence of complement control proteins CD55 and CD59 (Parker 2012). Of note, an anti-C5 antibody (eculizumab) is available as an efficient therapeutic option for PNH patients. In light of the functional C5-deficiency induced by eculizumab treatment requires prior vaccination against meningococci (Parker 2012). Regarding the cross-talk with immunoglobulins, the interaction of IC-bound C3b and its degradation products iC3b, C3dg, and C3d with distinct complement receptors expressed on phagocytes, B cells, and follicular dendritic cells (FDC) is of major importance. In this chapter, we will discuss the activation of complement by ICs, the role of C3b degradation products for the appropriate handling of such

ICs, and the role of complement in the development of B cell immunity. Further, we will focus on pathways initiated by the C5 fragment C5a and their cross-talk with IgG crystallizable fragment (Fc) receptors (Fig. 7.1).

7.2 Antibody-Mediated Activation of the Complement System

The activation of the complement system by antibodies (Ab) has been mainly studied in humans, mice, and rats. Among the different classes of immunoglobulins, IgM and IgG can activate complement either via the classical, the alternative, or the lectin pathway. IgA activates complement through the alternative or the lectin pathway. Activation of the classical pathway takes place when C1q recognizes IgG or IgM that form ICs with their cognate antigen. Similarly, unspecific aggregation of such Ab, for example, in individuals suffering from cryoglobulinemia, may activate C1q (Adu and Williams 1984). IgM is the most potent activator of complement. Even a single IgM pentamer can trigger the classical pathway of complement activation, whereas at least two IgG molecules are required for the same effect (Borsos and Rapp 1965). The IgM molecule exists in pentameric and hexameric forms, the latter of which is a 10–20 times more potent complement activator than its pentameric counterpart (Davis et al. 1988; Collins et al. 2002). The IgG subclasses differ in their potencies to activate the complement system. Human IgG activates C1q in the order IgG3 > IgG1 > IgG2, whereas IgG4 shows none to minor reactivity towards C1q (Ishizaka et al. 1967). In mice, the IgG2a/c, IgG2b, and IgG3 subclasses activate complement, whereas IgG1 is not fixing complement via the classical pathway (Klaus et al. 1979). In rats, all four subclasses (IgG1, IgG2a, IgG2b, IgG2c) are able to activate the complement system. For rat IgG2c, it was first assumed that this subclass is not activating complement (Medgyesi et al. 1978), but this view was based on experiments in heterologous systems, in which rat immunoglobulins were tested for complement activation with guinea pig or human serum as a source for complement. The same phenomenon was observed when the complement fixing abilities of bovine IgG subclasses (IgG1 and IgG2) were investigated. Both of these subclasses were able to fix complement in a homologous system, but when guinea pig serum was used, only IgG1 was able to activate complement (McGuire et al. 1979).

The binding site for C1q has been located within the C_H2 domain of the Fc portion of IgG (Duncan et al. 1988). C1q is part of the C1 complex comprising one C1q molecule (C1q is a hexamer with six identical subunits) and the C1r2:C1s2 proenzyme complex. The tails of the C1q hexamer bind to the C1r2:C1s2 complex, whereas the C-terminal globular heads of the protein bind to the Fc part of the Ab. Because of its hexameric structure, the binding affinity of C1q to immunoglobulins increases markedly with the complexity of the IgG. Monomeric IgG binds to C1q with a 10.000-fold lower affinity as compared with IgG that forms ICs (Burton and Woof 1992). The binding of C1q to ICs results in a conformational change in the C1r2:C1s2 complex driving the autolytic activation of C1r, which in turn activates the C1s serine protease. The activated C1s enzyme then cleaves C4 resulting in two fragments, the large C4b and the small C4a molecule. C4b can covalently bind to cell surfaces via its labile thioester, driving the cleavage of C2 into the large C2a and the small C2b fragment. Unbound C4b is quickly inactivated. Within the C4bC2a complex, C2a is the active protease that can cleave C3 into the small C3a and the large C3b fragment, which can also covalently attach to cell surfaces. Importantly, bypass pathways exist, in which the C1 complex activation by ICs results in direct cleavage of C3 by the alternative pathway without C2 (Knutzen Steuer et al. 1989) or C4 and C2 cleavage (May and Frank 1973a, b, c). In line with its evolutionary appearance before the development of the adaptive immune system and immunoglobulins (Nonaka and Kimura 2006), C1q not only binds to Ab but also to cell wall components and membrane proteins of all kinds of microorganisms, the prion infectious agent (Klein et al. 2001) as well as fragments of cellular and subcellular membranes (e.g., mitochondrial membranes), and other modified host proteins and phospholipids (Ghebrehiwet et al. 1983; Peitsch et al. 1988; Storrs et al. 1983; Kovacsovics et al. 1985). Further, it can bind C-reactive protein (CRP) and serum amyloid P (SAP) protein of the pentraxin family (Bottazzi et al. 2010).

The marked difference of the IgG subclasses to activate complement is puzzling because the C1q-binding site within the C_H2 domain at positions Glu 318, Lys 320, and Lys 322 is highly conserved in IgG molecules of different species (Duncan et al. 1988). However, the structure of the hinge region, which separates the antibody binding part (Fab) from the Fc part, is remarkably diverse. More specifically, the length of the hinge regions between different IgG subclasses varies resulting in different Fab arm segmental flexibility. The most effective complement activators among human IgGs, the IgG3 and IgG1 isotypes, have the longest hinge regions of the four IgG subclasses. The hinge region of the non-complement-activating human IgG4 isotype and its murine IgG1 counterpart display only limited flexibility as compared with other IgG subclasses, which results in limited movement of their Fab arms and less accessibility of this region for the C1q heads. The importance of the hinge region was strengthened by the observation that naturally occurring human IgG1 molecules with short and rigid hinges showed reduced ability to fix complement, as it is the case for Dob and Mcg human IgG1 myeloma proteins (Klein et al. 1981; Guddat et al. 1993). However, antibody-engineering studies carried out by several groups (Tan et al. 1990; Norderhaug et al. 1991; Tao et al. 1993) suggest that the difference in the hinge region is only one important factor enabling IgG molecules to activate complement. In fact, differences within the hinge region do not fully explain the different potencies of IgG subclasses to activate C1q, as the available data show no strict correlation between segmental flexibility and the ability to activate complement. A modified IgG3 with the rigid hinge of IgG4 exhibits significant complement activation ability and an IgG4 with the hinge of IgG3, although as flexible as IgG3, displays no detectable complement activity (Tan et al. 1990; Norderhaug et al. 1991). In addition to the hinge region, the inter-H-chain disulfide seems to be important for complement activation. Schur et al. (Schur and Christian 1964) showed that reduction of the inter-H-chain disulfide bridges in rabbit and human antibodies destroys their ability to interact with complement. More specifically, the amino acid composition of the C_H2 domain seems to be responsible for the isotype variation. Changing the C_H2 part from a human IgG1 to an IgG4 molecule created an IgG4 mutant that is almost as potent in activating complement as IgG1 (Tao et al. 1991; Greenwood et al. 1993). A Ser331Pro replacement improves complement activation of IgG4, suggesting that amino acids outside the C1q binding domain at positions 318, 320, and 322 contribute to classical pathway activation (Tao et al. 1993; Xu et al. 1994). Also, a fully functional and paired C_H3 domain is necessary for optimal complement activation (Utsumi et al. 1985). At this point, it is still unclear whether these structural properties are required for the appropriate orientation of the C_H2 domains or the functional integrity of the C1q-binding site (Brekke et al. 1995).

IgM mutant studies (Shulman et al. 1987) and X-ray analyses (Perkins et al. 1991) provided evidence that the C_H3 domain, which corresponds to the C_H2 domain in IgG, is critical for binding to C1q. Here, two clusters at the opposite ends of the C_H3 domain at positions Asp432, Pro434, Prolin436, Asp356, and Asp417 may form a binding site for C1q (Arya et al. 1994). These data suggest that the C1q-binding site for IgG and IgM is not identical. Taken together, several structural features within the IgG or IgM Ab promote the successful interaction of C1q with the antibody C_H2 and C_H3 regions.

In addition to the classical pathway, IgG molecules have been shown to activate complement via the alternative pathway, e.g., in the mouse model of K/BxN serum-transferred arthritis, which is mimicking human rheumatoid arthritis (Ji et al. 2002) or in patients with juvenile arthritis (Aggarwal et al. 2000). In the K/BxN model, FcγRIII- and C5a receptor (C5aR)-deficient mice are equally protected from disease development whereas C1q-deficient mice are not. Surprisingly, also factor B-deficient mice are protected, suggesting a crucial role for alternative pathway activation in this model. Notably, IgG1, which does not activate the classical pathway, is the dominant isotype in the serum of K/BxN mice (Kouskoff et al. 1996), and transfer of such K/BxN IgG1 antibodies induces arthritis (Ji et al. 2002) (although IgG2b is also present and contributes to the inflammatory response) (Nimmerjahn et al. 2010). Usually, the alternative pathway is associated with the detection of microbial surfaces, but what triggers this pathway in the arthritis model? It is well appreciated that spontaneous initiation of the alternative complement pathway takes place in the fluid phase resulting in continuous cleavage of C3 into C3a and C3b (Pangburn and Muller-Eberhard 1980; Pangburn et al. 1981). Such C3b has affinity for IgG molecules (Kulics et al. 1983) and can form $C3b_2$-IgG complexes that are partially protected from inactivation by factors H and I resulting in an increased half-life of $C3b_2$-IgG complexes. Further, they can bind properdin bivalently resulting in preferred sites for convertase formation (reviewed in (Lutz and Jelezarova 2006)). In contrast to C1q, which binds to the C_H2 domain, the docking site of nascent C3b is within the C_H1 domain of the IgG molecule (Shohet et al. 1993). Thus, certain IgG molecules can promote alternative pathway activation, which appears to be amplified by C5a-dependent neutrophil activation, resulting in C3, factor B, and properdin production (Schwaeble and Reid 1999). Of note, naturally occurring antibodies against yeast or bacterial polysaccharide structures preferentially form $C3b_2$-IgG complexes, suggesting that this pathway has primarily evolved as a defense mechanism that efficiently attacks microbial invaders in the circulation (Lutz and Jelezarova 2006).

Another important parameter for effective complement activation by IgG is the glycosylation of the Fc portion. As outlined in more detail in Chap. 6, IgG molecules are N-glycosylated in their C_H2 domain at Asn297. The N-glycans within the Fc portion of the IgG molecule are highly heterogeneous with regard to the composition of their terminal sugar residues, which can be mannose, N-acetyl-glucosamine (GlcNAc), bisecting GlcNAc, galactose, or sialic acid (Arnold et al. 2007a). Agalactosylated IgG molecules have been shown to either exert a reduced potency to activate complement or to lack it completely (Williams et al. 1973; Koide et al. 1977; Nose and Wigzell 1983; Tsuchiya et al. 1989). Further, when the terminal sugar is mannose, C1q binding and classical pathway activation is abrogated (Wright and Morrison 1994). This loss or decrease of C1q binding may result from allosteric changes of the C_H2 domain within the IgG molecule as the carbohydrate chains cover a hydrophobic patch in the C_H2 domain (Huber et al. 1976) that also involves regions between C_H1 and C_H2 (Silverton et al. 1977). As outlined above, the length and the accessibility of the hinge region is one important factor determining C1q binding. Notably, the terminal sugars in the biantennary glycan structure also define whether MBL can recognize the IgG molecule through its carbohydrate-recognition domains. Typically, MBL recognizes polysaccharide structures on bacteria, fungi, viruses, and parasites as well as on apoptotic host cells (Thiel and Gadjeva 2009). When the Fc glycan terminates with a GlcNAc residue, MBL can bind to IgG and activate the lectin pathway (Malhotra et al. 1995a). Human IgM comprises five N-linked glycosylation sites on each μ chain and one glycosylation site on the J chain. Some IgM glycoforms that express GlcNac-terminating structures can bind to MBL. However, antigen-bound IgM ICs do not bind to MBL as such target glycan structures become inaccessible upon antigen binding. Nevertheless, MBL-mediated activation of the lectin pathway may play a role in the handling of aggregated IgM molecules, i.e., in keeping such ICs soluble and remove them from the circulation. Rheumatoid arthritis patients express high levels of the agalactosyl-IgG-G0 glycoform, terminating in GlcNAc (Parekh et al. 1985) suggesting that the inflammatory response in the joint may result from undesired activation of complement by the lectin pathway (Malhotra et al. 1995a; Garred et al. 2000). Increased levels of the G0 glycoform have also been described in individuals suffering from systemic lupus erythematosus (SLE), Crohn's disease, or tuberculosis (Parekh et al. 1985, 1989; Bond et al. 1997). In addition to MBL, some reports demonstrated a positive correlation between the terminal Fc glycan galactose content and C1q binding in vitro (Raju 2008a; Hodoniczky et al. 2005; Boyd et al. 1995; Kobata 2008), suggesting that the adaptive immune system can fine-tune the activation of complement by either the classical or the lectin pathway at the level of IgG Fc glycan composition.

Other diseases, where lectin-dependent activation of complement may play an important role, are IgA-associated diseases such as primary IgA nephropathy and

Henoch-Schönlein purpura (reviewed in (Oortwijn et al. 2008; Floege 2011)). IgA can occur in monomeric and polymeric forms. The polymeric form contains a joining chain and may be associated with a secretory component. It can activate the alternative (Hiemstra et al. 1987) and the lectin (Roos et al. 2001) pathway. In patients with IgA nephropathy, mesangial deposition of IgA together with MBL, MASP-2, and C4 has been found, suggesting glomerular complement activation by the lectin pathway (Roos et al. 2006). At this point the glycan composition on polymeric IgA that triggers MBL binding and complement activation is unclear. Polymeric IgA heavy chains can be glycosylated via N-linked sugars or O-linked sugars (IgA1 heavy chain). Further, the joining chain and the secretory component may be glycosylated (Royle et al. 2003). Of note, IgA nephropathy is frequently associated with an elevation of circulating poorly galactosylated IgA1 O-glycoforms, in particular in patients with active nephropathy and proliferative mesangial changes (Floege 2011). It is tempting to speculate that such IgA molecules may be targets for MBL recognition, driving the activation of the lectin pathway and disease progression.

7.3 The Role of Complement in Handling of Immune Complexes

Complement activation by either the classical or the alternative pathway results in IC processing in several ways. First, complement is required to maintain the solubility of ICs (Schifferli et al. 1980) or to solubilize immune precipitates (Miller and Nussenzweig 1975). Second, C3b- or C4b-coated ICs can bind to complement receptor 1 (CR1, CD35) on erythrocytes (in humans) or platelets (in rodents), a process which is known as "immune adherence" (Nelson 1953). Third, C3b/C4b or iC3b binding to ICs allows these complexes to bind to CR1 or CR3 (CD11b/CD18) on phagocytes such as monocytes/macrophages or neutrophils, which facilitates phagocytosis of such ICs.

In 1941, Heidelberger was the first who noticed that addition of active complement delays precipitation of ICs (Heidelberger 1941). Several further quantitative chemical studies followed to understand this phenomenon in more detail (Heidelberger and Mayer 1942; Heidelberger et al. 1941a, b). When antibodies recognize their antigens, they form ICs. The formation of immune precipitates depends on the antigen valence, intermolecular Fc interactions, and the ratio of antigen and antibody. Indeed, with multivalent antigens and bivalent IgG molecules, an infinite lattice can develop either in the fluid phase or in tissues, when antibodies recognize a tissue-bound antigen, e.g., in the kidney. The deposition of immune precipitates in small vessels or directly in organs can result in devastating tissue inflammation. Thus, keeping ICs soluble or solubilizing immune precipitates and efficiently removing such ICs is of major importance to prevent undesired tissue inflammation. Today, we know that activation of the classical pathway promotes the inhibition of immune precipitation (Schifferli et al. 1980). As already mentioned above, Ig subclasses differ in their ability to activate complement. Thus, it is not surprising that IgG ICs are bound by erythrocytes and are gradually removed from the circulation, whereas monomeric IgA ICs bind very poorly to erythrocytes but disappear rapidly from the circulation, with increased deposition in the kidney, lung, and brain (Waxman et al. 1986). Under steady-state conditions, ICs are deposited continually but are removed by immune adherence and subsequent phagocytosis. The critical role for complement in this process is underscored by prolonged tissue deposition of ICs in hypocomplementemic patients (Manderson et al. 2004).

The exact mechanisms by which complement prevents immune precipitation are still incompletely understood. Possible explanations are that the complement components C1, C4, C2, and in particular C3b mask the epitopes within the antigen thereby reducing the number of available binding sites (Hong et al. 1984) or interfere with Fc-Fc interactions (Moller 1979; Moller and Steensgaard 1979). In addition to prevent immune precipitation, complement is capable to solubilize immune aggregates (Miller and Nussenzweig 1975). In contrast to the inhibition of immune precipitation, this effect is strictly dependent on the activation of the alternative pathway and covalent binding of C3b to the antigen/antibody complexes. As already outlined, such $C3b_2$-IgG complexes serve as a potent nucleus for alternative pathway convertase formation driving the massive production of the anaphylatoxins C3a and C5a (Schifferli et al. 1986). Thus, the price for efficient solubilization of immune aggregates is the induction of a proinflammatory scenario with the attraction and activation of neutrophils and monocytes.

When ICs are coated with C3b, they become targets for CR1, a single-chain integral membrane glycoprotein composed of 30 repeats of 59–72 amino acids (Krych-Goldberg and Atkinson 2001). CR1 can also bind C4b and iC3b; however, its highest affinity is towards C3b (Klickstein et al. 1987). Further, C1q (Klickstein et al. 1997) and MBL (Ghiran et al. 2000) can bind to CR1. It is widely expressed on erythrocytes, polymorphonuclear neutrophils (PMN), monocytes, B lymphocytes and follicular dendritic cells (see below), some T cell subsets, and glomerular podocytes. CR1 has several important functions: it has decay-accelerating properties for the C3 convertases and acts as a cofactor for factor I to cleave C3b into iC3b and further into C3dg which remains covalently attached to the ICs and C3c which is released from the complex. Similarly, CR1 also serves as a cofactor for factor I-mediated cleavage of C4 into iC4b and C4c. These properties of CR1 are important for the safe and efficient removal of soluble ICs from the circulation. In humans, the first step is the binding of C3b-coated ICs to erythrocytes. Notably, the affinity of CR1 for monomeric C3b is low (Arnaout et al. 1983). The efficient transportation of C3b-coated ICs is achieved through clustering of CR1 on erythrocytes (Paccaud et al. 1988; Chevalier and Kazatchkine 1989). In fact, 8–20 clusters are formed on the surface of each erythrocyte. Although CR1 is also expressed on neutrophils (even at higher levels), they bind C3b-coated ICs less efficiently as they are not clustered. This differential binding behavior may prevent accidental activation of PMN in the circulation (Paccaud et al. 1990). In rodents, which lack CR1 expression on erythrocytes, CR1-positive platelets mediate this transport and in this case, the immune adherence activity is promoted by platelet-associated factor H

(Miller et al. 1975; Alexander et al. 2001). As shown for the inhibition of immune precipitation, immune adherence is dependent on the activation of the classical pathway with regard to IC deposition (Paccaud et al. 1987; Taylor et al. 1983) or the activation of the lectin pathway, when mannose residues are recognized on microorganisms by MBL. The physiologic relevance of immune adherence is obvious with regard to immune tagging of microorganisms. Indeed, the studies by Nelson (1953) showed that several bacteria such as *Treponema pallidum* or *Streptococcus pneumoniae* bind to erythrocytes in a complement-dependent manner. Later it became clear that not only whole bacteria, but also fragments of bacteria together with the recognizing antibody form ICs, which also adhere to erythrocytes and are transported to the liver and spleen, where phagocytes can clear these opsonized ICs (Nardin et al. 1999) (Fig. 7.2). It should be mentioned at this point that certain viruses such as HIV or the hepatitis C virus may use erythrocytes as Trojan horses to deliver their cargo to spleen cells susceptible to viral infection (Hess et al. 2002).

The mechanisms promoting the transfer of C3b-coated ICs from clustered CR1 to macrophages in the liver and the spleen are complex and have puzzled researchers for decades. C3b-coated ICs bound to erythrocytes are cleared from the circulation within minutes and appear in the liver and spleen (Cornacoff et al. 1983). Due to the cofactor activity of CR1 for factor I, C3b is degraded to iC3b, C3dg, and C3c. C3dg is a ligand for CR2 (CD21) but not for CR1 or CR3 (CD11b/CD18). iC3b has a higher affinity for CR3 than for CR1, suggesting that iC3b-coated ICs may be transferred to CR3 on spleen macrophages, which express CR3 at high levels (Hess and Schifferli 2003). This scenario would result in erythrocytes that would be available for a new round of IC clearance (Fig. 7.2a). However, this model does not take into account that the factor I-mediated release of ICs from erythrocytes is much slower than what has been observed in vivo and that the effect depends on FcγRIII (Kimberly et al. 1989). A series of elegant in vitro and in vivo studies suggests a different model, in which engagement of CR1-bound ICs by macrophage FcγRs results in proteolytic cleavage of CR1 and shedding of the CR1/C3b-IC from the erythrocyte surface (Kuhn et al. 1998; Taylor et al. 1997; Reinagel and Taylor 2000) (Fig. 7.2b). The latter model is in line with the observation of reduced erythrocyte CR1 levels in patients with autoimmune diseases (Iida et al. 1982; Ross et al. 1985) or HIV infection (Tausk et al. 1986). In fact, SLE patients suffer from decreased opsonization of ICs leading to IC deposition in small vessels such as the capillaries of the glomerular tuft in the kidney (Hebert et al. 1994). Moreover, these individuals show an impaired clearance of apoptotic cells and bodies (Cohen et al. 2002; Ren et al. 2003). Botto et al. (1998) have shown that C1q-deficient mice develop a lupus-like phenotype and accumulate pathological levels of apoptotic bodies within the glomerular tuft of the kidneys which may also be relevant in human patients with C1q deficiency (Hess and Schifferli 2003b). Strikingly, C1q-deficient individuals have a 93 % risk to develop an SLE-like disease, followed by C4- (75 % risk) and C2-deficient individuals (10–30 %) (Agnello 1978; Pickering et al. 2000). Another receptor for C3b and iC3b expressed on resident tissue macrophages such as Kupffer cells in the liver is CRIg (Helmy et al. 2006), which is critical for

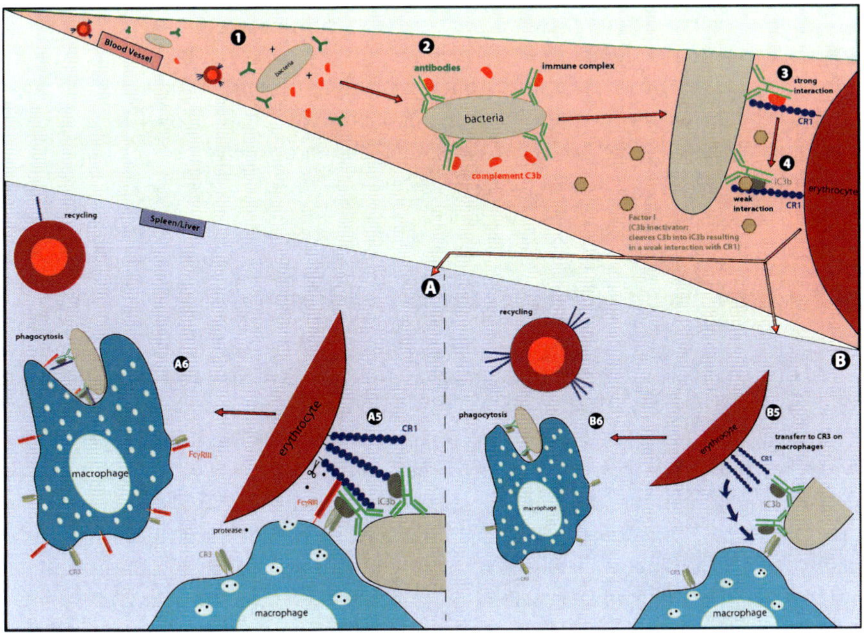

Fig. 7.2 *Complement-mediated clearance of ICs from the circulation*. (**1**) Pathogens, complement components, and antibodies are located in the circulation. (**2**) Antibodies recognize their antigens on bacteria, thereby activating the classical pathway of complement via C1q, leading to the covalent attachment of C3b. (**3**) C3b-tagged bacteria can bind to clustered CR1 on erythrocytes (immune adherence). CR1 serves as cofactor for factor I to promote the cleavage of C3b to iC3b. (**4**) iC3b has a weaker affinity for CR1 than C3b. Erythrocytes are transported to the liver or spleen, where ICs are safely cleared by macrophages. (**a**) and (**b**) depict two alternative mechanisms by which ICs are transferred to macrophages. (**a**) ICs bind to CR3 and FcγRIII on macrophages resulting in FcγR-dependent release of a macrophage-derived proteases (A5) that cleaves CR1, leaving the CR1 IC complex for phagocytosis and releasing the erythrocyte for recirculation (A6). In (**b**), the iC3b-tagged IC is transferred from CR1 on erythrocytes to CR3 on macrophages due to the higher affinity of iC3b for CR3 as compared with CR1 (B5). In this model, the erythrocyte is recycled with an intact CR1 (B6) (Modified from Hess and Schifferli 2003)

complement-mediated clearance of microbes, particles, or cells by Kupffer cells as they lack CR1 and only weakly express CR3 (Gregory et al. 2002). The absence of CrIg from inflammatory macrophages suggests that it plays an important role in clearing of iC3b-coated IC during steady-state conditions (Gorgani et al. 2008). CRIg also inhibits alternative pathway activation through binding to C3b and blockade of the C3 and C5 convertases (Wiesmann et al. 2006). A third function of CRIg, which is also termed V-set and Ig domain-containing 4 (VSIG4), is to suppress activation of T cells due to its function as a homolog of B7 costimulatory molecules (Vogt et al. 2006; Xu et al. 2010; Chen et al. 2010). Recent data suggest that this might be of importance for the regulation of autoreactive T cells in experimental type I diabetes (Fu et al. 2012).

Taken together, the complement system is critical in keeping antigen-antibody complexes small and soluble. Here, the complement system is not only able to prevent the formation of IC precipitates but also to solubilize preformed immune precipitates. Whereas the first mechanism, described in detail above, depends on activation of the classical pathway, the solubilization of preformed precipitates is exclusively dependent on the alternative pathway.

7.4 Complement Regulates the B Cell Response

Given that complement is part of the innate immune system that has evolved before the development of the adaptive immune system, it may not be surprising that complement instructs the development of an appropriate B cell response (Fearon and Locksley 1996). This has already been noted in the 1970s when C3 was found to bind to circulating B lymphocytes (Nussenzweig et al. 1971) and FDCs within lymphoid follicles (Papamichail et al. 1975). As already outlined above, C3 harbors a reactive thioester moiety that allows for covalent attachment to molecules or cellular surfaces. This is a crucial step within the complement activation, as it marks molecules or surfaces with a complement signature. Moreover, the tagged molecules or cells become targets for several complement receptors. With regard to B cell activation, CR1 and CR2 are of major importance. The cleavage fragments of iC3b, C3dg, and C3d bind to CR2. In addition to these ligands, CR1 also binds C3b and C4b (Molina et al. 1994). In mice, both receptors are splice products of the same locus (*Cr2*) that is co-expressed on chromosome 1 (Molina et al. 1990; Kurtz et al. 1990). In humans, they are encoded at separated loci (Carroll et al. 1988; Ahearn and Fearon 1989). Mature B cells, which are located in the B cell follicle, express CR1/CR2 as part of the B cell receptor complex along with CD19 and CD81 (Carter and Fearon 1992; Bradbury et al. 1992). The activation of the B cell receptor complex in the presence of CR1/CR2 ligands enhances B cell differentiation at several stages. Interaction of C3d-bound antigens with CR2 results in dramatic (up to 10.000-fold) reduction of the B cell activation threshold (Dempsey et al. 1996) and promotes the migration of the activated B cells to the T cell:B cell boundary. Here, B cell differentiation occurs including somatic cell hypermutation and class switch recombination (Fig. 7.3). In addition to the lowered threshold for B cell activation, CR1/CR2 co-receptor activation also rescues low-affinity B cells from caspase-dependent apoptosis (Barrington et al. 2005). Further, when B cells lack CR1/CR2, they fail to survive in germinal centers (GC) as it has been shown in an experimental system with lysozyme as a model antigen (Fischer et al. 1998). A prerequisite for the effective formation of GCs is an efficient antigen encounter of B cells. FDCs play a critical role in this regard as FDCs retain antigen on their surface via CR1 and CR2. Upon activation by FDCs, cognate B cells present their antigen to follicular T helper cells and differentiate into effector and memory B cells (Fig. 7.3). In the absence of efficient retention of antigen on FDCs through CR1/CR2, long-term antibody production as well as memory cells is markedly reduced (Barrington et al. 2002;

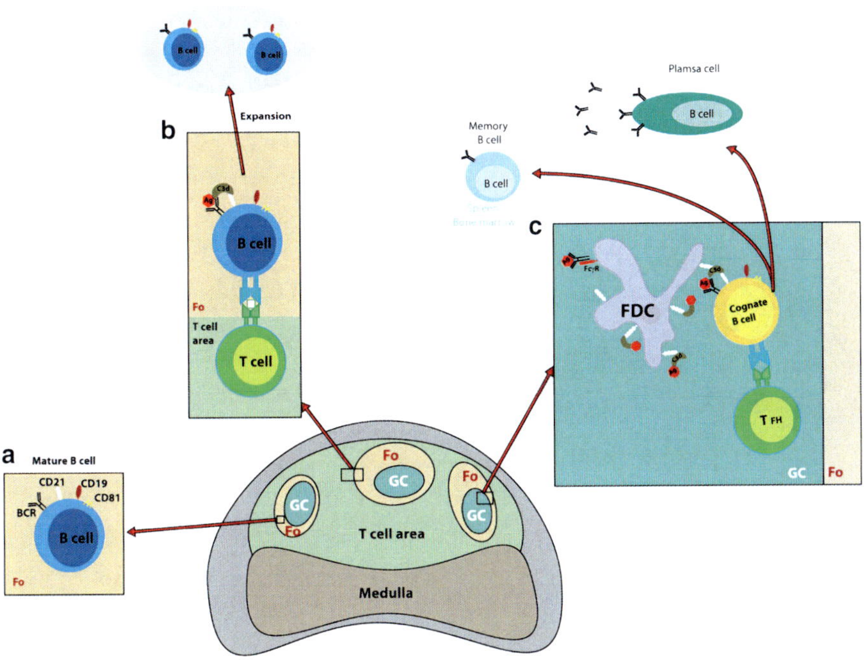

Fig. 7.3 *Role of complement in B cell differentiation within the lymph node*. (**a**) Mature B cells within the follicle (*Fo*) are positive for CD21, which is building a co-receptor complex with CD19 and CD81. (**b**) Binding of C3d to CD21 and the B cell receptor complex lowers the threshold for B cell activation – further, these events are directing the B cell to the B cell:T cell boundary. (**c**) Within the germinal center (*GC*), B cells acquire cognate antigen deposited on FDCs and present it to follicular T helper cells (T_{FH}). This leads to B cell differentiation into Ab-secreting plasma cells and memory B cells that reside in the spleen and the bone marrow (Adapted from Gonzalez et al. 2010b, 2011)

Brockman et al. 2006). Complement is also important for trafficking of lymph-born antigen into the B cell follicle. Recent data suggest that subcapsular sinus macrophages (SSM) capture C3b-coated antigens via CR3 (and FcγRIIB), which are widely expressed on SSMs (Phan et al. 2007). In a next step, noncognate naïve B cells take up the C3b (or C3d)-ICs from the SSMs and transfer such complexes to FDCs by a CR2-dependent process (Fig. 7.4) (Phan et al. 2009). At this point, it is still unclear what drives the transfer of complement-tagged ICs from naïve B cells to FDCs as both cells express CR1/CR2 on their surface. Further, it remains elusive, how large protein antigens are captured by SSMs prior to a specific immune response. Most likely, natural IgM or other innate sensors of the fluid phase such as C1q, MBL, ficolins, pentraxins, and C-reactive protein will bind to the antigen and activate complement. It has also been shown that marginal zone B cells can capture ICs by a CR1/CR2-dependent mechanism and transport them to FDCs within the B cell follicle (Youd et al. 2002; Pozdnyakova et al. 2003; Ferguson et al. 2004). Of course, multiple pathways exist to transport antigens into the follicles that are not

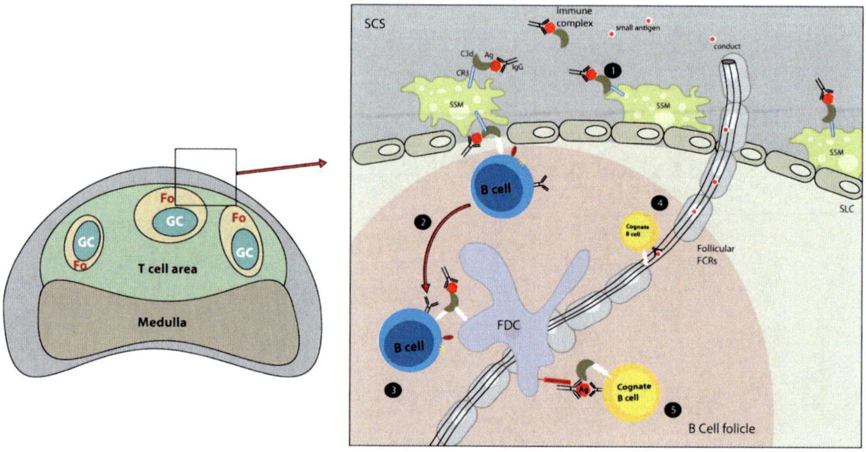

Fig. 7.4 *Transport of antigen in the lymph node to the site of B cell recognition*. (**1**) C3-coated ICs bind to CR3 on the surface of subcapsular sinus macrophages (*SSM*). (**2**) C3d-coated ICs are transported by naïve B cells from the subcapsular sinus (SCS) to FDCs. (**3**) ICs are transferred from the B cell to the FDC via a CR2-mediated mechanism. (**4**) Cognate B cells can capture antigens in different ways: small antigen is taken directly from the follicular conduits; large antigen complexes are taken from the surface of FDCs, (**5**) which are associated with FcγR or CR2 (Adapted from Gonzalez et al. 2011)

dependent on the activation of complement. For example, particulate antigens such as vesicular stomatitis virus (Junt et al. 2007) or protein-coated beads (Carrasco and Batista 2007) are rapidly phagocytosed by SSMs independent of C3. Of note, capture of influenza virus is mediated by a distinct pathway that involves MBL-dependent uptake by SSMs, transfer to medullary macrophages, and recognition by resident dendritic cells through the C-type lectin receptor SIGN-R1 (Gonzalez et al. 2010a). Taken together, signalling downstream of the complement receptors CR1, CR2, and CR3 on distinct and highly specialized immune cells within the peripheral lymphoid compartment builds a complex network that defines the activation and the outcome of the humoral immune response. The mechanisms and the pathways underlying such regulation have been outlined in detail in excellent recent reviews (Gonzalez et al. 2010b, 2011).

Further, mice with disrupted *Cr2* locus are prone to develop autoimmunity as a result of defective elimination of self-reactive B cells. In addition to the regulatory effects on T cell-dependent antigens, an increasing body of data suggests a critical role for CR2 in shaping the natural antibody repertoire (reviewed in (Holers 2005)). Natural antibodies are mainly produced by a subset of long-lived, self-replenishing B cells, i.e., B-1 cells. As germline-encoded molecules, they serve as low-affinity pattern recognition receptors that recognize pathogen as well as tissue-derived patterns. Importantly, *Cr2-/-* mice lack a subset of the natural antibody repertoire that is critical to recognize hypoxia-related damage on endothelial cells. On the one hand, this can be beneficial, as *Cr2-/-* mice are protected from intestinal ischemia/reperfusion injury. However, the impaired ability to recognize damaged cells may on the other hand lead to defective clearance and promote autoimmune responses. Collectively, these data demonstrate that the complement system is an integral part of the B cell compartment that shapes and regulates both the innate and the adaptive part of the B cell response.

7.5 Bidirectional Cross-Talk Between Complement and IgG Fc Receptors

Immune complexes drive the proinflammatory effector functions in autoimmune diseases including SLE, rheumatoid arthritis, and Goodpasture's syndrome (Madaio 1999; Korganow et al. 1999; Bolton 1996) among others. The inflammatory properties of ICs have been extensively studied in the Arthus reaction, serving as a model of a type III hypersensitivity reaction in which soluble ICs are formed after repeated administration of foreign antigens. The Arthus reaction is characterized by infiltration of PMNs, hemorrhage, and plasma exudation. Originally, horse serum has been repeatedly injected into the skin of rabbits, resulting in inflammation and neutrophil infiltration into the skin (Arthus 1903). For better standardization, the reverse passive model is frequently used, in which a specific antibody is injected at the desired site of inflammation, e.g., the peritoneum, the skin, or the lung, followed by intravenous application of the corresponding antigen. The activation of complement has initially been considered as the main effector mechanism promoting the inflammatory response underlying the Arthus reaction (Colten 1994). Indeed, classical pathway activation results in the production of C5a, which leads to increased vascular permeability and promotes migration and activation of immune cells (Gerard and Gerard 1994). Several studies from the Ravetch lab published during the past 15 years have shown that the presence of activating FcγR is crucial, since *Fcerg1*$^{-/-}$ mice, deficient in the common γ-chain (i.e., lacking all activating FcγR), were protected from inflammation in different models of IC diseases (Clynes et al. 1998, 1999; Park et al. 1998). Based on such data, the authors came to the conclusion that the initiation of the inflammatory cascade depended entirely on FcγR-triggered activation of inflammatory cells such as mast cells (Sylvestre and Ravetch 1994). In support of this view, they found that complement depletion by cobra venom factor (CVF) or genetic deficiency of C3 and C4 had at best minor effect on inflammation. This view was first challenged by data from a cutaneous Arthus reaction model, pointing towards complement-dependent and FcγR-dependent pathways that were depending on the genetic background of the mice used for the study (Hazenbos et al. 1996). In support for the impact of the genetic background, two studies demonstrated that CVF treatment had strong protective effects in peritoneal and cutaneous Arthus reaction models with C57BL/6 but not with a mixed C57BL/6/129 background (Heller et al. 1999; Baumann et al. 2000). Also, the missing anti-inflammatory effect of C3 or C4 deficiency does not take into account the bypass pathways, in which ICs can drive direct cleavage of C3 by the alternative pathway without C2 (Knutzen Steuer et al. 1989) and/or C4 cleavage (May and Frank 1973a, b, c). Finally, an emerging paradigm suggests that tissue-specific complement activation by resident and inflammatory cells promotes the production of several complement factors including C3a and C5a that are either activated locally by one of the three complement pathways or directly cleaved by proteases of cellular (Colten and Strunk 1993; Huber-Lang et al. 2002) or humoral origin (Huber-Lang et al. 2006, 2012; Amara et al. 2010). The importance of C5a for IC-mediated inflammation was clearly demonstrated in several models of pulmonary, peritoneal, and cutaneous Arthus reactions, in which C5aR-deficient mice were protected from IC injury (Bozic et al. 1996; Höpken et al. 1997). Further, in a model of collagen-induced arthritis, in which type II collagen-specific Ab were injected, C5aR-deficient mice were completely protected from disease development (Grant et al. 2002). Similarly, C5aR-deficient mice were protected in the K/BxN serum-induced arthritis model. Notably, FcγRIII-deficient mice were also protected in this model, pointing towards substantial cross-talk between the complement network and activating IgG Fc receptors (Ji et al. 2002).

First mechanistic insights into this cross-talk were provided using alveolar macrophages from C5aR-deficient mice. Treatment of such macrophages with C5a in vitro and in vivo resulted in strong transcriptional upregulation of FcγRIII (and the common γ-chain) and downregulation of the inhibitory FcγRIIB resulting in a decreased ratio of activating to inhibitory FcγRs (A/I ratio) (Shushakova et al. 2002). This threshold regulation of FcγRs has also been shown with peritoneal macrophages (Godau et al. 2004) and Kupffer cells (Kumar et al. 2006). Subsequent studies showed that $G\alpha_{i2}$- and PI3Kγ-dependent signalling downstream of the G-protein-coupled C5aR is critical for the regulation of FcγR expression in macrophages and that cooperative signalling between activating FcγRs and C5aR drives the production of cytokines and chemokines that eventually recruit and activate inflammatory neutrophils (Skokowa et al. 2005; Konrad et al. 2008). Of note, a feedback loop has been described between the upregulation of activating FcγRIII on Kupffer cells by C5a and the local production of complement. In a mouse model of autoimmune hemolytic anemia, an Ab-dependent type II autoimmune disease, activation of the C5a-C5aR axis first upregulates activating FcγRs followed by binding of IgG-autoantibodies to such IgG Fc receptors which promotes the production of C5 and its proteolytic cleavage into C5a. Eventually these events further enhance the expression of activating FcγRs (Fig. 7.5) (Kumar et al. 2006; Atkinson 2006). In addition to FcγRIII, FcγRIV also seems to contribute to this vicious inflammatory circle between complement and IgG Fc receptors, at least in an acute IC pulmonary hypersensitivity model (Syed et al. 2009). Taken together, activation of C5 by systemic or local complement activation or direct proteolytic cleavage through humoral- or cellular-derived proteases sets the threshold for immunoglobulin-mediated activation of immune cells by FcγRs. This cross-talk, including the FcγR amplification loop that generates more C5a, is desirable in an infection to ward off invaders by a fully armed immune response. As discussed above, complement receptors and FcγRs play critical roles in immune adherence of microorganisms to eliminate them from the bloodstream and to promote their subsequent elimination by phagocytosis. The flip side is that in autoimmune disease or allergies, such

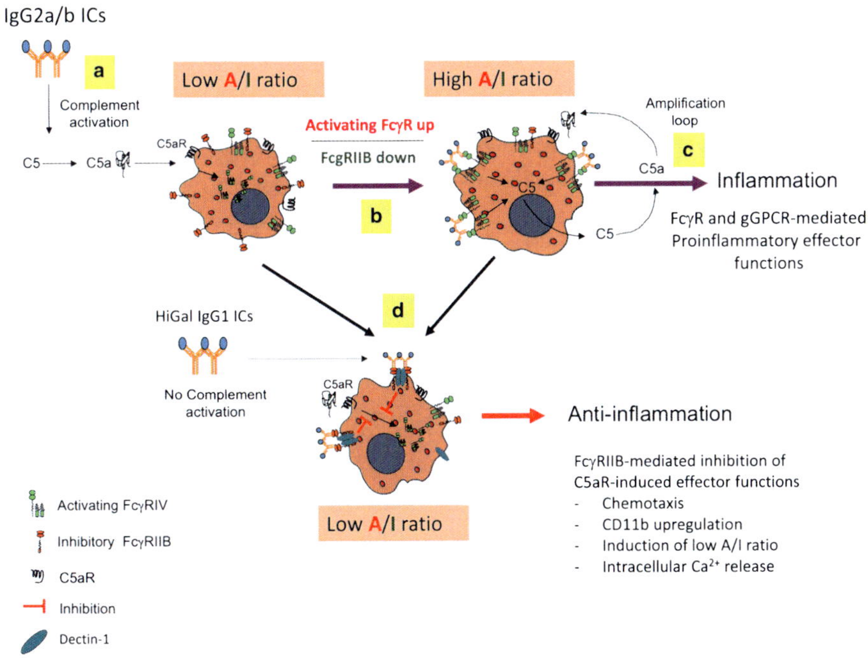

Fig. 7.5 *Bidirectional cross-talk between C5a receptor and IgG Fc receptor signalling*. (**a**) In the mouse system, IgG2a or IgG2b ICs activate complement, resulting in the generation of C5a. (**b**) Activation of the C5aR on macrophages results in the upregulation of activating FcγRs and down-regulation of the inhibitory FcγRIIB thereby reducing the threshold for FcγR-mediated stimulation of macrophages. (**c**) IC-triggered ligation of activating FcγRs promotes the production of C5 and activates proteases that cleave C5 into C5a, thereby amplifying the C5a-induced inflammation. (**d**) Highly galactosylated IgG1-ICs bind to FcγRIIB and associate FcγRIIB with Dectin-1. This heterodimer formation induces an anti-inflammatory pathway blocking C5aR-mediated effector functions

full-blown inflammation may drive undesired tissue damage. Indeed, activating and inhibitory FcγRs reciprocally promote or suppress pulmonary allergy in experimental asthma (Bandukwala et al. 2007; Dharajiya et al. 2010) and autoimmune models (Nimmerjahn and Ravetch 2006). Several reports suggest that this regulation of the A/I ratio depends on sialylated IgG molecules that bind to the C-type lectin receptor SIGNR1 on macrophages in the splenic marginal zone (Anthony et al. 2008). This interaction is supposed to promote the production of a yet unknown soluble factor that upregulates FcγRIIB on effector macrophages. Of note, the subset of highly sialylated IgGs has been suggested to be the active part within intravenous IgG (IVIG), which is used at high concentrations to treat several autoimmune diseases including immune thrombocytopenia and the Kawasaki syndrome as well as rheumatoid arthritis, SLE, and multiple sclerosis. An alternative mechanism for the effect of IVIG has been suggested by Basta et al., who found that IVIG is capturing

C3a and C5a thereby neutralizing their proinflammatory effects (Basta et al. 2003). In light of the regulatory impact of C5a on the A/I ratio, it has been shown that IVIG treatment protects macrophages from the C5a-mediated upregulation of FcγRIII and suppression of FcγRIIB resulting in sustained high FcγRIIB:FcγRIII ratio (Konrad et al. 2006).

Recent evidence suggests that that the cross-talk between C5aR and FcγR signalling is not uni- but bidirectional. In light of the strong direct and indirect proinflammatory properties of C5a through the ligation of its G-protein-coupled C5aR, it is surprising that our knowledge about the regulation of the C5a-C5aR axis is relatively scarce. In the fluid phase, the biologic activity of C5a is controlled by carboxypeptidases that rapidly cleave off a C-terminal arginine residue resulting in C5adesArg, which retains only 1–10 % of the inflammatory activity of C5a (Bokisch and MÅller-Eberhard 1970; Matthews et al. 2004). At the receptor level, TLR4 signalling has been shown to upregulate C5aR (Riedemann et al. 2002) and C5L2 expression (Raby et al. 2011). However, a detailed understanding of how the immune system regulates the effector responses of C5a has been lacking. A recent study provides evidence that IgG1-ICs can suppress C5aR-mediated effector functions in neutrophils by a novel anti-inflammatory pathway that involves the inhibitory FcγRIIB (Fig. 7.6) (Karsten et al. 2012). More specifically, IgG1-ICs block C5a-mediated ERK1/2 phosphorylation and the phospholipase C/inositol phosphate pathway, both of which are critical for C5a-mediated proinflammatory effector functions. Notably, the effect was independent of activating FcγRs, suggesting that it is not initiated by the classical ITAM/ITIM pairing of activating FcγR with FcγRIIB. In search for a potential partner that may promote the activation of FcγRIIB in the absence of the ITAM-bearing FcγR and that is expressed on neutrophils, we focused on the ITAM-like expressing C-type-lectin receptor Dectin-1. Our experiments revealed that IgG1-ICs engaged Dectin-1, resulting in Syk phosphorylation, which in turn promoted the phosphorylation of the ITIM within FcγRIIB and subsequent SHIP phosphorylation (Fig. 7.6). This pathway then blocked C5aR-mediated ERK1/2 phosphorylation and the phospholipase C/inositol phosphate pathway driven neutrophilic inflammation in vitro and in vivo in a model of IC peritonitis and of epidermolysis bullosa acquisita, an autoimmune skin blistering disease. Surprisingly, this suppressive effect of IgG1-ICs was strictly dependent on high galactosylation of the Fc glycan portion. Agalactosylated IgG1-ICs exerted no inhibitory effect. Dectin-1 is well characterized as a receptor that recognizes ß-1,3 glucans expressed by a broad range of fungal pathogens including *Candida albicans* and *Aspergillus* spp. (Geijtenbeek and Gringhuis 2009). However, no interaction with any Fc glycans has yet been reported. On the other hand, the N-glycan at Asn297 of the Fc moiety has been shown to promote IgG binding to FcγRs (Nimmerjahn and Ravetch 2005), the serum collection MBL (Malhotra et al. 1995), and C1q (Raju 2008). Our findings demonstrate that high N-glycan galactosylation of IgG1 molecules promotes cooperative signalling of the FcγRIIB with Dectin-1 resulting in an inhibitory signalling pathway that blocks proinflammatory effector functions of the C5aR. This mechanism is distinct from the

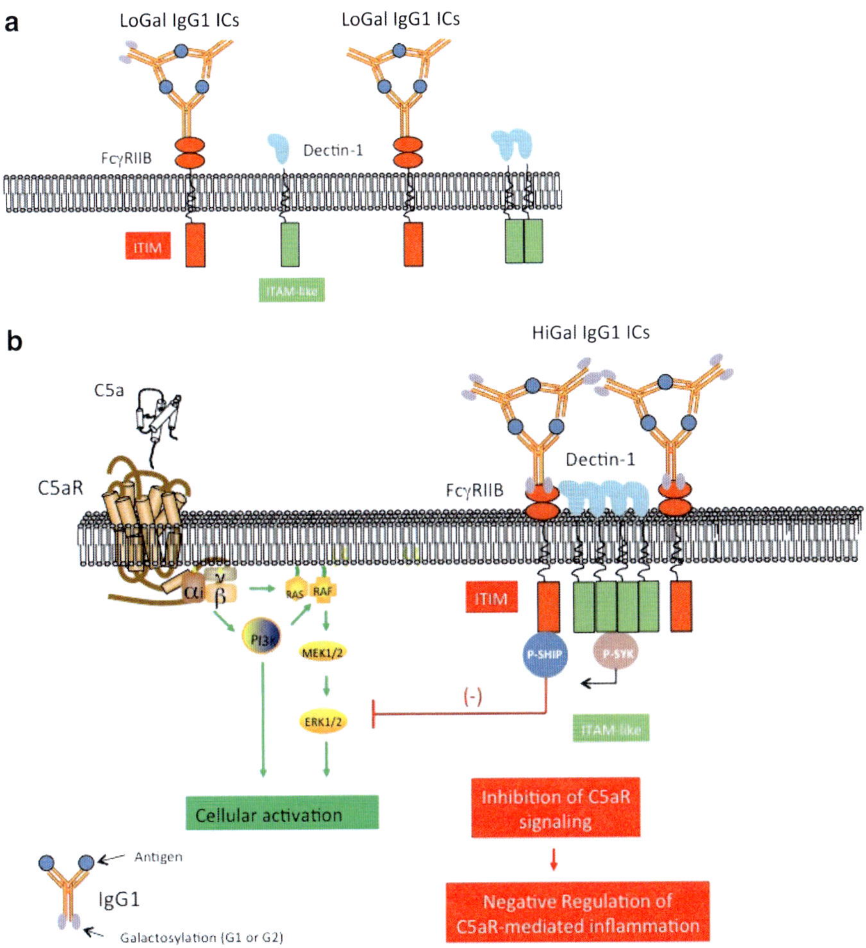

Fig. 7.6 *Two-step model by which highly galactosylated IgG1-ICs associate FcγRIIB with Dectin-1 and inhibit C5aR signalling*. (**a**) IgG1-ICs bind to FcγRIIB independent of their N-glycan galactosylation. Low or agalactosylated IgG1-ICs do not drive association of FcγRIIB with Dectin-1. (**b**) Highly galactosylated IgG1-ICs link FcγRIIB to Dectin-1 resulting in tyrosine phosphorylation of the ITAM-like motif downstream of Dectin-1 and transient phosphorylation of Syk. Further, HiGal-IgG1-IC drives the association of SHIP with the ITIM within FcγRIIB and its phosphorylation. This pathway inhibits C5a-mediated ERK1/2 phosphorylation and several cellular effector functions mediated by C5aR signalling (From Karsten et al. 2012)

anti-inflammatory properties mediated by sialylation of Fc-linked N-glycans (Anthony et al. 2008). About 25–35 % of Fc glycans from serum IgG are agalactosylated under steady-state conditions. This frequency markedly increases in experimental and human autoimmune conditions including rheumatoid arthritis, SLE, inflammatory bowel disease, and in infections (Arnold et al. 2007). Thus, the inhibitory effect of highly galactosylated IgG1-ICs may serve as a feedback loop to control complement and chemokine-mediated inflammation in autoimmunity and infection before IgGs can switch to the harmful agalactosyl glycoform. This effect may also contribute to the increased incidence of autoimmune phenomena in patients with hypogammaglobulinemia or common variable immunodeficiency (Agarwal and Cunningham-Rundles 2009).

7.6 Conclusions

Complement has developed before the appearance of adaptive immunity including the production of immunoglobulins. As outlined above, complement factors, cleavage fragments, and receptors interact with immunoglobulins in many ways. Cross-talk between Abs and complement protects the body from microbial invaders, apoptotic bodies, and other potentially harmful cells. Further, complement prevents immune precipitation and subsequent deposition of ICs in small vessels or tissues through defined and elaborated pathways of immune adherence and IC clearance, thereby maintaining tissue hemostasis and avoiding tissue damage. Whereas the first mechanism depends on activation of the classical pathway, the solubilization of preformed precipitates is exclusively dependent on the alternative pathway. Although several pathways of IC clearance have been elucidated, how exactly complement and IgG Fc receptors communicate to safely remove ICs under steady-state and inflammatory conditions is still incompletely understood and awaits further research. Another fascinating topic is the regulatory impact of complement on the B cell response and the shaping of the natural and induced antibody repertoire. The contribution of C3 cleavage fragments and the interaction with CR1 and CR2 on different types of B cells and FDCs has clearly been shown. At this point it is unclear whether other fragments of the complement system and their receptors may also contribute to B cell regulation. In particular, it will be of interest to delineate the molecular pathways underlying complement-mediated regulation of the innate B1 cell compartment, which is important for the control of infections with encapsulated bacteria. Finally, a bidirectional cross-talk between the C5a anaphylatoxin receptor with IgG Fc receptors exists that regulates IC-mediated inflammation in different ways. Under conditions, where IgG IC strongly activate complement, such activation will reduce the threshold for IC-mediated activation of inflammatory effector cells including macrophages and neutrophils. This way, pathogens will be synergistically and most efficiently attacked by humoral and cellular effector systems. The flipside of such cross-talk is deleterious inflammatory responses in case of autoimmune diseases. Notably, an anti-inflammatory pathway has evolved that can control the vicious circle between C5aR and activating FcγR-mediated inflammation at the level of IgG glycosylation. More research is required to provide a comprehensive view of the pathways that control Fc glycan composition. A detailed understanding of the mechanisms that control and regulate IgG, IgG Fc receptor, and complement protein and complement receptor glycosylation may be key to uncover novel pathways of complement and immunoglobulin interaction.

References

Adu D, Williams DG (1984) Complement activating cryoglobulins in the nephritis of systemic lupus erythematosus. Clin Exp Immunol 55:495–501

Agarwal S, Cunningham-Rundles C (2009) Autoimmunity in common variable immunodeficiency. Curr Allergy Asthma Rep 9:347–352

Aggarwal A, Bhardwaj A, Alam S, Misra R (2000) Evidence for activation of the alternate complement pathway in patients with juvenile rheumatoid arthritis. Rheumatology (Oxford) 39:189–192

Agnello V (1978) Association of systemic lupus erythematosus and SLE-like syndromes with hereditary and acquired complement deficiency states. Arthritis Rheum 21:S146–S152

Ahearn JM, Fearon DT (1989) Structure and function of the complement receptors, CR1 (CD35) and CR2 (CD21). Adv Immunol 46:183–219

Alexander JJ, Hack BK, Cunningham PN, Quigg RJ (2001) A protein with characteristics of factor H is present on rodent platelets and functions as the immune adherence receptor. J Biol Chem 276:32129–32135

Amara U et al (2010) Molecular intercommunication between the complement and coagulation systems. J Immunol 185:5628–5636

Anthony RM, Wermeling F, Karlsson MC, Ravetch JV (2008) Identification of a receptor required for the anti-inflammatory activity of IVIG. Proc Natl Acad Sci USA 105:19571–19578

Arnaout MA, Dana N, Melamed J, Medicus R, Colten HR (1983) Low ionic strength or chemical cross-linking of monomeric C3b increases its binding affinity to the human complement C3b receptor. Immunology 48:229–237

Arnold JN, Wormald MR, Sim RB, Rudd PM, Dwek RA (2007) The impact of glycosylation on the biological function and structure of human immunoglobulins. Annu Rev Immunol 25:21–50

Arthus M (1903) Inections repetees de serum de cheval chez le lapin. C R Soc Biol 55:817

Arya S et al (1994) Mapping of amino acid residues in the C mu 3 domain of mouse IgM important in macromolecular assembly and complement-dependent cytolysis. J Immunol 152:1206–1212

Atkinson JP (2006) C5a and Fcgamma receptors: a mutual admiration society. J Clin Invest 116:304–306

Bandukwala HS et al (2007) Signaling through Fc gamma RIII is required for optimal T helper type (Th)2 responses and Th2-mediated airway inflammation. J Exp Med 204:1875–1889

Barrington RA, Pozdnyakova O, Zafari MR, Benjamin CD, Carroll MC (2002) B lymphocyte memory: role of stromal cell complement and FcgammaRIIB receptors. J Exp Med 196:1189–1199

Barrington RA et al (2005) CD21/CD19 coreceptor signaling promotes B cell survival during primary immune responses. J Immunol 175:2859–2867

Basta M et al (2003) F(ab)′2-mediated neutralization of C3a and C5a anaphylatoxins: a novel effector function of immunoglobulins. Nat Med 9:431–438

Baumann U et al (2000) A codominant role of Fc gamma RI/III and C5aR in the reverse Arthus reaction. J Immunol 164:1065–1070

Bokisch VA, MÅller-Eberhard HJ (1970) Anaphylatoxin inactivator of human plasma: its isolation and characterization as a carboxypeptidase. J Clin Invest 49:2427–2436

Bolton WK (1996) Goodpasture's syndrome. Kidney Int 50:1753–1766

Bond A et al (1997) A detailed lectin analysis of IgG glycosylation, demonstrating disease specific changes in terminal galactose and N-acetylglucosamine. J Autoimmun 10:77–85

Borsos T, Rapp HJ (1965) Complement fixation on cell surfaces by 19S and 7S antibodies. Science 150:505–506

Bottazzi B, Doni A, Garlanda C, Mantovani A (2010) An integrated view of humoral innate immunity: pentraxins as a paradigm. Annu Rev Immunol 28:157–183

Botto M et al (1998) Homozygous C1q deficiency causes glomerulonephritis associated with multiple apoptotic bodies. Nat Genet 19:56–59
Boyd PN, Lines AC, Patel AK (1995) The effect of the removal of sialic acid, galactose and total carbohydrate on the functional activity of Campath-1H. Mol Immunol 32:1311–1318
Bozic CR, Lu B, Hopken UE, Gerard C, Gerard NP (1996) Neurogenic amplification of immune complex inflammation. Science 273:1722–1725
Bradbury LE, Kansas GS, Levy S, Evans RL, Tedder TF (1992) The CD19/CD21 signal transducing complex of human B lymphocytes includes the target of antiproliferative antibody-1 and Leu-13 molecules. J Immunol 149:2841–2850
Brekke OH, Michaelsen TE, Sandlie I (1995) The structural requirements for complement activation by IgG: does it hinge on the hinge? Immunol Today 16:85–90
Brockman MA, Verschoor A, Zhu J, Carroll MC, Knipe DM (2006) Optimal long-term humoral responses to replication-defective herpes simplex virus require CD21/CD35 complement receptor expression on stromal cells. J Virol 80:7111–7117
Burton DR, Woof JM (1992) Human antibody effector function. Adv Immunol 51:1–84
Carrasco YR, Batista FD (2007) B cells acquire particulate antigen in a macrophage-rich area at the boundary between the follicle and the subcapsular sinus of the lymph node. Immunity 27:160–171
Carroll MC et al (1988) Organization of the genes encoding complement receptors type 1 and 2, decay-accelerating factor, and C4-binding protein in the RCA locus on human chromosome 1. J Exp Med 167:1271–1280
Carter RH, Fearon DT (1992) CD19: lowering the threshold for antigen receptor stimulation of B lymphocytes. Science 256:105–107
Chen M, Muckersie E, Luo C, Forrester JV, Xu H (2010) Inhibition of the alternative pathway of complement activation reduces inflammation in experimental autoimmune uveoretinitis. Eur J Immunol 40:2870–2881
Chevalier J, Kazatchkine MD (1989) Distribution in clusters of complement receptor type one (CR1) on human erythrocytes. J Immunol 142:2031–2036
Clynes R, Dumitru C, Ravetch JV (1998) Uncoupling of immune complex formation and kidney damage in autoimmune glomerulonephritis. Science 279:1052–1054
Clynes R et al (1999) Modulation of immune complex-induced inflammation in vivo by the coordinate expression of activation and inhibitory Fc receptors. J Exp Med 189:179–185
Cohen PL et al (2002) Delayed apoptotic cell clearance and lupus-like autoimmunity in mice lacking the c-mer membrane tyrosine kinase. J Exp Med 196:135–140
Collins C, Tsui FW, Shulman MJ (2002) Differential activation of human and guinea pig complement by pentameric and hexameric IgM. Eur J Immunol 32:1802–1810
Colten HR (1994) Immunology. Drawing a double-edged sword. Nature 371:474–475
Colten HR, Strunk RC (1993) Synthesis of complement components in liver and at extrahepatic sites. In: Whaley K, Loos M, Weiler JM (eds) Complement in health and disease, vol 2. Kluwer Academic, Dordrecht, pp 127–158
Cornacoff JB et al (1983) Primate erythrocyte-immune complex-clearing mechanism. J Clin Invest 71:236–247
Davis AC, Roux KH, Shulman MJ (1988) On the structure of polymeric IgM. Eur J Immunol 18:1001–1008
Dempsey PW, Allison ME, Akkaraju S, Goodnow CC, Fearon DT (1996) C3d of complement as a molecular adjuvant: bridging innate and acquired immunity. Science 271:348–350
Dharajiya N et al (2010) FcgammaRIIb inhibits allergic lung inflammation in a murine model of allergic asthma. PLoS One 5:e9337
Duncan AR, Woof JM, Partridge LJ, Burton DR, Winter G (1988) Localization of the binding site for the human high-affinity Fc receptor on IgG. Nature 332:563–564
Fearon DT, Locksley RM (1996) The instructive role of innate immunity in the acquired immune response. Science 272:50–53
Ferguson AR, Youd ME, Corley RB (2004) Marginal zone B cells transport and deposit IgM-containing immune complexes onto follicular dendritic cells. Int Immunol 16:1411–1422

Fischer MB et al (1998) Dependence of germinal center B cells on expression of CD21/CD35 for survival. Science 280:582–585

Floege J (2011) The pathogenesis of IgA nephropathy: what is new and how does it change therapeutic approaches? Am J Kidney Dis 58:992–1004

Fu W, Wojtkiewicz G, Weissleder R, Benoist C, Mathis D (2012) Early window of diabetes determinism in NOD mice, dependent on the complement receptor CRIg, identified by noninvasive imaging. Nat Immunol 13:361–368

Fujita T, Matsushita M, Endo Y (2004) The lectin-complement pathway – its role in innate immunity and evolution. Immunol Rev 198:185–202

Garred P et al (2000) Two edged role of mannose binding lectin in rheumatoid arthritis: a cross sectional study. J Rheumatol 27:26–34

Geijtenbeek TB, Gringhuis SI (2009) Signalling through C-type lectin receptors: shaping immune responses. Nat Rev Immunol 9:465–479

Gerard C, Gerard NP (1994) C5A anaphylatoxin and its seven transmembrane-segment receptor. Annu Rev Immunol 12:775–808

Ghebrehiwet B, Randazzo BP, Dunn JT, Silverberg M, Kaplan AP (1983) Mechanisms of activation of the classical pathway of complement by Hageman factor fragment. J Clin Invest 71:1450–1456

Ghiran I et al (2000) Complement receptor 1/CD35 is a receptor for mannan-binding lectin. J Exp Med 192:1797–1808

Godau J et al (2004) C5a initiates the inflammatory cascade in immune complex peritonitis. J Immunol 173:3437–3445

Gonzalez SF et al (2010a) Capture of influenza by medullary dendritic cells via SIGN-R1 is essential for humoral immunity in draining lymph nodes. Nat Immunol 11:427–434

Gonzalez SF et al (2010b) Complement-dependent transport of antigen into B cell follicles. J Immunol 185:2659–2664

Gonzalez SF et al (2011) Trafficking of B cell antigen in lymph nodes. Annu Rev Immunol 29:215–233

Gorgani NN et al (2008) Complement receptor of the Ig superfamily enhances complement-mediated phagocytosis in a subpopulation of tissue resident macrophages. J Immunol 181:7902–7908

Grant EP et al (2002) Essential role for the C5a receptor in regulating the effector phase of synovial infiltration and joint destruction in experimental arthritis. J Exp Med 196:1461–1471

Greenwood J, Clark M, Waldmann H (1993) Structural motifs involved in human IgG antibody effector functions. Eur J Immunol 23:1098–1104

Gregory SH et al (2002) Complementary adhesion molecules promote neutrophil-Kupffer cell interaction and the elimination of bacteria taken up by the liver. J Immunol 168:308–315

Gros P, Milder FJ, Janssen BJ (2008) Complement driven by conformational changes. Nat Rev Immunol 8:48–58

Guddat LW, Herron JN, Edmundson AB (1993) Three-dimensional structure of a human immunoglobulin with a hinge deletion. Proc Natl Acad Sci USA 90:4271–4275

Hazenbos WL et al (1996) Impaired IgG-dependent anaphylaxis and Arthus reaction in Fc gamma RIII (CD16) deficient mice. Immunity 5:181–188

Hebert LA et al (1994) Effect of chronically increased erythrocyte complement receptors on immune complex nephritis. Kidney Int 45:493–499

Heidelberger M (1941) Quantitative chemical studies on complement or alexin: I. A method. J Exp Med 73:681–694

Heidelberger M, Mayer M (1942) Quantitative chemical studies on complement or alexin: IV. Addition of human complement to specific precipitates. J Exp Med 75:285–295

Heidelberger M, Rocha ESM, Mayer M (1941a) Quantitative chemical studies on complement or alexin: III. Uptake of complement nitrogen under varying experimental conditions. J Exp Med 74:359–367

Heidelberger M, Weil AJ, Treffers HP (1941b) Quantitative chemical studies on complement or alexin: II. The interrelation of complement with antigen-antibody compounds and with sensitized red cells. J Exp Med 73:695–709

Heller T et al (1999) Selection of a C5a receptor antagonist from phage libraries attenuating the inflammatory response in immune complex disease and ischemia/reperfusion injury. J Immunol 163:985–994
Helmy KY et al (2006) CRIg: a macrophage complement receptor required for phagocytosis of circulating pathogens. Cell 124:915–927
Hess C, Schifferli JA (2003) Immune adherence revisited: novel players in an old game. News Physiol Sci 18:104–108
Hess C et al (2002) Association of a pool of HIV-1 with erythrocytes in vivo: a cohort study. Lancet 359:2230–2234
Hiemstra PS, Gorter A, Stuurman ME, Van Es LA, Daha MR (1987) Activation of the alternative pathway of complement by human serum IgA. Eur J Immunol 17:321–326
Hodoniczky J, Zheng YZ, James DC (2005) Control of recombinant monoclonal antibody effector functions by Fc N-glycan remodeling in vitro. Biotechnol Prog 21:1644–1652
Holers VM (2005) Complement receptors and the shaping of the natural antibody repertoire. Springer Semin Immunopathol 26:405–423
Hong K et al (1984) Inhibition of immune precipitation by complement. J Immunol 133:1464–1470
Höpken UE, Lu B, Gerard NP, Gerard C (1997) Impaired inflammatory responses in the reverse arthus reaction through genetic deletion of the C5a receptor. J Exp Med 186:749–756
Huber R, Deisenhofer J, Colman PM, Matsushima M, Palm W (1976) Crystallographic structure studies of an IgG molecule and an Fc fragment. Nature 264:415–420
Huber-Lang M et al (2002) Generation of C5a by phagocytic cells. Am J Pathol 161:1849–1859
Huber-Lang M et al (2006) Generation of C5a in the absence of C3: a new complement activation pathway. Nat Med 12:682–687
Huber-Lang M et al (2012) Cathepsin D is released after severe tissue trauma in vivo and is capable of generating C5a in vitro. Mol Immunol 50:60–65
Iida K, Mornaghi R, Nussenzweig V (1982) Complement receptor (CR1) deficiency in erythrocytes from patients with systemic lupus erythematosus. J Exp Med 155:1427–1438
Ishizaka T, Ishizaka K, Salmon S, Fudenberg H (1967) Biologic activities of aggregated gamma-globulin. 8. Aggregated immunoglobulins of different classes. J Immunol 99:82–91
Janssen BJ et al (2005) Structures of complement component C3 provide insights into the function and evolution of immunity. Nature 437:505–511
Ji H et al (2002) Arthritis critically dependent on innate immune system players. Immunity 16:157–168
Junt T et al (2007) Subcapsular sinus macrophages in lymph nodes clear lymph-borne viruses and present them to antiviral B cells. Nature 450:110–114
Karsten CM et al (2012) Anti-inflammatory activity of IgG1 mediated by Fc galactosylation and association of FcγRIIB and dectin-1. Nat Med 18:1401–1406
Kimberly RP et al (1989) In vivo handling of soluble complement fixing Ab/dsDNA immune complexes in chimpanzees. J Clin Invest 84:962–970
Klaus GG, Pepys MB, Kitajima K, Askonas BA (1979) Activation of mouse complement by different classes of mouse antibody. Immunology 38:687–695
Klein M et al (1981) Expression of biological effector functions by immunoglobulin G molecules lacking the hinge region. Proc Natl Acad Sci USA 78:524–528
Klein MA et al (2001) Complement facilitates early prion pathogenesis. Nat Med 7:488–492
Klickstein LB et al (1987) Human C3b/C4b receptor (CR1). Demonstration of long homologous repeating domains that are composed of the short consensus repeats characteristics of C3/C4 binding proteins. J Exp Med 165:1095–1112
Klickstein LB, Barbashov SF, Liu T, Jack RM, Nicholson-Weller A (1997) Complement receptor type 1 (CR1, CD35) is a receptor for C1q. Immunity 7:345–355
Knutzen Steuer KL et al (1989) Lysis of sensitized sheep erythrocytes in human sera deficient in the second component of complement. J Immunol 143:2256–2261
Kobata A (2008) The N-linked sugar chains of human immunoglobulin G: their unique pattern, and their functional roles. Biochim Biophys Acta 1780:472–478

Köhl J (2006) The role of complement in danger sensing and transmission. Immunol Res 34:157–176
Koide N, Nose M, Muramatsu T (1977) Recognition of IgG by Fc receptor and complement: effects of glycosidase digestion. Biochem Biophys Res Commun 75:838–844
Konrad S, Baumann U, Schmidt RE, Gessner JE (2006) Intravenous immunoglobulin (IVIG)-mediated neutralisation of C5a: a direct mechanism of IVIG in the maintenance of a high Fc gammaRIIB to Fc gammaRIII expression ratio on macrophages. Br J Haematol 134:345–347
Konrad S et al (2008) Phosphoinositide 3-kinases gamma and delta, linkers of coordinate C5a receptor-Fcgamma receptor activation and immune complex-induced inflammation 99. J Biol Chem 283:33296–33303
Korganow AS et al (1999) From systemic T cell self-reactivity to organ-specific autoimmune disease via immunoglobulins. Immunity 10:451–461
Kouskoff V et al (1996) Organ-specific disease provoked by systemic autoimmunity. Cell 87:811–822
Kovacsovics T, Tschopp J, Kress A, Isliker H (1985) Antibody-independent activation of C1, the first component of complement, by cardiolipin. J Immunol 135:2695–2700
Krych-Goldberg M, Atkinson JP (2001) Structure-function relationships of complement receptor type 1. Immunol Rev 180:112–122
Kuhn SE, Nardin A, Klebba PE, Taylor RP (1998) Escherichia coli bound to the primate erythrocyte complement receptor via bispecific monoclonal antibodies are transferred to and phagocytosed by human monocytes in an in vitro model. J Immunol 160:5088–5097
Kulics J, Rajnavolgyi E, Fust G, Gergely J (1983) Interaction of C3 and C3b with immunoglobulin G. Mol Immunol 20:805–810
Kumar V et al (2006) Cell-derived anaphylatoxins as key mediators of antibody-dependent type II autoimmunity in mice. J Clin Invest 116:512–520
Kurtz CB, O'Toole E, Christensen SM, Weis JH (1990) The murine complement receptor gene family. IV. Alternative splicing of Cr2 gene transcripts predicts two distinct gene products that share homologous domains with both human CR2 and CR1. J Immunol 144:3581–3591
Lutz HU, Jelezarova E (2006) Complement amplification revisited. Mol Immunol 43:2–12
Madaio MP (1999) The role of autoantibodies in the pathogenesis of lupus nephritis. Semin Nephrol 19:48–56
Malhotra R et al (1995) Glycosylation changes of IgG associated with rheumatoid arthritis can activate complement via the mannose-binding protein. Nat Med 1:237–243
Manderson AP, Botto M, Walport MJ (2004) The role of complement in the development of systemic lupus erythematosus. Annu Rev Immunol 22:431–456
Matthews KW, Mueller-Ortiz SL, Wetsel RA (2004) Carboxypeptidase N: a pleiotropic regulator of inflammation. Mol Immunol 40:785–793
May JE, Frank MM (1973a) A new complement-mediated cytolytic mechanism – the C1-bypass activation pathway. Proc Natl Acad Sci USA 70:649–652
May JE, Frank MM (1973b) Hemolysis of sheep erythrocytes in guinea pig serum deficient in the fourth component of complement. I. Antibody and serum requirements. J Immunol 111:1671–1677
May JE, Frank MM (1973c) Hemolysis of sheep erythrocytes in guinea pig serum deficient in the fourth component of complement. II. Evidence for involvement of C1 and components of the alternate complement pathway. J Immunol 111:1668–1676
McGuire TC, Musoke AJ, Kurtti T (1979) Functional properties of bovine IgG1 and IgG2: interaction with complement, macrophages, neutrophils and skin. Immunology 38:249–256
Medgyesi GA, Fust G, Gergely J, Bazin H (1978) Classes and subclasses of rat immunoglobulins: interaction with the complement system and with staphylococcal protein A. Immunochemistry 15:125–129
Miller GW, Nussenzweig V (1975) A new complement function: solubilization of antigen-antibody aggregates. Proc Natl Acad Sci USA 72:418–422
Miller GW, Steinberg AD, Green I, Nussenzweig V (1975) Complement-dependent alterations in the handling of immune complexes by NZB/W mice. J Immunol 114:1166–1170

Molina H, Kinoshita T, Inoue K, Carel JC, Holers VM (1990) A molecular and immunochemical characterization of mouse CR2. Evidence for a single gene model of mouse complement receptors 1 and 2. J Immunol 145:2974–2983
Molina H, Kinoshita T, Webster CB, Holers VM (1994) Analysis of C3b/C3d binding sites and factor I cofactor regions within mouse complement receptors 1 and 2. J Immunol 153:789–795
Moller NP (1979) Fc-mediated immune precipitation. I. A new role of the Fc-portion of IgG. Immunology 38:631–640
Moller NP, Steensgaard J (1979) Fc-mediated immune precipitation. II. Analysis of precipitating immune complexes by rate-zonal ultracentrifugation. Immunology 38:641–648
Nardin A, Lindorfer MA, Taylor RP (1999) How are immune complexes bound to the primate erythrocyte complement receptor transferred to acceptor phagocytic cells? Mol Immunol 36:827–835
Nelson RA Jr (1953) The immune-adherence phenomenon; an immunologically specific reaction between microorganisms and erythrocytes leading to enhanced phagocytosis. Science 118:733–737
Nimmerjahn F, Ravetch JV (2005) Divergent immunoglobulin g subclass activity through selective Fc receptor binding. Science 310:1510–1512
Nimmerjahn F, Ravetch JV (2006) Fcgamma receptors: old friends and new family members. Immunity 24:19–28
Nimmerjahn F et al (2010) FcgammaRIV deletion reveals its central role for IgG2a and IgG2b activity in vivo. Proc Natl Acad Sci USA 107:19396–19401
Nonaka M, Kimura A (2006) Genomic view of the evolution of the complement system. Immunogenetics 58:701–713
Norderhaug L et al (1991) Chimeric mouse human IgG3 antibodies with an IgG4-like hinge region induce complement-mediated lysis more efficiently than IgG3 with normal hinge. Eur J Immunol 21:2379–2384
Nose M, Wigzell H (1983) Biological significance of carbohydrate chains on monoclonal antibodies. Proc Natl Acad Sci USA 80:6632–6636
Nussenzweig V, Bianco C, Dukor P, Eden A. Receptors for C3 on B lymphocytes: possible role in the immune system, vol 59 (ed. Amos, B.) 73–81 (Academic Press, 1971)
Oortwijn BD et al (2008) The role of secretory IgA and complement in IgA nephropathy. Semin Nephrol 28:58–65
Paccaud JP, Steiger G, Sjoholm AG, Spaeth PJ, Schifferli JA (1987) Tetanus toxoid-anti-tetanus toxoid complexes: a potential model to study the complement transport system for immune complex in humans. Clin Exp Immunol 69:468–476
Paccaud JP, Carpentier JL, Schifferli JA (1988) Direct evidence for the clustered nature of complement receptors type 1 on the erythrocyte membrane. J Immunol 141:3889–3894
Paccaud JP, Carpentier JL, Schifferli JA (1990) Difference in the clustering of complement receptor type 1 (CR1) on polymorphonuclear leukocytes and erythrocytes: effect on immune adherence. Eur J Immunol 20:283–289
Pangburn MK, Muller-Eberhard HJ (1980) Relation of putative thioester bond in C3 to activation of the alternative pathway and the binding of C3b to biological targets of complement. J Exp Med 152:1102–1114
Pangburn MK, Schreiber RD, Muller-Eberhard HJ (1981) Formation of the initial C3 convertase of the alternative complement pathway. Acquisition of C3b-like activities by spontaneous hydrolysis of the putative thioester in native C3. J Exp Med 154:856–867
Papamichail M et al (1975) Complement dependence of localisation of aggregated IgG in germinal centres. Scand J Immunol 4:343–347
Parekh RB et al (1985) Association of rheumatoid arthritis and primary osteoarthritis with changes in the glycosylation pattern of total serum IgG. Nature 316:452–457
Parekh R et al (1989) A comparative analysis of disease-associated changes in the galactosylation of serum IgG. J Autoimmun 2:101–114

Park SY et al (1998) Resistance of Fc receptor- deficient mice to fatal glomerulonephritis. J Clin Invest 102:1229–1238
Parker CJ (2012) Paroxysmal nocturnal hemoglobinuria. Curr Opin Hematol 19:141–148
Peitsch MC, Tschopp J, Kress A, Isliker H (1988) Antibody-independent activation of the complement system by mitochondria is mediated by cardiolipin. Biochem J 249:495–500
Perkins SJ, Nealis AS, Sutton BJ, Feinstein A (1991) Solution structure of human and mouse immunoglobulin M by synchrotron X-ray scattering and molecular graphics modelling. A possible mechanism for complement activation. J Mol Biol 221:1345–1366
Phan TG, Grigorova I, Okada T, Cyster JG (2007) Subcapsular encounter and complement-dependent transport of immune complexes by lymph node B cells. Nat Immunol 8:992–1000
Phan TG, Green JA, Gray EE, Xu Y, Cyster JG (2009) Immune complex relay by subcapsular sinus macrophages and noncognate B cells drives antibody affinity maturation. Nat Immunol 10:786–793
Pickering MC, Botto M, Taylor PR, Lachmann PJ, Walport MJ (2000) Systemic lupus erythematosus, complement deficiency, and apoptosis. Adv Immunol 76:227–324
Pozdnyakova O, Guttormsen HK, Lalani FN, Carroll MC, Kasper DL (2003) Impaired antibody response to group B streptococcal type III capsular polysaccharide in C3- and complement receptor 2-deficient mice. J Immunol 170:84–90
Raby AC et al (2011) TLR activation enhances C5a-induced pro-inflammatory responses by negatively modulating the second C5a receptor, C5L2. Eur J Immunol 41(9):2741–2752
Raju TS (2008) Terminal sugars of Fc glycans influence antibody effector functions of IgGs. Curr Opin Immunol 20:471–478
Reinagel ML, Taylor RP (2000) Transfer of immune complexes from erythrocyte CR1 to mouse macrophages. J Immunol 164:1977–1985
Ren Y et al (2003) Increased apoptotic neutrophils and macrophages and impaired macrophage phagocytic clearance of apoptotic neutrophils in systemic lupus erythematosus. Arthritis Rheum 48:2888–2897
Ricklin D, Hajishengallis G, Yang K, Lambris JD (2010) Complement: a key system for immune surveillance and homeostasis. Nat Immunol 11:785–797
Riedemann NC et al (2002) Increased C5a receptor expression in sepsis. J Clin Invest 110:101–108
Roos A et al (2001) Human IgA activates the complement system via the mannan-binding lectin pathway. J Immunol 167:2861–2868
Roos A et al (2006) Glomerular activation of the lectin pathway of complement in IgA nephropathy is associated with more severe renal disease. J Am Soc Nephrol 17:1724–1734
Ross GD et al (1985) Disease-associated loss of erythrocyte complement receptors (CR1, C3b receptors) in patients with systemic lupus erythematosus and other diseases involving autoantibodies and/or complement activation. J Immunol 135:2005–2014
Royle L et al (2003) Secretory IgA N- and O-glycans provide a link between the innate and adaptive immune systems. J Biol Chem 278:20140–20153
Sahu A, Kozel TR, Pangburn MK (1994) Specificity of the thioester-containing reactive site of human C3 and its significance to complement activation. Biochem J 302(Pt 2):429–436
Schifferli JA, Bartolotti SR, Peters DK (1980) Inhibition of immune precipitation by complement. Clin Exp Immunol 42:387–394
Schifferli JA, Steiger G, Paccaud JP (1986) Complement mediated inhibition of immune precipitation and solubilization generate different concentrations of complement anaphylatoxins (C4a, C3a, C5a). Clin Exp Immunol 64:407–414
Schur PH, Christian GD (1964) The role of disulfide bonds in the complement-fixing and precipitating properties of 7s rabbit and sheep antibodies. J Exp Med 120:531–545
Schwaeble WJ, Reid KB (1999) Does properdin crosslink the cellular and the humoral immune response? Immunol Today 20:17–21
Shohet JM, Pemberton P, Carroll MC (1993) Identification of a major binding site for complement C3 on the IgG1 heavy chain. J Biol Chem 268:5866–5871

Shulman MJ, Collins C, Pennell N, Hozumi N (1987) Complement activation by IgM: evidence for the importance of the third constant domain of the mu heavy chain. Eur J Immunol 17:549–554

Shushakova N et al (2002) C5a anaphylatoxin is a major regulator of activating versus inhibitory FcgammaRs in immune complex-induced lung disease. J Clin Invest 110:1823–1830

Silverton EW, Navia MA, Davies DR (1977) Three-dimensional structure of an intact human immunoglobulin. Proc Natl Acad Sci USA 74:5140–5144

Skattum L, van Deuren M, van der Poll T, Truedsson L (2011) Complement deficiency states and associated infections. Mol Immunol 48:1643–1655

Skokowa J et al (2005) Macrophages induce the inflammatory response in the pulmonary Arthus reaction through G alpha i2 activation that controls C5aR and Fc receptor cooperation. J Immunol 174:3041–3050

Storrs SB, Kolb WP, Olson MS (1983) C1q binding and C1 activation by various isolated cellular membranes. J Immunol 131:416–422

Syed SN et al (2009) Both FcgammaRIV and FcgammaRIII are essential receptors mediating type II and type III autoimmune responses via FcRgamma-LAT-dependent generation of C5a. Eur J Immunol 39:3343–3356

Sylvestre DL, Ravetch JV (1994) Fc receptors initiate the Arthus reaction: redefining the inflammatory cascade. Science 265:1095–1098

Tack BF, Harrison RA, Janatova J, Thomas ML, Prahl JW (1980) Evidence for presence of an internal thiolester bond in third component of human complement. Proc Natl Acad Sci USA 77:5764–5768

Tan LK, Shopes RJ, Oi VT, Morrison SL (1990) Influence of the hinge region on complement activation, C1q binding, and segmental flexibility in chimeric human immunoglobulins. Proc Natl Acad Sci USA 87:162–166

Tao MH, Canfield SM, Morrison SL (1991) The differential ability of human IgG1 and IgG4 to activate complement is determined by the COOH-terminal sequence of the CH2 domain. J Exp Med 173:1025–1028

Tao MH, Smith RI, Morrison SL (1993) Structural features of human immunoglobulin G that determine isotype-specific differences in complement activation. J Exp Med 178:661–667

Tausk FA, McCutchan A, Spechko P, Schreiber RD, Gigli I (1986) Altered erythrocyte C3b receptor expression, immune complexes, and complement activation in homosexual men in varying risk groups for acquired immune deficiency syndrome. J Clin Invest 78:977–982

Taylor RP, Burge J, Horgan C, Shasby DM (1983) The complement-mediated binding of soluble antibody/dsDNA immune complexes to human neutrophils. J Immunol 130:2656–2662

Taylor RP et al (1997) Bispecific monoclonal antibody complexes facilitate erythrocyte binding and liver clearance of a prototype particulate pathogen in a monkey model. J Immunol 159:4035–4044

Thiel S, Gadjeva M (2009) Humoral pattern recognition molecules: mannan-binding lectin and ficolins. Adv Exp Med Biol 653:58–73

Tsuchiya N et al (1989) Effects of galactose depletion from oligosaccharide chains on immunological activities of human IgG. J Rheumatol 16:285–290

Utsumi S, Okada M, Udaka K, Amano T (1985) Preparation and biologic characterization of fragments containing dimeric and monomeric C gamma 2 domain of rabbit IgG. Mol Immunol 22:811–819

Vogt L et al (2006) VSIG4, a B7 family-related protein, is a negative regulator of T cell activation. J Clin Invest 116:2817–2826

Waxman FJ et al (1986) Differential binding of immunoglobulin A and immunoglobulin G1 immune complexes to primate erythrocytes in vivo. Immunoglobulin A immune complexes bind less well to erythrocytes and are preferentially deposited in glomeruli. J Clin Invest 77:82–89

Wiesmann C et al (2006) Structure of C3b in complex with CRIg gives insights into regulation of complement activation. Nature 444:217–220

Williams RC Jr, Osterland CK, Margherita S, Tokuda S, Messner RP (1973) Studies of biologic and serologic activities of rabbit-IgG antibody depleted of carbohydrate residues. J Immunol 111:1690–1698
Wright A, Morrison SL (1994) Effect of altered CH2-associated carbohydrate structure on the functional properties and in vivo fate of chimeric mouse-human immunoglobulin G1. J Exp Med 180:1087–1096
Xu Y, Oomen R, Klein MH (1994) Residue at position 331 in the IgG1 and IgG4 CH2 domains contributes to their differential ability to bind and activate complement. J Biol Chem 269:3469–3474
Xu S et al (2010) Induction of T cells suppression by dendritic cells transfected with VSIG4 recombinant adenovirus. Immunol Lett 128:46–50
Youd ME, Ferguson AR, Corley RB (2002) Synergistic roles of IgM and complement in antigen trapping and follicular localization. Eur J Immunol 32:2328–2337

Chapter 8
Regulation of Immunological Responses by the Neonatal Fc Receptor for IgG, FcRn

Kristi Baker, Timo Rath, and Richard S. Blumberg

Abstract Immunoglobulins (Ig) are the hallmark of the humoral immune system and can be found, in varying concentrations, in all tissues of the body as well as throughout the circulatory system. Of the various isotypes, IgG is by far the most abundant in serum due to its extremely long half-life of 7–23 days. Indeed, while other Ig isotypes are known for very specialized roles in allergy (IgE) and pathogen neutralization (IgA), IgG is the main driver of many of the functions originally attributed to humoral immunity including opsonization, complement activation, and antibody-dependent cellular cytotoxicity (ADCC). Increasingly, however, it is being recognized that the functions of IgG extend beyond simply that of a humoral immune system component to that of a potent integrator of the humoral and cellular arms of a coordinated adaptive immune response.

Keywords Immunoglobulins • Humoral immune system • IgG • Immunological responses • Fc receptor • Antibody-dependent cellular cytotoxicity • FcRn • Intracellular trafficking • Regulation of IgG

Immunoglobulins (Ig) are the hallmark of the humoral immune system and can be found, in varying concentrations, in all tissues of the body as well as throughout the circulatory system. Of the various isotypes, IgG is by far the most abundant in

K. Baker, Ph.D. (✉) • T. Rath, M.D.
Division of Gastroenterology, Hepatology and Endoscopy, Brigham and Women's Hospital and Harvard Medical School, 75 Francis Street, Thorn 1415, 02115 Boston, MA, USA
e-mail: kdbaker@partners.org; trath@partners.org

R.S. Blumberg, MD
Division of Gastroenterology, Hepatology and Endoscopy, Harvard Medical School and Brigham and Women's Hospital, 75 Francis Street, Thorn 1415, 02115 Boston, MA, USA
e-mail: rblumberg@partners.org

F. Nimmerjahn (ed.), *Molecular and Cellular Mechanisms of Antibody Activity*, DOI 10.1007/978-1-4614-7107-3_8, © Springer Science+Business Media New York 2013

serum due to its extremely long half-life of 7–23 days (Spiegelberg 1989, 1968; Spiegelberg and Weigle 1965; Morell et al. 1970). Indeed, while other Ig isotypes are known for very specialized roles in allergy (IgE) and pathogen neutralization (IgA), IgG is the main driver of many of the functions originally attributed to humoral immunity including opsonization, complement activation, and antibody-dependent cellular cytotoxicity (ADCC) (Schroeder and Cavacini 2010). Increasingly, however, it is being recognized that the functions of IgG extend beyond simply that of a humoral immune system component to that of a potent integrator of the humoral and cellular arms of a coordinated adaptive immune response.

The ability of IgG to bridge the gap between divergent branches of such a sophisticated response network depends in large part on the interaction of IgG with its various receptors. The most widely appreciated of these fall into the Fcγ receptor (FcγR) family which is a class of cell surface receptors widely expressed by cells of the hematopoietic lineage. Two broad classes of human FcγR exist, those which transmit activating signals (FcγRI, FcγRIIa, FcγRIIc, FcγRIII, FcγRIV) through the SYK tyrosine kinase and those which transmit inhibitory signals (FcγRIIb) via an ITIM motif in their cytoplasmic tail (Nimmerjahn and Ravetch 2010). Homologous structures exist in mice except for the absence of an FcγRIIa homologue (Nimmerjahn and Ravetch 2010). These receptors bind to the conserved Fc region of IgG with varying degrees of affinity determined by the IgG subclass and the valence of the IgG-containing immune complex (IgG IC). Consequently, IgG exerts a substantial part of its actions via converting antigen recognition of soluble and particulate antigen into cellular signals which are tunable based on the composition of the ligand. FcγR binding by IgG and IgG IC is known to induce both signaling and internalization of the ligand, thereby delivering it into the endolysosomal system. Given that FcγR only bind to IgG at neutral pH and that the compartments of the endosomal systems undergo gradual acidification as they mature, it is likely that the IgG or IgG IC dissociates from internalizing FcγR in these vesicles (Baker et al. 2011; Russell 2007). Until recently, it was unknown whether active trafficking of the IgG or IgG IC proceeded intracellularly following this event.

The identification and characterization of a largely intracellular trafficking receptor for IgG, first postulated to exist by Brambell in 1966, has recently shed light on how IgG and IgG IC behave within cells (Brambell 1966). The neonatal Fc receptor for IgG (FcRn), which binds to IgG and albumin in a non-mutually exclusive manner, is known to traffic to, and bind ligand at, the cell surface but is in fact a predominantly intracellularly distributed receptor (Chaudhury et al. 2003, 2006). FcRn was first described as a transporter of IgG within the intestinal epithelial cells (IEC) of neonatal rodents, where it drove absorption of maternal IgG in suckling pups, and thus enabled the transfer of passive immunity from mother to offspring (Simister and Rees 1985; Simister and Mostov 1989a). While FcRn is dramatically down-regulated in the IEC of rodents to nearly nondetectable levels upon weaning, it is expressed throughout life in the IEC of many other animals, including humans, and in all other cell types in which it has been described, such as placental syncytiotrophoblasts (Leach et al. 1996), endothelial cells (Borvak et al. 1998; Antohe et al. 2001;

Ober et al. 2004a), pulmonary epithelial cells (Sakagami et al. 2006), mammary epithelial cells (Cianga et al. 2003), kidney podocytes (Akilesh et al. 2008), hepatocytes (Blumberg et al. 1995), and a host of hematopoietic cells including monocytes, macrophages, dendritic cells (DC), polymorphonuclear leukocytes (Zhu et al. 2001), and B cells (Mi et al. 2008). The ubiquity of FcRn expression across tissues reflects its importance in the maintenance of organismal homeostasis. In particular, the contribution of FcRn within antigen-presenting cells (APC), such as macrophages and DC, to intracellular routing of IgG IC has been found to be critically important in directing these complexes into processing pathways by which epitopes are generated for loading onto MHC class I and class II (Baker et al. 2011; Qiao et al. 2008). Thus, FcRn within cells of the innate branch of the immune system represents a critical component of the integration machinery by which IgG translates an antigen-specific humoral response into a specific cellular response involving both $CD4^+$ and $CD8^+$ T cells. This chapter will therefore focus on the immunobiology of FcRn.

8.1 The Physiology of FcRn

8.1.1 Molecular Structure of FcRn

FcRn is an MHC class I-related molecule encoded by the *Fcgrt* gene and in which the peptide-binding groove has become collapsed and cannot functionally be loaded with peptide (Simister and Mostov 1989a). Similar to MHC class I and other members of this family such as CD1d, FcRn requires heterodimeric association with species-matched β2-microglobulin (β2M) in order to acquire the appropriate glycosylation for maturation, trafficking to the cell surface, and efficient ligand binding (Claypool et al. 2002; Praetor and Hunziker 2002). Indeed, despite sharing only 22–29 % sequence homology with MHC class I, many of the β2M contact sites remain conserved in FcRn, supporting the idea of a common ancestry between the two groups of molecules (Simister and Ahouse 1996). FcRn itself consists of three extracellular domains with varying degrees of glycosylation depending upon the species of origin, a single-pass transmembrane domain, and a short cytoplasmic tail (Martin et al. 2001; Burmeister et al. 1994a, b).

FcRn remains functionally conserved across a wide range of species including humans, rodents (Catunda Lemos et al. 2012; Ober et al. 2001), sheep (Mayer et al. 2002), cattle (Kacskovics 2004; Kacskovics et al. 2000, 2006a; Lu et al. 2007; Mayer et al. 2004), swine (Schnulle and Hurley 2003; Stirling et al. 2005), Egyptian water buffalo (Sayed-Ahmed et al. 2010), and camels (Kacskovics et al. 2006b). An FcRn orthologue, IgYR, is also expressed in avian and reptilian species (West et al. 2004; He and Bjorkman 2011; Tesar et al. 2008). Despite functional conservation of FcRn itself across many species, considerable molecular divergence is evident in FcRn molecules from different species. The amino acid chains of mouse and rat

FcRn are 91 % identical, but human FcRn shares only 68 % homology with the rat molecule (West and Bjorkman 2000; Ahouse et al. 1993). Although many of the amino acid alterations across species are conservative changes, some result in differential FcRn glycosylation patterns which can alter the subcellular distribution and trafficking of the molecule (Martin et al. 2001; Vaughn and Bjorkman 1998; Kuo et al. 2009). For example, the human FcRn molecule (hFcRn), which is predominantly located along the basolateral surface of IEC, possesses a single carbohydrate side chain in the α2 domain. In contrast, rodent FcRn has a total of four N-glycan-linked side chains spread across the α1, α2, and α3 domains and is found predominantly at the apical surface of intestinal epithelial cells. Modification of hFcRn to generate a "rodentized" form with additional carbohydrate side chains results in redistribution of the hFcRn to the apical cell surface (Kuo et al. 2009). Critically, these modifications also result in a reversal of the dominant FcRn trafficking direction within these cells, as will be discussed further below. One additional point to consider regarding FcRn in different species is the degree of "promiscuity" of the receptor with respect to ligand binding. Whereas murine FcRn (mFcRn) binds well to IgG from a range of different species including humans, rabbits, and cattle, hFcRn exhibits a much more restricted binding pattern and is only known to interact with IgG from rabbits and guinea pigs (Ober et al. 2001; Zhou et al. 2005a; Andersen et al. 2010). Such potential ligand–receptor mismatches, particularly of the human receptor with the murine ligand, are a particularly important consideration when designing experimental systems in which to study the function and kinetics of hFcRn and will be addressed in a subsequent section.

8.1.2 Interaction of FcRn with IgG

Binding of FcRn to IgG occurs in a strict 2:1 stoichiometric ratio with each of two FcRn molecules binding to the C_H2–C_H3 domain interface of one of the heavy chains in a single IgG molecule (West and Bjorkman 2000; Martin and Bjorkman 1999). Numerous studies inducing amino acid substitutions in the IgG–Fc region have demonstrated that there are three critical residues in the ligand molecule required for interaction with FcRn: I253, H310, and H435 (Fig. 8.1) (Martin et al. 2001; Raghavan et al. 1993, 1995; Medesan et al. 1997; Kim et al. 1999; Jin-Kyoo et al. 1994; Wang et al. 2000; Porollo and Meller 2007). Simultaneous mutation of all of these three amino acids to alanine abrogates the IgG–FcRn interaction without affecting the ability of IgG to bind to FcγR (Qiao et al. 2008). Within FcRn, the specific amino acid residues which contact IgG vary across species, but all include conserved histidine residues (such as H250/H251 in rat FcRn and H248/H249 in human FcRn) whose charge represents a critical feature for ligand binding (Martin et al. 2001; West and Bjorkman 2000; Raghavan et al. 1993, 1995; Kuo and Aveson 2011).

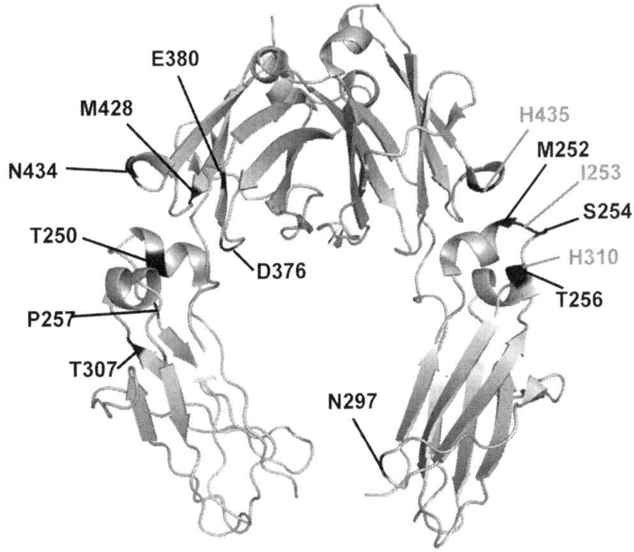

Fig. 8.1 *Mutations in the human IgG1 Fc domain known to enhance (black labels) or inhibit (gray labels) IgG–FcRn binding*. Various mutations, singly or in combination, in the indicated amino acids of the hIgG–Fc region are known to alter the binding of IgG to FcRn (Kuo and Aveson 2011) (Diagram created using POLYVIEW-3D based on the crystal structure of the Fc region of hIgG1 (Protein Data Bank ID: 3DO3) based on Kuo and Aveson (2011), (Porollo and Meller 2007; Kuo and Aveson 2011))

Somewhat counterintuitively for a receptor directed towards extracellular ligands, FcRn only binds to its ligands at acidic pH ≤ 6.5 (Vaughn and Bjorkman 1998; Raghavan et al. 1993, 1995; Rodewald 1976). This strict pH dependence is due not to pH-dependent conformation changes in the receptor but rather to the pH-dependent titration of several salt bridges critical for the FcRn–IgG interaction (Chaudhury et al. 2006; Vaughn and Bjorkman 1998; Raghavan et al. 1993, 1995). At a neutral pH ≥ 7.0, the acidic histidine residues of the Fc domain of IgG which interact with FcRn become deprotonated, thereby disabling binding and allowing for ligand dissociation. In contrast, at more acidic pH in the range of 4–6.5, protonation of these same residues at the FcRn–IgG interface stabilizes ligand interaction and permits efficient binding of IgG to FcRn (Baker et al. 2011). The strict pH-dependent ligand binding exhibited by FcRn is well suited to its subcellular distribution since the pH range over which FcRn efficiently binds IgG overlaps almost entirely with that found within the endolysosomal system where the receptor is predominantly situated. Hence, the particular characteristics which have evolved for receptor binding during the FcRn–IgG interaction enable FcRn to function as an exquisitely efficient mediator of ligand trafficking.

8.1.3 Intracellular Trafficking of FcRn

Given that FcRn was first described as a transporter of IgG across the polarized epithelial cells of the intestinal epithelium, the vast majority of studies to date on the intracellular distribution and trafficking of FcRn have been conducted in these types of cells. Under steady-state conditions, FcRn is observed in subapical vesicles of primary IECs and throughout the endolysosomal system in transfected Madin–Darby canine kidney (MDCK) cells although with somewhat of an apical predominance (Shah et al. 2003; Claypool et al. 2004). In MDCK cells, this distribution is not altered by the presence of IgG (Claypool et al. 2004). Of the fraction of FcRn which is located at the cell surface, it is predominantly apically distributed in neonatal rodents but basolaterally distributed in humans of all ages (McCarthy et al. 2000, 2001). As discussed below, this difference likely reflects the major role played by FcRn in these organisms at different stages of development since the predominant direction of FcRn transcytosis in neonatal rodents is apical → basolateral whereas in humans it is basolateral → apical. The differential distribution of FcRn homologues in neonatal rodents and humans has been attributed both to cytoplasmic tail variations and to differential glycosylation patterns between species (Kuo et al. 2009; McCarthy et al. 2001).

The lack of FcγR expression by epithelial cells under steady-state conditions has led to the proposition of two major mechanisms for IgG uptake by these cells. The first of these relies upon pinocytic or endocytic uptake, as occurs in endothelial cells, and it is unclear to what extent this process might play a role at an epithelial boundary (Ober et al. 2004a, b). The second proposed mechanism of uptake involves receptor-mediated internalization of IgG by FcRn itself. Although the extracellular environment is typically at a neutral pH that would inhibit IgG binding by FcRn, the actions of a sodium–hydrogen exchange pump at the apical membrane of IEC is posited to create an acidic environment at the luminal surface of the epithelium which would generate permissive conditions for FcRn–IgG binding (Claypool et al. 2004; Hattori et al. 2001). While direct experimental evidence to support this mechanism is lacking, it is very clear that FcRn is capable of actively retrieving IgG from the luminal environment of the gut. Indeed, in the jejunum of neonatal rodents, where the luminal environment is known to be at an acidic pH ≈ 6.0, gold-labeled IgG has been demonstrated to enter into IEC within apical clathrin-coated pits (He et al. 2008). Although FcRn was not directly shown to be bound to the IgG at this point, these data support the notion that FcRn itself might be directly involved in the uptake of IgG under permissive binding conditions. Studies of FcRn exocytosis likewise provide a clue as to how FcRn might first be internalized by IEC. Single-cell fluorescence analysis has indicated that multiple mechanisms exist for FcRn-mediated exocytosis of its cargo. While complete fusion events between the cell membrane and compartments containing IgG-bound FcRn were observed, partial fusion events dubbed the "prolonged release" were also documented. These events were characterized by only partial mixing of the membrane material and the temporary distribution of FcRn into the cell membrane before its return to the partially fused vesicle (Ober et al. 2004b; Lencer and Blumberg 2005). Should the pH of this vesicle remain acidic during this partial fusion event, it is conceivable that FcRn would be surrounded by an immediate microenvironment conducive to ligand binding and that this would permit its direct involvement in ligand uptake.

Subsequent to internalization, it is abundantly clear that the transcytosis of IgG across IEC is directly driven by FcRn. Critically, FcRn-enabled transcytosis is bidirectional, proceeding with near-equal efficiency in both the apical → basolateral and basolateral → apical directions despite the fact that one direction or the other tends to predominate under a given set of circumstances and via a process yet to be defined (Claypool et al. 2004; McCarthy et al. 2000; Tesar and Björkman 2010). The bidirectional transcytosis of IgG by FcRn has been observed in many different types of polarized epithelia including that from intestine, placenta, and kidney (Leach et al. 1996; Antohe et al. 2001; Akilesh et al. 2008; Claypool et al. 2004; Dickinson et al. 1999; Ellinger et al. 1999; Firan et al. 2001; Gafencu et al. 2003; Kim et al. 2009; Simister 2003; Kobayashi et al. 2002). This unique ability of FcRn to both deliver IgG into the intestinal lumen and then retrieve and transport it back into the lamina propria is in stark contrast to the receptor for the classic mucosal immunoglobulin, IgA. The polymeric IgA receptor (pIgR) unidirectionally transports IgA into the intestinal lumen, and while its trafficking route does partially intermix with that of FcRn, it is efficiently removed from early, recycling, and sorting endosomes (Jerdeva et al. 2010). In contrast, FcRn accumulates in these compartments, as well as in multivesicular bodies and a complex series of tubules which likely serve as connections between the vesicles (He et al. 2008). Furthermore, while FcRn is distributed throughout the endolysosomal system of IEC, pIgR was found to concentrate in the apical region of the cells. The mechanisms governing the bidirectional transport function of FcRn are not fully understood, as noted, but it is clear that the cytoplasmic domain plays a dominant role. In particular, the cytoplasmic tail of FcRn contains a tryptophan-based endocytosis sorting signal, a dileucine-based internalization motif, a calmodulin binding site required for transcytosis, and two serine phosphorylation sites, one of which is required for diversion of FcRn away from an apical recycling pathway and into a transcytotic one (McCarthy et al. 2001; Wernick et al. 2005; Wu and Simister 2001; Dickinson et al. 2008). While other direct binding partners for FcRn remain to be identified, certain intracellular components are known to affect FcRn-mediated trafficking despite not being known to directly bind FcRn. In particular, myosin Vb and Rab25 have been implicated in the diversion of IgG-bound FcRn away from a recycling pathway and into a transcytotic one (Tzaban et al. 2009). Rab11a, a known component of recycling endosomes, however, directs IgG–FcRn towards the basolateral membrane of IEC. While the details of FcRn trafficking in cell types other than IEC remain to be clarified, it is clear from the studies on IEC that intracellular FcRn movement is not stochastic but rather proceeds by carefully orchestrated and separable mechanisms which, collectively, enable FcRn to carry out a multitude of functions.

8.2 Regulation of IgG by FcRn

Long before FcRn was discovered, a receptor capable of regulating various aspects of IgG biology, namely, maternofetal transfer, anaphylactic sensitization, and systemic catabolism, was proposed to exist (Brambell 1966; Brambell et al. 1958). Experimental evidence now clearly indicates that FcRn is the dominant player in each of these processes.

8.2.1 FcRn-Mediated Transcytosis Drives Prenatal and Neonatal Delivery of IgG from Mother to Offspring

The immune system of neonates is incompletely developed, and in humans, it can take up to 1–2 years before an individual is capable of mounting a full range of immunological responses (Conroy et al. 2009; Kelly et al. 2007). Throughout this critical stage of immune system development, a major source of immunological protection comes from passively acquired immunoglobulins transferred from mother to offspring during either the gestational or suckling periods of life. FcRn is an integral component in both of these processes.

During the antenatal period, IgG is transferred across the placenta from the mother's circulation into the developing fetus. FcRn expression has been documented in the syncytiotrophoblasts of the human placenta, where it mediates transfer of IgG to the fetus beginning in the second trimester of pregnancy and peaking in the third trimester (Simister 2003; Story et al. 1994; Szlauer et al. 2009). Following uptake of IgG by these cells, which is believed to occur passively by pinocytosis, it becomes bound to FcRn in the early endosomes and is then transited across the cell and released during an exocytosis event at the neutral pH of the basolateral side of the cell. While FcRn has not been identified in the rodent placenta, it is clearly expressed in the endodermal cells of the yolk sac, from which it has been shown to mediate maternofetal IgG transfer (Kim et al. 2009; Mohanty et al. 2010). Although it is likely that additional Fc receptors participate in this process, experimental evidence clearly indicates that in the absence of FcRn, fetuses contain negligible quantities of IgG (Kim et al. 2009; Mohanty et al. 2010).

FcRn continues to enable maternal transfer or immune protection during the early postnatal period. Expression of FcRn in the mammary glands of both lactating humans and mice allows for transfer of IgG into breast milk, while FcRn within the duodenal and jejunal IEC of neonates, where acidic luminal conditions persist subsequent to exit from the stomach, permits the uptake and transcytosis of maternal IgG into the circulation of the suckling offspring (Simister and Rees 1985; Cianga et al. 2003; Cianga et al. 1999; Jones and Waldmann 1972; Simister and Mostov 1989b; Israel et al. 1997). Indeed, FcRn-mediated transfer of allergen-specific IgG1 from allergen-exposed mothers to their nursing offspring has been shown to protect the neonates from the subsequent development of allergic responses

(Matson et al. 2010). Furthermore, it has been demonstrated that neonatal FcRn can transport not only maternal IgG across the intestinal epithelium but also IgG-containing immune complexes which are present in mother's milk. Indeed, the transfer of such IgG IC was found not only to drive the development of oral tolerance but also to protect offspring from later acquiring allergic diseases such as asthma (Verhasselt et al. 2008; Mosconi et al. 2010). It is therefore obvious that the actions of FcRn during neonatal life contribute significantly to both the immediate protection of the immunodeficient offspring as well as to the establishment of baseline immune responses which may persist throughout a significant portion of the progeny's life.

8.2.2 FcRn Maintains Systemic IgG Homeostasis

In light of the central role played by IgG in humoral immune responses such as opsonization, complement activation, and ADCC, it is not surprising that it would be advantageous for an organism to possess a substantial baseline pool of IgG from which to draw in times of need. However, continual synthesis of large quantities of new protein is a highly energy-intensive process, thereby generating selective pressure for the evolution of an IgG salvage process which protects existing IgG molecules from rapid degradation. The broad temporal and tissue distribution of FcRn reflects the dominant role it plays in this process. Indeed, it has been demonstrated that four times as much IgG can be rescued by FcRn-mediated recycling of circulating IgG than can be produced by the body (Anderson et al. 2006). Mice deficient in either FcRn or β2M are severely systemically hypogammaglobulinemic due to an increased rate of IgG catabolism and have IgG half-lives of approximately 1.4 days compared to 9 days for wild-type (WT) mice (Chaudhury et al. 2003; Roopenian et al. 2003; Kim et al. 2008). The consequence of such low serum IgG levels is highlighted by an equivalent human syndrome, familial hypercatabolic hypoproteinemia. This disorder has been documented in a pair of siblings in whom a mutation in the β2M gene leading to a nonfunctional protein was discovered (Waldmann and Terry 1990; Wani et al. 2006). The rate of IgG catabolism in these patients was increased by fivefold compared to that in healthy controls, and FcRn expression levels were decreased by greater than 80 %. While the patients did suffer from chemical diabetes and skeletal abnormality, it remains unclear whether this was causally related to the β2M deficiency.

Nevertheless, while FcRn's potency at protecting IgG from entering a catabolic pathway is beneficial in an energetic sense, it also acts as a double-edged sword in that FcRn-mediated protection does not discriminate between protective and pathogenic IgG. A wide variety of autoimmune diseases including arthritis, myasthenia gravis, epidermolysis bullosa, pemphigus, and, possibly, lupus can be attributed at least partially to the presence of pathogenic circulating autoantibodies which target self-antigens and lead to the disease-inducing destruction of otherwise healthy tissue (Akilesh et al. 2004; Liu et al. 2007; Csorba et al. 2010; Sesarman et al. 2008,

2010; Aoyama 2010; Li et al. 2005; Denny et al. 2006; Zhou et al. 2005b). FcRn's role in perpetuating these illnesses by enabling autoantibody persistence has been demonstrated experimentally. *Fcgrt-/-* mice are protected from disease development in the K/BxN serum transfer model of autoimmune arthritis unless extremely high doses of pathogenic IgG are transferred in (Akilesh et al. 2004). Likewise, high-dose administration of intravenous immunoglobulin (IVIg) or administration of IgG engineered for high affinity FcRn binding, both of which saturate available FcRn and thus disable its ability to protect pathogenic IgG, was able to protect WT mice from disease development (Akilesh et al. 2004; Petkova et al. 2006; Patel et al. 2011). Epidermolysis bullosa acquisita, a skin blistering disease attributed to the pathogenic effects of anti-collagen VII IgG, is also enabled by the ability of FcRn to protect these antibodies since *Fcgrt-/-* mice are protected from disease development (Csorba et al. 2010; Sesarman et al. 2008, 2010). Similar FcRn-driven pathology has been seen in pemphigus, another skin blistering disease associated with pathogenic anti-desmoglein IgG (Aoyama 2010; Li et al. 2005). Thus, whether FcRn contributes in a beneficial or harmful way to the development of IgG-mediated diseases depends largely on the specificity of the IgG being protected and whether it is predominantly pathogenic or protective in nature.

It is unlikely that a single tissue site contributes to FcRn-mediated protection of IgG. Initial studies using bone marrow-chimeric mice demonstrated an approximately 50 % reduction in serum IgG levels in both WT mice chimerized with *Fcgrt-/-* bone marrow and *Fcgrt-/-* mice chimerized with WT bone marrow, indicating that FcRn expression within the hematopoietic system likely accounts for approximately half of the total IgG protection within an organism (Qiao et al. 2008). The generation of a mouse model with a floxed FcRn gene and its cross with Tie2-Cre mice to generate an endothelial-specific knockdown of FcRn subsequently support the role of endothelial FcRn in contributing to the majority of the remaining half of the IgG recycling and protection function (Montoyo et al. 2009). In contrast, the epithelial component accounts for a near-negligible fraction of this protection (Yoshida et al. 2006). Following IgG internalization by an endothelial cell or hematopoietic cell, FcRn appears to first bind its ligand within the early endosomal compartments. This appears to protect IgG from being sorted into a degradation-promoting pathway and instead routes it towards a sorting endosome from which tubular extensions containing FcRn have been shown to protrude in the direction of the cell surface (Ober et al. 2004a). At the neutral pH at the cell surface, IgG would be expected to rapidly dissociate from FcRn and return to the circulation. While the ability of FcRn to protect monomeric IgG from catabolism is beyond dispute, it has been demonstrated that polymeric IgG in the form of immune complexes (IC) is routed quite differently by FcRn within hematopoietic cells and is actually targeted towards a degradative pathway (Qiao et al. 2008). Such differential treatment of its ligand implies that FcRn is capable of discriminating between monovalent and polyvalent ligands, a property which is critical for its ability to regulate diverse immune responses.

8.3 Coordination of Complex Immune Responses by FcRn

8.3.1 Epithelial FcRn and the Regulation of Mucosal Immune Responses in Adult Life

The ability of FcRn to drive bidirectional transport of IgG across polarized epithelial barriers makes it an important contributor not only to the steady-state distribution of IgG throughout an organism but also to the delivery of specific IgG from one tissue site to another. This is particularly critical in enabling the efficient priming of immune responses, which often occur at sites removed from the intended target of the response. FcRn's ability to transcytose not only IgG but IgG-complexed antigens is thus a decisive factor in the coordination of immune responses in highly compartmentalized areas. This is particularly true along the interfaces of the body which enter into direct contact with exogenously acquired substances and organisms such as the intestines. The mucosal surfaces of the body represent the largest such interface, and while IgA has classically been associated with such surfaces, there is an increasing appreciation for the role of mucosal IgG in driving both physiological and pathological immune responses within the mucosae.

8.3.1.1 Transcytosis of IgG-Complexed Antigen and Neutralization of Luminal Infections

Under steady-state conditions, the distribution of FcRn in intestinal epithelial cells (IEC) reflects its predominant physiological role in the organism. In neonatal rodents, where FcRn drives the uptake of maternal IgG, the receptor is predominantly located on the apical surface of the epithelial cells and the predominant direction of FcRn transport is apical → basolateral (Kuo et al. 2009). In contrast, in humans of all ages where FcRn is predominantly basolaterally localized, the predominant route of FcRn transport is basolateral → apical (Claypool et al. 2004). Thus, the considerable amount of IgG present in the intestinal lumen under steady-state conditions, which can reach a concentration of 800 μg/ml IgG in the rectal lumen, can largely be attributed to epithelial FcRn-mediated transport of IgG from the lamina propria into the lumen (Claypool et al. 2004; Kozlowski et al. 1997). Abundant evidence exists, however, that significant FcRn-mediated retrieval of IgG from the intestinal lumen also occurs and that, in many instances, this IgG is bound to luminal antigens (He et al. 2008; Jerdeva et al. 2010; Yoshida et al. 2004). As outlined in Fig. 8.2, successful transport of the intact IgG IC across IEC represents a significant source of antigenic delivery into the lamina propria and, from there, into the mesenteric lymph nodes (MLN) where a large proportion of mucosal immune responses are known to be primed (Yoshida et al. 2006). Thus, FcRn within mucosal epithelial cells is an integral component of a first-line mucosal immune monitoring system.

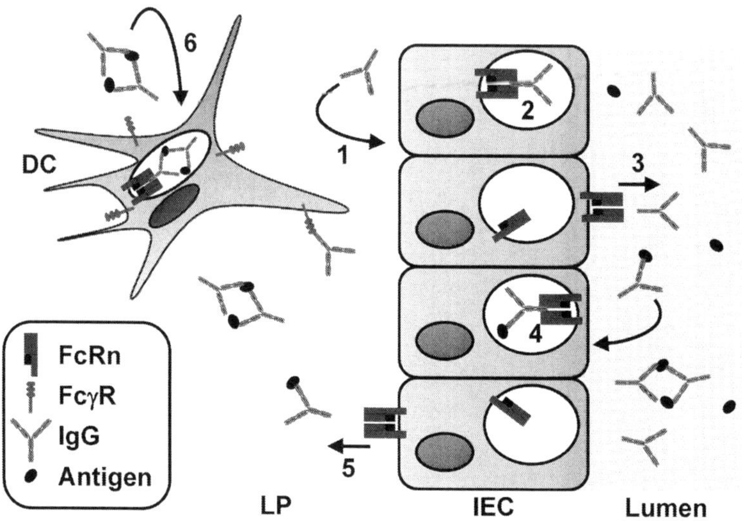

Fig. 8.2 *Sites of FcRn-driven immune coordination within the mucosal immune system*. FcRn is functionally expressed in two critical sites of the mucosal immune system: intestinal epithelial cells (*IEC*) and antigen-presenting cells such as dendritic cells (*DC*) and macrophages. FcRn-mediated coordination of mucosal immunity begins within the IEC where, subsequent to macropinocytic IgG uptake from the intestinal lamina propria (*LP*) (*1*), FcRn within the acidic endolysosomal system of IEC (*2*) directs the basolateral → apical transcytosis of IgG. At the relatively neutral pH of the apical surface, IgG is then released into the lumen (*3*), where it can bind to luminal antigen. Following reuptake, FcRn within IEC is then able to retrieve the IgG immune complex (*IgG IC*) and retrograde transport it from the apical to basal surface by transcytosis (*4*). Subsequent to its pH-dependent release in the LP (*5*), the IgG IC can be retrieved by FcγR on DC of the LP which migrate to the MLN. Here, FcRn actively routes the IgG IC into highly efficient antigen-processing pathways (*6*) which generate epitopes for loading onto MHC class I and class II. Thus, the actions of FcRn within multiple intestinal cell types enable the coordinated delivery of luminal antigens into the mucosal immune system and the priming of cognate cell-mediated immune responses

The important contribution of FcRn to coordinating mucosal immune responses was first demonstrated in the gastrointestinal tract. When fed ovalbumin (OVA), mice transgenic for hFcRn and hβ2M, in which functional hFcRn is expressed in adult life in the IEC as is found in humans, were able to activate OVA-specific $CD4^+$ T cells in their MLN only when also given anti-OVA IgG systemically (Yoshida et al. 2004). *Fcgrt-/-* mice were unable to prime such an immune response even when administered an equivalent dose of anti-OVA IgG. Furthermore, luminal antigen taken up as an IgG IC in an FcRn-dependent manner was observed in $CD11c^+$ MLN DC, confirming successful delivery of the antigen in cells capable of direct $CD4^+$ T cell activation. The physiological importance of this process in conferring protection against an epithelial-specific intestinal pathogen, *Citrobacter rodentium*, a murine infectious agent equivalent to enterohemorrhagic or enteropathogenic *Escherichia coli* in humans, was subsequently demonstrated (Yoshida et al. 2006). Mice transgenic for IEC-specific mFcRn (IFABP-mFcRn) expression and administered anti-*C. rodentium* IgG systemically were protected from infection upon challenge with OVA-expressing *C. rodentium*. Specifically, these mice experienced far less extreme body weight loss, decreased intestinal pathology, far lower levels of *C. rodentium* CFU/mg feces, and increased survival rates compared to *Fcgrt*$^{-/-}$ mice. Furthermore, OVA-specific CD4^{+} T cells adoptively transferred into the IFABP-mFcRn transgenic mice proliferated more significantly in the MLN and secreted greater quantities of IFNγ and IL-4 upon ex vivo restimulation than did CD4^{+} T cells from *Fcgrt*$^{-/-}$ mice. Recently, FcRn-mediated protection against a gastric epithelial-specific pathogen has also been demonstrated. The gastric epithelium of *Fcgrt*$^{-/-}$ mice chronically infected with *Helicobacter heilmanii* was found to have significant *H. heilmanii* invasion deep into gastric pits and developed significant quantities of gastric lymphoid follicles (Ben Suleiman et al. 2012). WT mice, in contrast, had only minor surface infection of the gastric epithelium by the pathogen. This correlated with the presence of elevated anti-*H. heilmanii* IgG in gastric secretions in WT compared to *Fcgrt*$^{-/-}$ mice despite equivalent serum levels of bacteria-specific IgG. Similarly, far greater colonization of the gastric epithelium was seen *Fcgrt*$^{-/-}$ mice 8 weeks after infection with *Helicobacter pylori*. Thus, these studies clearly demonstrate that the presence of FcRn within epithelial cells along mucosal boundaries can be protective against epithelial-specific bacterial invasion.

Very recently, the importance of epithelial FcRn in protecting mucosal surfaces has been extended to a different form of pathogen and to two additional mucosal sites. A transcytosis assay using a rat FcRn (rFcRn)-transfected epithelial cell line demonstrated that apically applied influenza virus was efficiently neutralized when an anti-influenza hemagglutinin (HA) antibody was present in the basolateral chamber of the assay but not when an irrelevant IgG was present (Bai et al. 2011). In the former condition, viral replication in the epithelial cells was significantly inhibited, reducing the yield by approximately 100-fold relative the latter condition after a 24 h incubation. Confocal microscopy of the rFcRn containing epithelial cells indicated that in the presence of the influenza neutralizing IgG, viral particles were prevented from entering into the cell nucleus, possibly due to increased delivery of the IgG-complexed viral particles to LAMP-2-positive compartments in which viral degradation could occur. The physiological importance of this process was evident in the fact that WT mice but not *Fcgrt*$^{-/-}$ mice treated prophylactically with the anti-HA IgG were protected upon viral challenge and experienced reduced mortality and decreased lung pathology. A related study demonstrated that FcRn mediates transport of IgG across cells of the female genital tract and that systemically administered IgG could be detected in the vaginal lumen of WT but not *Fcgrt*$^{-/-}$ mice (Li et al. 2011). Passive immunization of WT mice with IgG against herpes simplex virus-2 (HSV-2) was further shown to confer protection upon intravaginal viral challenge, decreasing mortality compared to similarly treated *Fcgrt*$^{-/-}$ mice or to un-immunized WT mice. Collectively, these data demonstrate the importance of efficient FcRn function within mucosal epithelial cells in protecting against pathogen-induced epithelial cell damage capable of compromising the defensive function of the barrier.

8.3.1.2 FcRn-Targeted Mucosal Antivirus Vaccines

The potential to exploit FcRn's transcytotic function at mucosal barriers for therapeutic purposes is enormous, as outlined in Fig. 8.3. The first step in this direction was the demonstration that airway epithelial cells expressing FcRn are capable of retrieving an intranasally administered fusion protein between erythropoietin and the Fc region of IgG (Epo–Fc) and delivering it into systemic circulation (Spiekermann et al. 2002). If mutations were introduced into the Fc region of Epo–Fc which disabled its ability to bind FcRn, no fusion protein was detected in the circulation of WT mice. This finding was then extended to both non-human primates and to humans themselves, all of which are known to express FcRn in epithelial cells of the central and upper airways (Bitonti et al. 2004; Dumont et al. 2005). Approximately 70 % of the administered fusion protein was delivered safely and effectively to the central region of the lungs of the human volunteers and the fusion protein was confirmed to retain bioactivity by the demonstration of increased levels of circulating reticulocytes following airway Epo–Fc administration (Dumont et al. 2005). Importantly, the circulating half-life of the airway-administered Epo–Fc fusion protein was greatly extended when compared to erythropoietin delivered via a standard route, persisting for 2–10-fold longer in circulation due to FcRn-mediated protection from catabolism of the Fc-fused molecule (Bitonti et al. 2004). A direct demonstration of the potential for FcRn-mediated delivery to drive therapeutic protection has recently been reported in two related papers. In each of these, fusion proteins were created between a viral protein and the Fc region of IgG, and the chimeric molecules were administered intranasally, in combination with an adjuvant, to mice which were subsequently subjected to vaginal challenge by either HSV-2 or HIV Gag-expressing vaccinia virus (Ye et al. 2011; Lu et al. 2011). In both instances, immunized WT mice were found to develop strong antiviral responses capable of protecting the mice from virus-induced pathology and/or mortality whereas *Fcgrt*$^{-/-}$ mice failed to develop protective immunity. Efficient production of antiviral IgG in addition to the activation of virus-specific CD4$^+$ and CD8$^+$ T cells was observed in vaccinated WT mice. Consistent with previous reports that immune activation at one mucosal site can confer protection to distant mucosal sites, virus-specific IgG was found in vaginal washes of the immunized WT mice, and activated T and B cells were localized to vaginal tissues, indicating the potency of FcRn's ability to coordinate multiple branches of the immune system and to function in both the induction and effector phases of vaccine-induced protection (Mestecky 1987).

8.3.1.3 Diseases Mediated by Immune Complex Deposition

Even in the absence of prior vaccination, the localized transcytotic actions of FcRn can exert potent protective effects against IgG IC-driven disease. In the kidney, FcRn within podocytes clears IgG from the glomerular basement membrane and prevents the deposition of IgG at this filtration barrier (Akilesh et al. 2008;

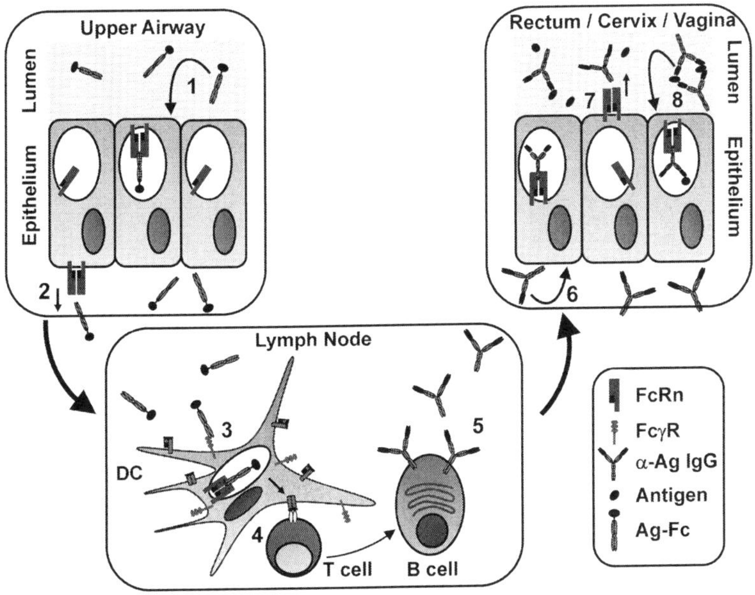

Fig. 8.3 *Mechanism of FcRn-targeted mucosal vaccination strategy*. The generation of effective mucosal immune responses is particularly important for protection against diseases transmitted at mucosal sites, such as HIV. As such, vaccines aimed at instilling virus-neutralizing protection are likely to be far more effective at preventing initial infection when they stimulate the generation of a localized immune response at the site of initial infection. Recognition of the important role played by FcRn in coordinating immune responses across diverse mucosal sites has stimulated the development of novel vaccine strategies which exploit the functions of FcRn within both epithelial and hematopoietic cells to drive virus-neutralizing immunity. Intranasal administration of chimeric Fc-fused therapeutic antigens leads to the delivery of the antigen to the upper airways. FcRn within airway epithelial cells binds to the Fc portion of the chimeric molecules in the presence of adjuvant and mediates its apical → basolateral transcytosis (*1*). At the basolateral surface of the epithelial cells, the therapeutic molecule is released in the neutral pH environment and taken up by local antigen-presenting cells and/or enters the systemic circulation (*2*). The Fc domain of the chimeric molecule enables efficient FcγR-mediated uptake (*3*) and delivery into the FcRn-driven antigen presentation pathway known to generate $CD4^{+}$ and $CD8^{+}$ T cell activation which is promoted by the presence of adjuvants (*4*). Signals from helper T cells activated in a FcRn-dependent manner facilitate the development of B cells into plasma cells which secrete IgG specific for the antigenic component of the therapeutic molecule (*5*). FcRn-mediated basolateral → apical transcytosis (*6*) across epithelial cells of the cervical, vaginal, or rectal wall enables the subsequent delivery of antigen-neutralizing IgG directly into the primary sites of infection (*7*). Retrieval of luminal IgG-complexed antigen and FcRn-driven retrograde transcytosis of the IgG IC into the local mucosal immune system then likely serves as a positive feedback loop to reinforce the neutralizing response without the need for additional doses of the vaccine (*8*). Exploitation of the immune-regulating functions of FcRn thus represents an attractive and adaptable therapeutic target for the treatment of many mucosae-centered diseases

Sarav et al. 2009). As *Fcgrt*$^{-/-}$ mice age, they accumulate glomerular IgG deposits in their kidneys resembling the phenotype seen in many human glomerular diseases (Akilesh et al. 2008). Similarly, FcRn within cells of the blood–brain barrier contributes to the clearance of IgG IC containing β-amyloid protein from the brain and therefore prevents the buildup of β-amyloid plaques characteristic of Alzheimer's disease (Deane et al. 2005, 2009; Schlachetzki et al. 2002). This ability of FcRn to transport both monomeric IgG and polymeric IgG IC confers upon it a very unique ability not only to prevent deposition of harmful complexes and shuttle IgG-opsonized pathogens across tissue barriers but, by virtue of its expression within hematopoietic cells, also to act as a central driver of immune responses against IgG-complexed antigens.

8.3.2 FcRn Within Antigen-Presenting Cells and the Intracellular Processing of IgG-Complexed Antigen

The documentation of FcRn expression within cells of the hematopoietic system opened the door to the possibility that FcRn might regulate trafficking of IgG IC within cells known to be potent immune response orchestrators (Zhu et al. 2001). In particular, APC such as DC, macrophages, and B cells are all known to express FcRn (Zhu et al. 2001; Mi et al. 2008). While the functions of FcRn within epithelial cells have long been the focus of FcRn-centered research, appreciation for the ability of FcRn to contribute directly to immune regulation is growing.

8.3.2.1 FcRn Is Required by APC for the Presentation of Epitopes from IgG IC on MHC Class II

When antigen was first shown to be delivered to MLN DC following FcRn-mediated IgG IC uptake from the intestinal lumen, it was unknown whether this antigen remained complexed with IgG and, if so, whether FcRn within these DC contributed to activation of local T cells in response to the IgG IC (Yoshida et al. 2004). Subsequent experiments demonstrated conclusively that FcRn within DC is indeed a decisive factor in routing IgG-complexed antigen into a degradation pathway enabling the generation and loading of epitopes onto MHC class II (Qiao et al. 2008). DC from WT mice pulsed with IgG IC were found to efficiently activate $CD4^+$ T cell proliferation and cytokine secretion whereas IgG IC-treated DC from *Fcgrt*$^{-/-}$ mice elicited minimal activation. Similar results were seen when WT DC were pulsed either with IgG IC or with IHH-IgG IC, in which three point mutations (I253A, H310A, and H435A) had been introduced to the IgG–Fc region, thereby disabling its ability to bind FcRn but not FcgR. WT DC pulsed with IgG IC induced strong $CD4^+$ T cell proliferation, whereas WT DC pulsed with IHH-IgG IC were activated only at tenfold greater antigen concentrations. IgG IC, but not IHH-IgG IC, within WT DC were found to be directed into LAMP-1-positive compartments, consistent with the immune complexes being directed by FcRn into a degradative antigen-processing pathway. Indeed, while FcRn was found to prolong the persistence of monomeric IgG, polymeric IgG IC was found to be rapidly cleared from circulation in WT mice at a rate similar to that seen for IgG monomers in *Fcgrt*$^{-/-}$ mice. Reciprocal bone marrow chimeras confirmed that the rapid degradation of IgG IC was conferred by FcRn function within the hematopoietic and not the parenchymal cell compartment, consistent with FcRn facilitating the degradation of IgG IC within APC. Indirect evidence hinted at the physiological importance of FcRn-mediated antigen presentation since protection of WT mice from *C. rodentium* infection, previously demonstrated to be dependent on FcRn expression, was also found to depend upon the activation of antigen-specific IFNγ-producing CD4^{+} T cells (Shiomi et al. 2010).

The ability of FcRn to promote IgG IC degradation resulting in antigen presentation has also been demonstrated in macrophages, where acidification of the phagosomes was found to be a critical factor, consistent with the acidic requirements for FcRn–IgG interaction (Liu et al. 2011). While the details of how FcRn routes IgG IC towards eventual degradation, and indeed how it is even able to sense the valency of its ligand, remain unknown, an intriguing possibility has been raised by the demonstration that FcRn interacts directly with the invariant chain (Ii) which is commonly associated with immature MHC class II molecules (Ye et al. 2008). That the distribution of FcRn to early and late endosomes was dependent upon Ii expression indicates that subcellular movements of FcRn may be coordinated with those of other molecules involved in antigen processing and presentation, thereby promoting its direct participation in these processes.

8.3.2.2 FcRn Drives the Cross-Presentation of Antigen from IgG IC on MHC Class I in Dendritic Cells

FcRn's ability to facilitate MHC class II-mediated antigen presentation suggested that it might play a broader role in the intracellular coordination of IgG IC trafficking within DC and, thus, that it might also enable the cross-presentation of IgG-complexed antigens on MHC class I molecules (Fig. 8.4). While the contribution of FcRn to IgG IC cross-presentation by CD8^{+}CD11b^{-} DC, which are the most potent cross-presenters of soluble antigen, was found to be negligible, FcRn was found to potently drive the cross-presentation of IgG IC in CD8^{-}CD11b^{+} DC subset (Baker et al. 2011). In contrast to CD8^{+}CD11b^{-} DC, in which phagosomal pH is buffered to near-neutral levels subsequent to antigen uptake, CD8^{-}CD11b^{+} DC rapidly acidify their phagosomes to a pH which is permissive to the IgG–FcRn interaction (Baker et al. 2011; Russell 2007; Savina et al. 2006). Importantly, FcγR was found to be critical for FcRn-mediated cross-presentation, supporting the conclusion that the functions served by FcRn within CD8^{-}CD11b^{+} DC are restricted to IgG IC intracellular trafficking rather than internalization. Thus, following FcγR-mediated internalization, FcRn binds its ligand in pH permissive vesicles and actively routes its ligand into processing pathways which enable particularly efficient antigen

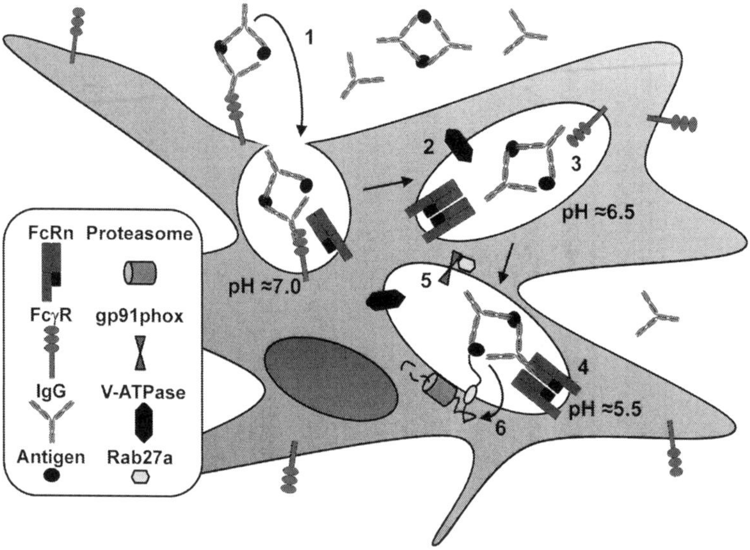

Fig. 8.4 *Mechanism of FcRn-driven cross-presentation of IgG-complexed antigen within DC.* Lifelong FcRn expression within antigen-presenting cells such as DC enables the coordination of humoral IgG and cell-mediated CD8+ T cell responses by FcRn. IgG-complexed antigen is initially captured and internalized in an FcγR-dependent manner in DC (*1*). As the pH of the IgG IC containing endosomal/phagosomal compartment begins to acidify through the recruitment of V-ATPase components (*2*), the IgG IC is released from FcγR which only bind at neutral pH (*3*). Continued acidification of the endosomal/phagosomal compartment generates a permissive pH for FcRn–IgG binding (*4*). As yet undefined signals subsequently trigger recruitment of the machinery known to be critical for the cross-presentation of soluble antigens, such as gp91phox and Rab27a, to intracellular compartments containing IgG IC-cross-linked FcRn (*5*). The IgG-complexed antigen is then shuttled into the proteasome dependent cytosolic cross-presentation pathway (*6*) which leads to the efficient generation of epitopes for loading onto MHC class I. Intracellular FcRn thus controls the delivery of IgG-opsonized antigens into a processing pathway generating cytotoxic CD8+ T cell-mediated immunity

processing. Specifically, IgG IC were found to enter into the cytosolic cross-presentation pathway only when able to bind FcRn within acidifying phagosomes to which the molecular machinery known to drive the cross-presentation of soluble antigen was recruited. FcRn-driven cross-presentation was demonstrated to drive potent CD8+ T cell activation and proliferation both in vitro and in vivo, thereby demonstrating the physiological importance of this process and implicating FcRn in the coordination of multiple aspects of cell-mediated immunity.

8.3.2.3 FcRn-Dependent Antigen-Presentation and Cross-Presentation Exacerbate Inflammation in Response to IgG IC

Perhaps nowhere in the body is the balance between homeostasis and inflammation as dependent upon the actions of FcRn as the mucosae, where foreign antigen is abundant in very close proximity to the host and IgG is present in large quantities even in the absence of inflammation. The ability of FcRn to bridge the gap between humoral and cell-mediated immunity during the course of disease has been elegantly shown by the demonstration that IgG specific for luminal bacterial antigens greatly increases the severity of intestinal inflammation but only in the presence of FcRn (Kobayashi et al. 2009). WT mice possessing large quantities of circulating anti-flagellin IgG due to prior immunization with this antigen developed severe colitis subsequent to breakdown of the intestinal epithelial cell barrier. In contrast, unimmunized WT mice or immunized *Fcgrt*$^{-/-}$ mice were protected from intestinal inflammation. Reciprocal bone marrow chimeras demonstrated that colitis was driven by the actions of FcRn within hematopoietic cells and thus likely depended upon the presentation and cross-presentation of bacterial antigens to $CD4^+$ and $CD8^+$ T cells responsible for driving the inflammation. These data place into context the significance of antibacterial IgG commonly found in the serum of patients with inflammatory bowel disease (IBD) and infectious enterocolitis and imply that, due to the actions of FcRn, such pathogenic IgG may be an important factor contributing to ongoing intestinal inflammation (Bossuyt 2006; Adams et al. 2008; Furrie et al. 2004; Lodes et al. 2004).

8.4 FcRn as a Therapeutic Target

In light of the large number of diseases associated with FcRn and our detailed understanding of the nature of its ligand, one can easily understand why FcRn is an attractive therapeutic target (Kuo et al. 2010; Roopenian et al. 2010; Roopenian and Sun 2010; Ward et al. 2009). When delivering IgG-based therapeutics for the treatment of disease, increasing the IgG–FcRn interaction by modulation of the Fc region of the drug may be beneficial in multiple respects. The first of these is that increased FcRn binding to the Fc domain of an IgG-based pharmacotherapeutic will prolong its half-life and theoretically enable its transport across a polarized epithelial barrier. On the other hand, blocking IgG–Fc–FcRn interactions will increase endogenous IgG catabolism as might be desired in an autoimmune disease. In a similar manner, it can be anticipated that increasing the binding of an IgG Fc region to FcRn will promote entry of the IgG Fc-containing ligand into antigen-processing pathways within DC if the IgG is present within an immune complex.

8.4.1 Therapies Which Block the Binding of Endogenous IgG to FcRn

In diseases driven by the presence of excessive autoantibodies, such as rheumatoid arthritis, myasthenia gravis, epidermolysis bullosa, pemphigus, and, possibly, lupus, the aim of therapy is to prevent IgG-mediated destruction of the affected tissues (Akilesh et al. 2004; Liu et al. 2007; Csorba et al. 2010; Sesarman et al. 2008, 2010;

Aoyama 2010; Li et al. 2005; Denny et al. 2006; Zhou et al. 2005b). One means to achieve this is to administer a therapy that will saturate available FcRn sites, thereby blocking the ability of FcRn to protect the autoantibodies from degradation. Traditionally, this has been done in a relatively nonspecific manner by the administration of high-dose IVIg with the aim of flooding the patients' FcRn with innocuous IgG and thus preventing rescue of the destructive autoantibodies (Akilesh et al. 2004; Li et al. 2005; Crow et al. 2011; Hansen and Balthasar 2002). Recently, however, FcRn is being targeted more specifically for the treatment of autoimmune diseases through the use of modified IgG having altered binding properties. The first of these to be tested were dubbed "AbDegs" for their ability to promote the degradation of endogenous IgG (Vaccaro et al. 2005). AbDegs consist of IgG engineered to possess a series of mutations in sites known to influence IgG–FcRn interactions (M252Y, S254T, T256E, H433K, and N434F) and which confer upon the AbDeg not only a 23-fold increase in FcRn binding affinity at pH 6.0 but which enable continued binding of the engineered IgG at pH 7.0. Thus, systemic FcRn is quickly saturated following administration of a small dose of AbDegs, and prolonged reduction in serum levels can be observed in mice treated with this therapeutic. Treatment of arthritic mice with a single dose of AbDegs has been shown to induce similar therapeutic benefits as a 25- to 50-fold higher dose of IVIg, strongly supporting it as a candidate for human therapeutic development (Patel et al. 2011).

The introduction of different single or combined mutations in the IgG–Fc region of potential therapeutics has been reported, most of which increase binding affinity at pH 6.0 but not in the neutral pH range (Yeung et al. 2009; Hinton et al. 2004, 2006). While these engineered IgG molecules exhibit increased in vivo half-life, their therapeutic potential in humans remains to be directly assessed but is currently under development. These include a humanized anti-CD4 antibody with an FcRn-binding enhancing mutation and anti-VEGF molecules containing similar mutations and which lead to greater reductions in tumor growth compared to the parent molecule (Peters et al. 2010; Dumont et al. 2012; Suzuki et al. 2010; Zheng et al. 2011; Yeung et al. 2010; Zalevsky et al. 2010). Similarly, the increased efficacy of several different Fc-conjugated pharmacotherapies, which are capable of prolonged systemic persistence thanks to FcRn-mediated rescue from catabolism, has already been demonstrated. These include Fc-fusion proteins currently in clinical use such as etanercept as well as more experimental compounds such as Fc-conjugated Factors VIII and IX for the treatment of hemophilia (Kuo et al. 2010; Peters et al. 2010; Dumont et al. 2012; Suzuki et al. 2010; Zheng et al. 2011; Yeung et al. 2010; Zalevsky et al. 2010; Popivanova et al. 2008).

The ability of FcRn to drive immune system activation is highly desirable for the induction of cell-mediated immunity against an invading pathogen, but, as has been demonstrated in the context of IBD, the actions of FcRn within DC can lead to persistent and pathological inflammation in the presence of large quantities of IgG IC where the IgG itself is not autoreactive (Baker et al. 2011; Kobayashi et al. 2009). Development of drugs capable of inhibiting the FcRn–IgG interaction within APC is thus also a valuable pursuit, and a number of small molecular inhibitors of FcRn have already been created for this purpose. Administration of these inhibitors, such as cyclic peptides which disrupt FcRn binding to IgG, is capable of decreasing serum IgG levels by almost 80 % in cynomolgus monkeys (Mezo et al. 2008a). Further refinements and improvements to these inhibitors are currently underway, including PEGylation, which improve peptide half-life, but experimental evidence of their efficacy in treating IgG immune complex-mediated diseases has yet to be assessed (Mezo et al. 2008b, 2011; Low and Mezo 2009). Certainly, these approaches have a significant likelihood of success given the demonstration that anti-FcRn antibodies ameliorate myasthenia gravis in a rodent model (Liu et al. 2007).

8.4.2 *Therapies Which Direct an Antigen for FcRn-Mediated Degradation*

In contrast to diseases ameliorated by prolonging the activity of a therapeutic molecule, some illnesses can be benefitted by administering an FcRn-targeted therapy designed to be shuttled into the FcRn-mediated degradation pathway in DC and which promotes the development of a coordinated cell-mediated immune response against the FcRn-targeted antigen. Indeed, transgenic animal models have already demonstrated that increased FcRn expression boosts the yields of antigen-specific IgG generated in response to vaccination (Catunda Lemos et al. 2012; Cervenak et al. 2010). In humans, the alternate strategy of modifying an antigen to increase its ability to bind endogenous FcRn is of particular interest in the development of vaccines against defined antigens of bacterial or viral origin, as has recently been demonstrated against HSV-2 and HIV-1 Gag-expressing vaccinia virus (Bai et al. 2011; Lu et al. 2011). This strategy is likely to be aided by the creation of multimeric vaccinations capable of evading the FcRn-mediated salvage pathway in other cell types and instead being rapidly targeted for degradation by DC. In addition, the introduction of mutations enabling FcRn binding in the vaccine Fc regions will promote stronger FcRn binding and a greater likelihood of diversion of the vaccine molecule towards a degradative pathway known to activate $CD4^+$ and $CD8^+$ T cells (Baker et al. 2011; Qiao et al. 2008; Zalevsky et al. 2010). In combination with the ability to mucosally administer such molecules for noninvasive FcRn-mediated uptake, these developing strategies have enormous potential for the treatment of a wide array of diseases, particularly those occurring at mucosal sites.

8.4.3 *Caveats to FcRn-Targeted Therapy*

The obvious attractiveness of FcRn-targeted therapies should not overshadow the fact that FcRn itself does not operate independently. This is particularly true in the case of that FcRn-driven antigen presentation, which relies upon initial IgG IC internalization by surface FcγR (Baker et al. 2011). This is an important consideration brought to light by a series of recent genetic studies which have revealed that, while clear functional evidence exists for the association of FcRn with the course of numerous diseases, little evidence was found for FcRn polymorphisms as a genetic disease risk factor. A single study having examined polymorphisms in the *Fcgrt* gene itself identified two non-synonymous variations (R210Q and S297T) for which the Japanese population was heterozygous but did not observe any impact of these variations on FcRn localization or recycling efficiency (Ishii-Watabe et al. 2010). Several studies have also examined the impact of variable number tandem repeats (VNTR) located within the FcRn promoter on rates of maternal–fetal IgG transfer, lupus, and common variable immunodeficiency (CVID) patients treated with IVIg (Freiberger et al. 2010a; Zhou et al. 2009; Freiberger et al. 2010b). Of these, the only impact identified for the FcRn promoter polymorphisms was an increase in the extent of lung fibrosis and bronchiectasis seen in CVID patients having lower FcRn expression levels (Freiberger et al. 2010b).

The advantages afforded by ongoing development in FcRn-targeted therapies bring to the forefront the important point that reliable testing models are needed in order to properly validate the likely usefulness of any one strategy in clinical use. The value of standard murine models for this process is complicated by the pleiotropic effects of FcRn, the differences in cross-species human and murine FcRn–IgG binding, and the slightly different tissue expression pattern of epithelial FcRn in humans and mice (Leach et al. 1996; Ober et al. 2001; Zhou et al. 2005a; Andersen et al. 2010; Tiwari and Junghans 2005). A cautionary note is to be found in a preliminary pair of studies demonstrating that while IgG harboring mutations enabling FcRn binding at both acidic and neutral pH were more effectively transported by hFcRn across epithelial boundaries than were non-engineered IgG in in vitro assays, the increased rate of in vivo catabolism afforded by the strong mFcRn binding negated any potentially therapeutic effects for this molecule (Mi et al. 2008; Vaccaro et al. 2006). The authors observed that this discrepant finding resulted from the fact that affinity increases across the pH spectrum had a much greater effect for murine FcRn binding to IgG than for human FcRn binding. The development of mouse models expressing both human FcRn and human β2M under their endogenous promoters and on an *mFcgrt*$^{-/-}$ background offers an important and alternative tool for the validation of newly developed therapeutics (Roopenian et al. 2003; Yoshida et al. 2004; Roopenian et al. 2010). However, this hFcRn/hβ2M transgenic model comes with an important limitation of its own in that murine IgG is not efficiently bound by the human receptor and so the mice are mildly hypogammaglobulinemic. While this may be an advantage for examining the kinetics of exogenously administered IgG subsequent to drug treatment, it is a factor that must be taken into consideration when studying immune complex-mediated diseases. Although the current limitations in model systems with which to study new FcRn-targeted therapies is an undeniable hindrance, it is by no means an insurmountable impediment and should therefore not discourage the development of such potentially valuable therapeutic molecules.

8.5 Concluding Remarks

Optimal immunological responses involve not only many branches of the immune system but also contributing signals derived from many non-hematopoietic parenchymal cell types such as the epithelium and endothelium. The distribution of FcRn across all of these cell types may be seen as an important indicator of its central role in coordinating immune responses. First described as a transporter of monomeric IgG across epithelial boundaries, FcRn is also critical for the maintenance of systemic IgG levels, the delivery of specific antigen into cells lacking antigen-binding receptors, and the ensuing initiation of a potent cell-mediated immune response. FcRn's capability for intracellular transport and trafficking of intact IgG and IgG IC across epithelial boundaries delivers the first line of immunological defense to fetuses and neonates. Subsequently, this lifelong transcytotic function keeps the host in constant communication with antigens lining the mucosal surfaces while its recycling function regulates systemic levels of IgG. FcRn within APC directs responses to potential threats signaled by IgG IC-mediated antigen delivery and enables the routing of IgG IC into processing pathways generating $CD4^+$ and $CD8^+$ T cell priming. From humble beginnings as a simple transport molecule, FcRn is now emerging as a central player in the regulation of many branches of mucosal and systemic immunity. A greater understanding of the complexities of FcRn-driven immune regulation will thus place us in a better position from which to both exploit its beneficial functions and inhibit its harmful ones.

Acknowledgements This work was supported by the Canadian Institutes of Health Research, the Deutsche Forschungsgemeinschaft (RA 2040/1–1), the High Pointe Foundation, the National Institutes of Health Research (DK53056, DK053162, DK088199, and DK044319), and the Harvard Digestive Diseases Center (NIH P30DK034854).

References

Adams RJ, Heazlewood SP, Gilshenan KS, O'Brien M, McGuckin MA, Florin TH (2008) IgG antibodies against common gut bacteria are more diagnostic for Crohn's disease than IgG against mannan or flagellin. Am J Gastroenterol 103(2):386–396

Ahouse JJ, Hagerman CL, Mittal P, Gilbert DJ, Copeland NG, Jenkins NA et al (1993) Mouse MHC class I-like Fc receptor encoded outside the MHC. J Immunol 151(11):6076–6088

Akilesh S, Petkova S, Sproule TJ, Shaffer DJ, Christianson GJ, Roopenian D (2004) The MHC class I-like Fc receptor promotes humorally mediated autoimmune disease. J Clin Invest 113(9):1328–1333

Akilesh S, Huber TB, Wu H, Wang G, Hartleben B, Kopp JB et al (2008) Podocytes use FcRn to clear IgG from the glomerular basement membrane. Proc Natl Acad Sci USA 105(3):967–972

Andersen JT, Daba MB, Berntzen G, Michaelsen TE, Sandlie I (2010) Cross-species binding analyses of mouse and human neonatal Fc receptor (FcRn) show dramatic differences in immunoglobulin G (IgG) and albumin binding. J Biol Chem 285(7):4826–4836

Anderson CL, Chaudhury C, Kim J, Bronson CL, Wani MA, Mohanty S (2006) Perspective– FcRn transports albumin: relevance to immunology and medicine. Trends Immunol 27(7):343–348
Antohe F, Radulescu L, Gafencu A, Ghetie V, Simionescu M (2001) Expression of functionally active FcRn and the differentiated bidirectional transport of IgG in human placental endothelial cells. Hum Immunol 62(2):93–105
Aoyama Y (2010) What's new in i.v. immunoglobulin therapy and pemphigus: high-dose i.v. immunoglobulin therapy and its mode of action for treatment of pemphigus. J Dermatol 37(3):239–245
Bai Y, Ye L, Tesar DB, Song H, Zhao D, Bjorkman PJ et al (2011) Intracellular neutralization of viral infection in polarized epithelial cells mediated by neonatal Fc receptor (FcRn)-mediated IgG transport. Proc Natl Acad Sci USA 108(45):18406–18411
Baker K, Qiao S-W, Kuo TT, Aveson VG, Platzer B, Andersen J-T et al (2011) Neonatal Fc receptor for IgG (FcRn) regulates cross-presentation of IgG immune complexes by CD8-CD11b+dendritic cells. Proc Natl Acad Sci USA 108(24):9927–9932
Ben Suleiman Y, Yoshida M, Nishiumi S, Tanaka H, Mimura T, Nobutani K et al (2012) Neonatal Fc receptor for IgG (FcRn) expressed in the gastric epithelium regulates bacterial infection in mice. Mucosal Immunol 5(1):87–98
Bitonti AJ, Dumont JA, Low SC, Peters RT, Kropp KE, Palombella VJ et al (2004) Pulmonary delivery of an erythropoietin Fc fusion protein in non-human primates through an immunoglobulin transport pathway. Proc Natl Acad Sci USA 101(26):9763–9768
Blumberg RS, Koss T, Story CM, Barisani D, Polischuk J, Lipin A et al (1995) A major histocompatibility complex class I-related Fc receptor for IgG on rat hepatocytes. J Clin Invest 95(5):2397–2402
Borvak J, Richardson J, Medesan C, Antohe F, Radu C, Simionescu M et al (1998) Functional expression of the MHC class I-related receptor, FcRn, in endothelial cells of mice. Int Immunol 10(9):1289–1298
Bossuyt X (2006) Serologic markers in inflammatory bowel disease. Clin Chem 52(2):171–181
Brambell FWR (1966) The transmission of immunity from mother to young and the catabolism of immunoglobulins. Lancet 288(7473):1087–1093
Brambell FWR, Halliday R, Morris IG (1958) Interference by human and bovine serum and serum protein fractions with the absorption of antibodies by suckling rats and mice. Proc R Soc Lond B Biol Sci 149(934):1–11
Burmeister WP, Gastinel LN, Simister NE, Blum ML, Bjorkman PJ (1994a) Crystal structure at 2.2 A resolution of the MHC-related neonatal Fc receptor. Nature 372(6504):336–343
Burmeister WP, Huber AH, Bjorkman PJ (1994b) Crystal structure of the complex of rat neonatal Fc receptor with Fc. Nature 372(6504):379–383
Catunda Lemos AP, Cervenak J, Bender B, Hoffmann OI, Baranyi M, Kerekes A et al (2012) Characterization of the rabbit neonatal Fc receptor (FcRn) and analyzing the immunophenotype of the transgenic rabbits that overexpresses FcRn. PLoS One 7(1):e28869
Cervenak J, Bender B, Schneider Z, Magna M, Carstea BV, Liliom K et al (2010) Neonatal FcR overexpression boosts humoral immune response in transgenic mice. J Immunol 186(2):959–968
Chaudhury C, Mehnaz S, Robinson JM, Hayton WL, Pearl DK, Roopenian DC et al (2003) The major histocompatibility complex-related Fc receptor for IgG (FcRn) binds albumin and prolongs its lifespan. J Exp Med 197(3):315–322
Chaudhury C, Brooks CL, Carter DC, Robinson JM, Anderson CL (2006) Albumin binding to FcRn: distinct from the FcRn-IgG interaction. Biochemistry 45(15):4983–4990
Cianga P, Cianga C, Cozma L, Ward ES, Carasevici E (2003) The MHC class I related Fc receptor, FcRn, is expressed in the epithelial cells of the human mammary gland. Hum Immunol 64(12):1152–1159
Claypool SM, Dickinson BL, Yoshida M, Lencer WI, Blumberg RS (2002) Functional reconstitution of human FcRn in Madin-Darby canine kidney cells requires co-expressed human beta 2-microglobulin. J Biol Chem 277(31):28038–28050

Claypool SM, Dickinson BL, Wagner JS, Johansen FE, Venu N, Borawski JA et al (2004) Bidirectional transepithelial IgG transport by a strongly polarized basolateral membrane Fc gamma-receptor. Mol Biol Cell 15(4):1746–1759

Conroy ME, Shi HN, Walker WA (2009) The long-term health effects of neonatal microbial flora. Curr Opin Allergy Clin Immunol 9(3):197–201

Crow AR, Suppa SJ, Chen X, Mott PJ, Lazarus AH (2011) The neonatal Fc receptor (FcRn) is not required for IVIg or anti-CD44 monoclonal antibody-mediated amelioration of murine immune thrombocytopenia. Blood 118(24):6403–6406

Csorba K, Sesarman A, Oswald E, Feldrihan V, Fritsch A, Hashimoto T et al (2010) Cross-reactivity of autoantibodies from patients with epidermolysis bullosa acquisita with murine collagen VII. Cell Mol Life Sci 67(8):1343–1351

Deane R, Sagare A, Hamm K, Parisi M, LaRue B, Guo H et al (2005) IgG-assisted age-dependent clearance of Alzheimer's amyloid beta peptide by the blood–brain barrier neonatal Fc receptor. J Neurosci 25(50):11495–11503

Deane R, Bell RD, Sagare A, Zlokovic BV (2009) Clearance of amyloid-beta peptide across the blood–brain barrier: implication for therapies in Alzheimer's disease. CNS Neurol Disord Drug Targets 8(1):16–30

Denny MF, Chandaroy P, Killen PD, Caricchio R, Lewis EE, Richardson BC et al (2006) Accelerated macrophage apoptosis induces autoantibody formation and organ damage in systemic lupus erythematosus. J Immunol 176(4):2095–2104

Dickinson BL, Badizadegan K, Wu Z, Ahouse JC, Zhu X, Simister NE et al (1999) Bidirectional FcRn-dependent IgG transport in a polarized human intestinal epithelial cell line. J Clin Invest 104(7):903–911

Dickinson BL, Claypool SM, D'Angelo JA, Aiken ML, Venu N, Yen EH et al (2008) Ca2+–dependent calmodulin binding to FcRn affects immunoglobulin G transport in the transcytotic pathway. Mol Biol Cell 19(1):414–423

Dumont JA, Bitonti AJ, Clark D, Evans S, Pickford M, Newman SP (2005) Delivery of an erythropoietin-Fc fusion protein by inhalation in humans through an immunoglobulin transport pathway. J Aerosol Med 18(3):294–303

Dumont JA, Liu T, Low SC, Zhang X, Kamphaus G, Sakorafas P et al (2012) Prolonged activity of a recombinant factor VIII-Fc fusion protein in hemophilia A mice and dogs. Blood 119(13):3024–3030

Ellinger I, Schwab M, Stefanescu A, Hunziker W, Fuchs R (1999) IgG transport across trophoblast-derived BeWo cells: a model system to study IgG transport in the placenta. Eur J Immunol 29(3):733–744

Firan M, Bawdon R, Radu C, Ober RJ, Eaken D, Antohe F et al (2001) The MHC class I-related receptor, FcRn, plays an essential role in the maternofetal transfer of gamma-globulin in humans. Int Immunol 13(8):993–1002

Freiberger T, Ravcuková B, Grodecká L, Kurecová B, Jarkovský J, Bartonková D et al (2010a) No association of FCRN promoter VNTR polymorphism with the rate of maternal-fetal IgG transfer. J Reprod Immunol 85(2):193–197

Freiberger T, Grodecka L, Ravcukova B, Kurecova B, Postranecka V, Vlcek J et al (2010b) Association of FcRn expression with lung abnormalities and IVIG catabolism in patients with common variable immunodeficiency. Clin Immunol 136(3):419–425

Furrie E, Macfarlane S, Cummings JH, Macfarlane GT (2004) Systemic antibodies towards mucosal bacteria in ulcerative colitis and Crohn's disease differentially activate the innate immune response. Gut 53(1):91–98

Gafencu A, Heltianu C, Burlacu A, Hunziker W, Simionescu M (2003) Investigation of IgG receptors expressed on the surface of human placental endothelial cells. Placenta 24(6):664–676

Hansen RJ, Balthasar JP (2002) Intravenous immunoglobulin mediates an increase in anti-platelet antibody clearance via the FcRn receptor. Thromb Haemost 88(6):898–899

Hattori R, Otani H, Moriguchi Y, Matsubara H, Yamamura T, Nakao Y et al (2001) NHE and ICAM-1 expression in hypoxic/reoxygenated coronary microvascular endothelial cells. Am J Physiol Heart Circ Physiol 280(6):H2796–H2803

He Y, Bjorkman PJ (2011) Structure of FcRY, an avian immunoglobulin receptor related to mammalian mannose receptors, and its complex with IgY. Proc Natl Acad Sci USA 108(30):12431–12436

He W, Ladinsky MS, Huey-Tubman KE, Jensen GJ, McIntosh JR, Bjorkman PJ (2008) FcRn-mediated antibody transport across epithelial cells revealed by electron tomography. Nature 455(7212):542–546

Hinton PR, Johlfs MG, Xiong JM, Hanestad K, Ong KC, Bullock C et al (2004) Engineered human IgG antibodies with longer serum half-lives in primates. J Biol Chem 279(8):6213–6216

Hinton PR, Xiong JM, Johlfs MG, Tang MT, Keller S, Tsurushita N (2006) An engineered human IgG1 antibody with longer serum half-life. J Immunol 176(1):346–356

Ishii-Watabe A, Saito Y, Suzuki T, Tada M, Ukaji M, Maekawa K et al (2010) Genetic polymorphisms of FCGRT encoding FcRn in a Japanese population and their functional analysis. Drug Metab Pharmacokinet 25(6):578–587

Israel EJ, Taylor S, Wu Z, Mizoguchi E, Blumberg RS, Bhan A et al (1997) Expression of the neonatal Fc receptor, FcRn, on human intestinal epithelial cells. Immunology 92(1):69–74

Jerdeva GV, Tesar DB, Huey-Tubman KE, Ladinsky MS, Fraser SE, Bjorkman PJ (2010) Comparison of FcRn- and pIgR-mediated transport in MDCK cells by fluorescence confocal microscopy. Traffic 11(9):1205–1220

Jin-Kyoo K, May-Fang T, Victor G, Ward ES (1994) Localization of the site of the murine IgG1 molecule that is involved in binding to the murine intestinal Fc receptor. Eur J Immunol 24(10):2429–2434

Jones EA, Waldmann TA (1972) The mechanism of intestinal uptake and transcellular transport of IgG in the neonatal rat. J Clin Invest 51(11):2916–2927

Kacskovics I (2004) Fc receptors in livestock species. Vet Immunol Immunopathol 102(4):351–362

Kacskovics I, Wu Z, Simister NE, Frenyo LV, Hammarstrom L (2000) Cloning and characterization of the bovine MHC class I-like Fc receptor. J Immunol 164(4):1889–1897

Kacskovics I, Kis Z, Mayer B, West AP Jr, Tiangco NE, Tilahun M et al (2006a) FcRn mediates elongated serum half-life of human IgG in cattle. Int Immunol 18(4):525–536

Kacskovics I, Mayer B, Kis Z, Frenyo LV, Zhao Y, Muyldermans S et al (2006b) Cloning and characterization of the dromedary (Camelus dromedarius) neonatal Fc receptor (drFcRn). Dev Comp Immunol 30(12):1203–1215

Kelly D, King T, Aminov R (2007) Importance of microbial colonization of the gut in early life to the development of immunity. Mutat Res 622(1–2):58–69

Kim JK, Firan M, Radu CG, Kim CH, Ghetie V, Ward ES (1999) Mapping the site on human IgG for binding of the MHC class I-related receptor, FcRn. Eur J Immunol 29(9):2819–2825

Kim J, Bronson CL, Wani MA, Oberyszyn TM, Mohanty S, Chaudhury C et al (2008) Beta 2-microglobulin deficient mice catabolize IgG more rapidly than FcRn- alpha-chain deficient mice. Exp Biol Med (Maywood) 233(5):603–609

Kim J, Mohanty S, Ganesan LP, Hua K, Jarjoura D, Hayton WL et al (2009) FcRn in the yolk sac endoderm of mouse is required for IgG transport to fetus. J Immunol 182(5):2583–2589

Kobayashi N, Suzuki Y, Tsuge T, Okumura K, Ra C, Tomino Y (2002) FcRn-mediated transcytosis of immunoglobulin G in human renal proximal tubular epithelial cells. Am J Physiol Renal Physiol 282(2):F358–F365

Kobayashi K, Qiao SW, Yoshida M, Baker K, Lencer WI, Blumberg RS (2009) An FcRn-dependent role for anti-flagellin immunoglobulin G in pathogenesis of colitis in mice. Gastroenterology 137(5):1746–1756, e1

Kozlowski PA, Cu-Uvin S, Neutra MR, Flanigan TP (1997) Comparison of the oral, rectal, and vaginal immunization routes for induction of antibodies in rectal and genital tract secretions of women. Infect Immun 65(4):1387–1394

Kuo TT, Aveson VG (2011) Neonatal Fc receptor and IgG-based therapeutics. MAbs 3(5):422–430

Kuo TT, de Muinck EJ, Claypool SM, Yoshida M, Nagaishi T, Aveson VG et al (2009) N-glycan moieties in neonatal Fc receptor determine steady-state membrane distribution and directional transport of IgG. J Biol Chem 284(13):8292–8300

Kuo T, Baker K, Yoshida M, Qiao S-W, Aveson V, Lencer W et al (2010) Neonatal Fc receptor: from immunity to therapeutics. J Clin Immunol 30(6):777–789

Leach JL, Sedmak DD, Osborne JM, Rahill B, Lairmore MD, Anderson CL (1996) Isolation from human placenta of the IgG transporter, FcRn, and localization to the syncytiotrophoblast: implications for maternal-fetal antibody transport. J Immunol 157(8):3317–3322

Lencer WI, Blumberg RS (2005) A passionate kiss, then run: exocytosis and recycling of IgG by FcRn. Trends Cell Biol 15(1):5–9

Li N, Zhao M, Hilario-Vargas J, Prisayanh P, Warren S, Diaz LA et al (2005) Complete FcRn dependence for intravenous Ig therapy in autoimmune skin blistering diseases. J Clin Invest 115(12):3440–3450

Li Z, Palaniyandi S, Zeng R, Tuo W, Roopenian DC, Zhu X (2011) Transfer of IgG in the female genital tract by MHC class I-related neonatal Fc receptor (FcRn) confers protective immunity to vaginal infection. Proc Natl Acad Sci USA 108(11):4388–4393

Liu L, Garcia AM, Santoro H, Zhang Y, McDonnell K, Dumont J et al (2007) Amelioration of experimental autoimmune myasthenia gravis in rats by neonatal FcR blockade. J Immunol 178(8):5390–5398

Liu X, Lu L, Yang Z, Palaniyandi S, Zeng R, Gao L-Y et al (2011) The neonatal FcR-mediated presentation of immune-complexed antigen is associated with endosomal and phagosomal pH and antigen stability in macrophages and dendritic cells. J Immunol 186(8):4674–4686

Lodes MJ, Cong Y, Elson CO, Mohamath R, Landers CJ, Targan SR et al (2004) Bacterial flagellin is a dominant antigen in Crohn disease. J Clin Invest 113(9):1296–1306

Low SC, Mezo AR (2009) Inhibitors of the FcRn: IgG protein-protein interaction. AAPS J 11(3):432–434

Lu W, Zhao Z, Zhao Y, Yu S, Zhao Y, Fan B et al (2007) Over-expression of the bovine FcRn in the mammary gland results in increased IgG levels in both milk and serum of transgenic mice. Immunology 122(3):401–408

Lu L, Palaniyandi S, Zeng R, Bai Y, Liu X, Wang Y et al (2011) An FcRn-targeted mucosal vaccine strategy effectively induces HIV-1 antigen-specific immunity to genital infection. J Virol 85(20):10542–10553

Martin WL, Bjorkman PJ (1999) Characterization of the 2:1 complex between the class I MHC-related Fc receptor and its Fc ligand in solution. Biochemistry 38(39):12639–12647

Martin WL, West AP Jr, Gan L, Bjorkman PJ (2001) Crystal structure at 2.8 A of an FcRn/heterodimeric Fc complex: mechanism of pH-dependent binding. Mol Cell 7(4):867–877

Matson AP, Thrall RS, Rafti E, Lingenheld EG, Puddington L (2010) IgG transmitted from allergic mothers decreases allergic sensitization in breastfed offspring. Clin Mol Allergy 8:9

Mayer B, Zolnai A, Frenyo LV, Jancsik V, Szentirmay Z, Hammarstrom L et al (2002) Redistribution of the sheep neonatal Fc receptor in the mammary gland around the time of parturition in ewes and its localization in the small intestine of neonatal lambs. Immunology 107(3):288–296

Mayer B, Kis Z, Kajan G, Frenyo LV, Hammarstrom L, Kacskovics I (2004) The neonatal Fc receptor (FcRn) is expressed in the bovine lung. Vet Immunol Immunopathol 98(1–2):85–89

McCarthy KM, Yoong Y, Simister NE (2000) Bidirectional transcytosis of IgG by the rat neonatal Fc receptor expressed in a rat kidney cell line: a system to study protein transport across epithelia. J Cell Sci 113(Pt 7):1277–1285

McCarthy KM, Lam M, Subramanian L, Shakya R, Wu Z, Newton EE et al (2001) Effects of mutations in potential phosphorylation sites on transcytosis of FcRn. J Cell Sci 114(Pt 8):1591–1598

Medesan C, Matesoi D, Radu C, Ghetie V, Ward ES (1997) Delineation of the amino acid residues involved in transcytosis and catabolism of mouse IgG1. J Immunol 158(5):2211–2217

Mestecky J (1987) The common mucosal immune system and current strategies for induction of immune responses in external secretions. J Clin Immunol 7(4):265–276

Mezo AR, McDonnell KA, Hehir CA, Low SC, Palombella VJ, Stattel JM et al (2008a) Reduction of IgG in nonhuman primates by a peptide antagonist of the neonatal Fc receptor FcRn. Proc Natl Acad Sci USA 105(7):2337–2342

Mezo AR, McDonnell KA, Castro A, Fraley C (2008b) Structure-activity relationships of a peptide inhibitor of the human FcRn: human IgG interaction. Bioorg Med Chem 16(12):6394–6405

Mezo AR, Low SC, Hoehn T, Palmieri H (2011) PEGylation enhances the therapeutic potential of peptide antagonists of the neonatal Fc receptor, FcRn. Bioorg Med Chem Lett 21(21):6332–6335

Mi W, Wanjie S, Lo ST, Gan Z, Pickl-Herk B, Ober RJ et al (2008) Targeting the neonatal fc receptor for antigen delivery using engineered fc fragments. J Immunol 181(11):7550–7561

Mohanty S, Kim J, Ganesan LP, Phillips GS, Hua K, Jarjoura D et al (2010) IgG is transported across the mouse yolk sac independently of Fc[gamma]RIIb. J Reprod Immunol 84(2):133–144

Montoyo HP, Vaccaro C, Hafner M, Ober RJ, Mueller W, Ward ES (2009) Conditional deletion of the MHC class I-related receptor FcRn reveals the sites of IgG homeostasis in mice. Proc Natl Acad Sci USA 106(8):2788–2793

Morell A, Terry WD, Waldmann TA (1970) Metabolic properties of IgG subclasses in man. J Clin Invest 49(4):673–680

Mosconi E, Rekima A, Seitz-Polski B, Kanda A, Fleury S, Tissandie E et al (2010) Breast milk immune complexes are potent inducers of oral tolerance in neonates and prevent asthma development. Mucosal Immunol 3(5):461–474

Nimmerjahn F, Ravetch JV (2010) Antibody-mediated modulation of immune responses. Immunol Rev 236(1):265–275

Ober RJ, Radu CG, Ghetie V, Ward ES (2001) Differences in promiscuity for antibody-FcRn interactions across species: implications for therapeutic antibodies. Int Immunol 13(12):1551–1559

Ober RJ, Martinez C, Vaccaro C, Zhou J, Ward ES (2004a) Visualizing the site and dynamics of IgG salvage by the MHC class I-related receptor, FcRn. J Immunol 172(4):2021–2029

Ober RJ, Martinez C, Lai X, Zhou J, Ward ES (2004b) Exocytosis of IgG as mediated by the receptor, FcRn: an analysis at the single-molecule level. Proc Natl Acad Sci USA 101(30):11076–11081

Patel DA, Puig-Canto A, Challa DK, Montoyo HP, Ober RJ, Ward ES (2011) Neonatal Fc receptor blockade by Fc engineering ameliorates arthritis in a murine model. J Immunol 187(2):1015–1022

Peters RT, Low SC, Kamphaus GD, Dumont JA, Amari JV, Lu Q et al (2010) Prolonged activity of factor IX as a monomeric Fc fusion protein. Blood 115(10):2057–2064

Petkova SB, Akilesh S, Sproule TJ, Christianson GJ, Al Khabbaz H, Brown AC et al (2006) Enhanced half-life of genetically engineered human IgG1 antibodies in a humanized FcRn mouse model: potential application in humorally mediated autoimmune disease. Int Immunol 18(12):1759–1769

Cianga P, Medesan C, Richardson JA, Ghetie V, Ward ES (1999) Identification and function of neonatal Fc receptor in mammary gland of lactating mice. Eur J Immunol 29(8):2515–2523

Popivanova BK, Kitamura K, Yu W, Kondo T, Kagaya T, Kaneko S et al (2008) Blocking TNF-alpha in mice reduces colorectal carcinogenesis associated with chronic colitis. J Clin Invest 118(2):560–570

Porollo A, Meller J (2007) Versatile annotation and publication quality visualization of protein complexes using POLYVIEW-3D. BMC Bioinformatics 8:316

Praetor A, Hunziker W (2002) beta(2)-Microglobulin is important for cell surface expression and pH-dependent IgG binding of human FcRn. J Cell Sci 115(Pt 11):2389–2397

Qiao SW, Kobayashi K, Johansen FE, Sollid LM, Andersen JT, Milford E et al (2008) Dependence of antibody-mediated presentation of antigen on FcRn. Proc Natl Acad Sci USA 105(27):9337–9342

Raghavan M, Gastinel LN, Bjorkman PJ (1993) The class I major histocompatibility complex related Fc receptor shows pH-dependent stability differences correlating with immunoglobulin binding and release. Biochemistry 32(33):8654–8660

Raghavan M, Bonagura VR, Morrison SL, Bjorkman PJ (1995) Analysis of the pH dependence of the neonatal Fc receptor/immunoglobulin G interaction using antibody and receptor variants. Biochemistry 34(45):14649–14657

Rodewald R (1976) pH-dependent binding of immunoglobulins to intestinal cells of the neonatal rat. J Cell Biol 71(2):666–669

Roopenian D, Sun V (2010) Clinical ramifications of the MHC family Fc receptor FcRn. J Clin Immunol 30(6):790–797

Roopenian DC, Christianson GJ, Sproule TJ, Brown AC, Akilesh S, Jung N et al (2003) The MHC class I-like IgG receptor controls perinatal IgG transport, IgG homeostasis, and fate of IgG-Fc-coupled drugs. J Immunol 170(7):3528–3533

Roopenian DC, Christianson GJ, Sproule TJ (2010) Human FcRn transgenic mice for pharmacokinetic evaluation of therapeutic antibodies. Methods Mol Biol 602:93–104

Russell DG (2007) New ways to arrest phagosome maturation. Nat Cell Biol 9(4):357–359

Sakagami M, Omidi Y, Campbell L, Kandalaft LE, Morris CJ, Barar J et al (2006) Expression and transport functionality of FcRn within rat alveolar epithelium: a study in primary cell culture and in the isolated perfused lung. Pharm Res 23(2):270–279

Sarav M, Wang Y, Hack BK, Chang A, Jensen M, Bao L et al (2009) Renal FcRn reclaims albumin but facilitates elimination of IgG. J Am Soc Nephrol 20(9):1941–1952

Savina A, Jancic C, Hugues S, Guermonprez P, Vargas P, Moura IC et al (2006) NOX2 controls phagosomal pH to regulate antigen processing during cross-presentation by dendritic cells. Cell 126(1):205–218

Sayed-Ahmed A, Kassab M, Abd-Elmaksoud A, Elnasharty M, El-Kirdasy A (2010) Expression and immunohistochemical localization of the neonatal Fc receptor (FcRn) in the mammary glands of the Egyptian water buffalo. Acta Histochem 112(4):383–391

Schlachetzki F, Zhu C, Pardridge WM (2002) Expression of the neonatal Fc receptor (FcRn) at the blood–brain barrier. J Neurochem 81(1):203–206

Schnulle PM, Hurley WL (2003) Sequence and expression of the FcRn in the porcine mammary gland. Vet Immunol Immunopathol 91(3–4):227–231

Schroeder HW Jr, Cavacini L (2010) Structure and function of immunoglobulins. J Allergy Clin Immunol 125(Suppl 2):S41–S52

Sesarman A, Sitaru AG, Olaru F, Zillikens D, Sitaru C (2008) Neonatal Fc receptor deficiency protects from tissue injury in experimental epidermolysis bullosa acquisita. J Mol Med 86(8):951–959

Sesarman A, Vidarsson G, Sitaru C (2010) The neonatal Fc receptor as therapeutic target in IgG-mediated autoimmune diseases. Cell Mol Life Sci 67(15):2533–2550

Shah U, Dickinson BL, Blumberg RS, Simister NE, Lencer WI, Walker WA (2003) Distribution of the IgG Fc receptor, FcRn, in the human fetal intestine. Pediatr Res 53(2):295–301

Shiomi H, Masuda A, Nishiumi S, Nishida M, Takagawa T, Shiomi Y et al (2010) Gamma interferon produced by antigen-specific CD4+ T cells regulates the mucosal immune responses to Citrobacter rodentium infection. Infect Immun 78(6):2653–2666

Simister NE (2003) Placental transport of immunoglobulin G. Vaccine 21(24):3365–3369

Simister NE, Ahouse JC (1996) The structure and evolution of FcRn. Res Immunol 147(5):333–337

Simister NE, Mostov KE (1989a) An Fc receptor structurally related to MHC class I antigens. Nature 337(6203):184–187

Simister NE, Mostov KE (1989b) Cloning and expression of the neonatal rat intestinal Fc receptor, a major histocompatibility complex class I antigen homolog. Cold Spring Harb Symp Quant Biol 54(Pt 1):571–580

Simister NE, Rees AR (1985) Isolation and characterization of an Fc receptor from neonatal rat small intestine. Eur J Immunol 15(7):733–738

Spiegelberg HL (1989) Biological role of different antibody classes. Int Arch Allergy Appl Immunol 90(Suppl 1):22–27
Spiegelberg HL, Weigle WO (1965) Studies on the catabolism of gamma- G subunits and chains. J Immunol 95(6):1034–1040
Spiegelberg HL, Fishkin BG, Grey HM (1968) Catabolism of human gammaG-immunoglobulins of different heavy chain subclasses. I. Catabolism of gammaG-myeloma proteins in man. J Clin Invest 47(10):2323–2330
Spiekermann GM, Finn PW, Ward ES, Dumont J, Dickinson BL, Blumberg RS et al (2002) Receptor-mediated immunoglobulin G transport across mucosal barriers in adult life: functional expression of FcRn in the mammalian lung. J Exp Med 196(3):303–310
Stirling CM, Charleston B, Takamatsu H, Claypool S, Lencer W, Blumberg RS et al (2005) Characterization of the porcine neonatal Fc receptor–potential use for trans-epithelial protein delivery. Immunology 114(4):542–553
Story CM, Mikulska JE, Simister NE (1994) A major histocompatibility complex class I-like Fc receptor cloned from human placenta: possible role in transfer of immunoglobulin G from mother to fetus. J Exp Med 180(6):2377–2381
Suzuki T, Ishii-Watabe A, Tada M, Kobayashi T, Kanayasu-Toyoda T, Kawanishi T et al (2010) Importance of neonatal FcR in regulating the serum half-life of therapeutic proteins containing the Fc domain of human IgG1: a comparative study of the affinity of monoclonal antibodies and Fc-fusion proteins to human neonatal FcR. J Immunol 184(4):1968–1976
Szlauer R, Ellinger I, Haider S, Saleh L, Busch BL, Knofler M et al (2009) Functional expression of the human neonatal Fc-receptor, hFcRn, in isolated cultured human syncytiotrophoblasts. Placenta 30(6):507–515
Tesar DB, Björkman PJ (2010) An intracellular traffic jam: Fc receptor-mediated transport of immunoglobulin G. Curr Opin Struct Biol 20(2):226–233
Tesar DB, Cheung EJ, Bjorkman PJ (2008) The chicken yolk sac IgY receptor, a mammalian mannose receptor family member, transcytoses IgY across polarized epithelial cells. Mol Biol Cell 19(4):1587–1593
Tiwari B, Junghans RP (2005) Functional analysis of the mouse Fcgrt 5′ proximal promoter. Biochim Biophys Acta 1681(2–3):88–98
Tzaban S, Massol RH, Yen E, Hamman W, Frank SR, Lapierre LA et al (2009) The recycling and transcytotic pathways for IgG transport by FcRn are distinct and display an inherent polarity. J Cell Biol 185(4):673–684
Vaccaro C, Zhou J, Ober RJ, Ward ES (2005) Engineering the Fc region of immunoglobulin G to modulate in vivo antibody levels. Nat Biotechnol 23(10):1283–1288
Vaccaro C, Bawdon R, Wanjie S, Ober RJ, Ward ES (2006) Divergent activities of an engineered antibody in murine and human systems have implications for therapeutic antibodies. Proc Natl Acad Sci USA 103(49):18709–18714
Vaughn DE, Bjorkman PJ (1998) Structural basis of pH-dependent antibody binding by the neonatal Fc receptor. Structure 6(1):63–73
Verhasselt V, Milcent V, Cazareth J, Kanda A, Fleury S, Dombrowicz D et al (2008) Breast milk-mediated transfer of an antigen induces tolerance and protection from allergic asthma. Nat Med 14(2):170–175
Waldmann TA, Terry WD (1990) Familial hypercatabolic hypoproteinemia. A disorder of endogenous catabolism of albumin and immunoglobulin. J Clin Invest 86(6):2093–2098
Wang Y, Geer LY, Chappey C, Kans JA, Bryant SH (2000) Cn3D: sequence and structure views for Entrez. Trends Biochem Sci 25(6):300–302
Wani MA, Haynes LD, Kim J, Bronson CL, Chaudhury C, Mohanty S et al (2006) Familial hypercatabolic hypoproteinemia caused by deficiency of the neonatal Fc receptor, FcRn, due to a mutant beta2-microglobulin gene. Proc Natl Acad Sci USA 103(13):5084–5089
Ward ES, Ober RJ, Frederick WA (2009) Chapter 4 Multitasking by exploitation of intracellular transport functions: the many faces of FcRn. Adv Immunol 103:77–115, Academic Press
Wernick NL, Haucke V, Simister NE (2005) Recognition of the tryptophan-based endocytosis signal in the neonatal Fc receptor by the mu subunit of adaptor protein-2. J Biol Chem 280(8):7309–7316

West AP Jr, Bjorkman PJ (2000) Crystal structure and immunoglobulin G binding properties of the human major histocompatibility complex-related Fc receptor. Biochemistry 39(32):9698–9708

West AP Jr, Herr AB, Bjorkman PJ (2004) The chicken yolk sac IgY receptor, a functional equivalent of the mammalian MHC-related Fc receptor, is a phospholipase A2 receptor homolog. Immunity 20(5):601–610

Wu Z, Simister NE (2001) Tryptophan- and dileucine-based endocytosis signals in the neonatal Fc receptor. J Biol Chem 276(7):5240–5247

Ye L, Liu X, Rout SN, Li Z, Yan Y, Lu L et al (2008) The MHC class II-associated invariant chain interacts with the neonatal Fc gamma receptor and modulates its trafficking to endosomal/lysosomal compartments. J Immunol 181(4):2572–2585

Ye L, Zeng R, Bai Y, Roopenian DC, Zhu X (2011) Efficient mucosal vaccination mediated by the neonatal Fc receptor. Nat Biotech 29(2):158–163

Yeung YA, Leabman MK, Marvin JS, Qiu J, Adams CW, Lien S et al (2009) Engineering human IgG1 affinity to human neonatal Fc receptor: impact of affinity improvement on pharmacokinetics in primates. J Immunol 182(12):7663–7671

Yeung YA, Wu X, Reyes AE 2nd, Vernes JM, Lien S, Lowe J et al (2010) A therapeutic anti-VEGF antibody with increased potency independent of pharmacokinetic half-life. Cancer Res 70(8):3269–3277

Yoshida M, Claypool SM, Wagner JS, Mizoguchi E, Mizoguchi A, Roopenian DC et al (2004) Human neonatal Fc receptor mediates transport of IgG into luminal secretions for delivery of antigens to mucosal dendritic cells. Immunity 20(6):769–783

Yoshida M, Kobayashi K, Kuo TT, Bry L, Glickman JN, Claypool SM et al (2006) Neonatal Fc receptor for IgG regulates mucosal immune responses to luminal bacteria. J Clin Invest 116(8):2142–2151

Zalevsky J, Chamberlain AK, Horton HM, Karki S, Leung IWL, Sproule TJ et al (2010) Enhanced antibody half-life improves in vivo activity. Nat Biotech 28(2):157–159

Zheng Y, Scheerens H, Davis JC Jr, Deng R, Fischer SK, Woods C et al (2011) Translational pharmacokinetics and pharmacodynamics of an FcRn-variant anti-CD4 monoclonal antibody from preclinical model to phase I study. Clin Pharmacol Ther 89(2):283–290

Zhou J, Mateos F, Ober RJ, Ward ES (2005a) Conferring the binding properties of the mouse MHC class I-related receptor, FcRn, onto the human ortholog by sequential rounds of site-directed mutagenesis. J Mol Biol 345(5):1071–1081

Zhou J, Pop LM, Ghetie V (2005b) Hypercatabolism of IgG in mice with lupus-like syndrome. Lupus 14(6):458–466

Zhou XJ, Yu L, Zhu L, Hou P, Lv JC, Yu F et al (2009) Association between polymorphisms in the FCGRT gene and lupus nephritis in Chinese patients. Clin Exp Rheumatol 27(4):609–614

Zhu X, Meng G, Dickinson BL, Li X, Mizoguchi E, Miao L et al (2001) MHC class I-related neonatal Fc receptor for IgG is functionally expressed in monocytes, intestinal macrophages, and dendritic cells. J Immunol 166(5):3266–3276

Chapter 9
Antibody Mediated Regulation of Humoral Immunity

Birgitta Heyman

Abstract Antibodies can regulate the antibody response against the antigen they bind to. This phenomenon is called antibody feedback regulation and can be both positive and negative. Passively administered erythrocyte-specific IgG antibodies can completely prevent an antibody response against erythrocytes. This is used clinically in Rhesus prophylaxis. IgM can enhance the antibody response to large antigens and this effect is dependent on the ability of IgM to activate complement. Enhancement by IgG3 is also dependent on complement whereas IgG1, IgG2a, and IgG2b enhance antibody responses to protein antigens via IgG Fc receptors. Finally, IgE is a potent enhancer of both antibody and CD4 T cells responses and this effect is mediated by the low-affinity receptor for IgE, CD23.

Keywords Rhesus prophylaxis • Feedback regulation • CD23 • IgM • IgE • Complement receptor • CD21 • Fc receptors

9.1 Background

Antibodies passively administered together with their specific antigen can regulate the immune response to this antigen. This process is referred to as antibody feedback regulation. Not only passively administered but also endogenously produced antibodies have regulatory capacity. Antibody feedback regulation is potent, and microgram amounts can cause more than 99 % suppression or over a 100-fold enhancement of specific antibody responses. The ability of antibodies to regulate the production of themselves was first noticed by Emil von Behring who was awarded the first Nobel Prize for Physiology or Medicine in 1901 for his work on

B. Heyman, M.D., Ph.D (✉)
Department of Medical Biochemistry and Microbiology, Uppsala University, Box 852, Biomedical Center Husargatan 3, SE-751 23, Uppsala, Sweden
e-mail: birgitta.heyman@imbim.uu.se

F. Nimmerjahn (ed.), *Molecular and Cellular Mechanisms of Antibody Activity*,
DOI 10.1007/978-1-4614-7107-3_9,

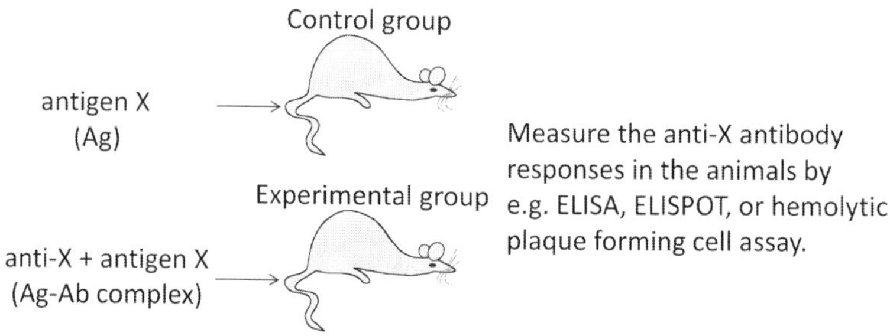

Fig. 9.1 *Protocol for studying antibody feedback regulation.* A mouse is immunized in the tail vein with purified specific antibody in physiologic buffer solution. Within minutes to hours, the antigen for which the antibody is specific, is also injected in the tail vein. Controls are immunized with antigen alone. After days to weeks, the active immune response in the two groups is compared. By using various monoclonal antibodies with known isotype, sometimes with point mutations that change the effector functions, and knockout mice lacking different Fc and complement receptors, the molecular mechanisms behind antibody feedback regulation can be analyzed

serum therapy, especially its application against diphtheria. In those days, diphtheria, caused by *Corynebacterium diphtheriae* producing a very potent toxin, was a disease with high mortality. Von Behring treated patients with antiserum against the diphtheria toxin and managed to lower the mortality dramatically (Grundbacher 1992). While experimenting to obtain high antitoxin titers in the experimental animals used as a source for the therapeutic antitoxin, he found that immunization with immune complexes could induce higher titers than immunization with the toxin alone (von Behring and Wernicke 1892). This led to the realization that antibodies could feedback regulate antibody production, and ever since, scientists have tried to elucidate the underlying mechanisms. A milestone in the understanding of antibody feedback regulation was when Claudia Henry and Niels Jerne found that IgG antibodies (then called 7S antigen receptors) suppressed, whereas IgM antibodies (19S antigen receptors) enhanced the antibody response to sheep erythrocytes (SRBC) (Henry and Jerne 1968). The early studies of antibody feedback regulation have been extensively reviewed by two pioneers in the field, Jonathan Uhr and Göran Möller (Uhr and Möller 1968). With the advent of more sophisticated methods for separation of antibodies into different classes and subclasses, monoclonal antibody technology, and gene targeted mice, the understanding of antibody feedback regulation has advanced but is still far from complete (reviewed in Heyman 2000, 2003; Hjelm et al. 2006; Nimmerjahn and Ravetch 2010).

The most common experimental protocol used to study antibody feedback regulation is passive administration of antigen and purified antigen-specific antibodies intravenously (iv). Both antigen and antibody are dissolved in physiological buffers and given within minutes or hours of one another (Fig. 9.1). Mice are by far the most frequently studied animals, but the relevance of this regulatory system for humans is illustrated by the successful use of IgG-mediated suppression in Rhesus prophylaxis in Rh negative women (Bowman 1988). The antibody responses are determined by measuring serum titers or single antibody-producing B cells.

Table 9.1 Overview of antibody feedback regulation

Regulating Ab isotype	Antigens	Regulatory effect on Ab response	CD4+ T cells	Suggested mechanism(s)
IgG1, 2a, 2b, 3	Erythrocytes (e.g., SRBC) proteins in adjuvants	Suppression	None	(Independent of complement and Fcγ receptors) – Epitope masking – Ag clearance/deviation
IgM	Erythrocytes malaria parasites KLH	Enhancement	Unknown	(Complement dependent) – Ag trapping on CR1/2$^+$ FDC in follicles – Ag transport to follicles by CR1/2$^+$ MZB
IgG1, 2a, 2b	Soluble proteins (e.g., OVA, BSA KLH)	Enhancement	Enhancement	(Fcγ-receptor dependent) – Ag presentation to CD4$^+$ – T cells by FcγR$^+$ APC
IgG3	Soluble proteins (e.g., OVA, BSA)	Enhancement	None	(Complement dependent) – Ag trapping on CR1/2$^+$ FDC in follicles – Ag transport to follicles by CR1/2$^+$ MZB
IgE	Soluble proteins (e.g., OVA, BSA)	Enhancement	Enhancement	(FcεRII/CD23 dependent) – Ag transport to follicles on CD23$^+$ B cells followed by uptake and Ag presentation to CD4$^+$ T cells by CD11c$^+$ APC

Feedback regulation is antigen specific. Thus, a regulating antibody will only affect the response to the antigen to which it binds. However, the response to all epitopes, and not only those recognized by the regulating antibody, is usually affected. An important factor in antibody-mediated regulation is localization of antigen in the body. Since no anatomical structure can be maintained in vitro, such studies are not always relevant to the in vivo situation. Therefore, the focus of this chapter will be on in vivo studies. An overview of antibody feedback regulation is presented in Table 9.1.

9.2 Suppression of Antibody Responses

Although antibodies of the IgM and IgE classes normally enhance antibody responses, very high doses of IgM, or IgM administered after the antigen, can induce suppression (Möller and Wigzell 1965; Wason 1973; Brüggemann and Rajewsky 1982) and IgE administered with a particulate antigen like SRBC can turn suppressive (Karlsson et al. 1999). However, IgG is by far the most important suppressive antibody class and the discussion below will be limited to this.

9.2.1 IgG Suppresses Responses to Particulate Antigens and Proteins in Adjuvants

Small amounts of IgG can suppress responses to proteins administered in adjuvants (Cerottini et al. 1969; Strannegård and Belin 1970; Krieger et al. 1983), but most of our knowledge comes from studies of responses to erythrocytes administered iv in salt solutions. Suppression is efficient, and often 90–99 % of a primary antibody response is inhibited (Henry and Jerne 1968; Möller and Wigzell 1965; Brüggemann and Rajewsky 1982; Karlsson et al. 1999; Heyman and Wigzell 1984, 1985; Quintana et al. 1987; Brinc et al. 2008a; Getahun and Heyman 2009). Induction of memory (priming) and secondary antibody responses can also be suppressed, but the effects are generally less pronounced than in a primary response (Karlsson et al. 1999; Heyman and Wigzell 1985; Brinc et al. 2008a; Nicholas 1969; Safford and Tokuda 1971; Enriquez-Rincon and Klaus 1984; Dresser 1990; Karlsson et al. 2001). High doses of IgG are more suppressive than low doses (Henry and Jerne 1968; Heyman and Wigzell 1984; Dresser 1990), strong responses are more difficult to suppress than weak responses, and high-affinity IgG is more suppressive than low-affinity IgG (Brüggemann and Rajewsky 1982; Dresser 1990; Walker and Siskind 1968). In most experiments, IgG is administered in close temporal relationship to the antigen. However, IgG administered up to 5 days after SRBC can inhibit an ongoing antibody response (Henry and Jerne 1968; Karlsson et al. 2001; Chan and Sinclair 1973). Studies using monoclonal IgG have shown that all murine subclasses, IgG1, IgG2a, IgG2b, and IgG3, are suppressive (Brüggemann and Rajewsky 1982; Karlsson et al. 1999; Heyman and Wigzell 1984; Wiersma et al. 1989).

The ability of IgG to suppress responses to erythrocytes has been used in clinical practice since the 1960s to prevent hemolytic disease of the fetus and newborn (Clarke et al. 1963). Rhesus blood group D-negative (RhD^-) women carrying RhD^+ fetuses are sometimes immunized against fetal erythrocytes entering the maternal circulation via transplacental hemorrhage (Urbaniak and Greiss 2000). The IgG anti-RhD produced by the mother will cross the placenta and destroy fetal erythrocytes which causes the hemolytic disease. Passive administration of human IgG anti-D antibodies to RhD^- women during pregnancy or immediately after delivery has been successful in preventing maternal immunization and has drastically reduced the frequency of hemolytic disease of the fetus and newborn (Bowman 1988). Although this therapy has been used for decades, the mechanism of action remains unresolved. It has been debated to which extent the suppression of responses to strongly immunogenic doses of xenogeneic erythrocytes in animal models reflects the suppression of responses to allogeneic RhD^+ erythrocytes in humans (Kumpel and Elson 2001; Brinc and Lazarus 2009). However, it seems unlikely that IgG-mediated suppression of antibody responses to foreign erythrocytes administered iv would follow completely different rules, depending on whether they are xenogeneic or allogeneic.

9.2.2 *Epitope-Specific Versus Non-epitope-Specific Suppression*

IgG antibodies mediate many of their effector functions via Fc receptors for IgG (FcγRs) and by activating complement. Both require an intact Fc portion. An indirect way of determining whether the Fc part is required for suppression has been to test whether IgG specific for one epitope can suppress the response also to other epitopes on the same antigen. Epitope specificity has been seen as an indication of Fc independence, whereas non-epitope specificity indicated Fc dependence. In animal studies, both epitope-specific (Brody et al. 1967; Möller 1985) and non-epitope-specific suppression (Brüggemann and Rajewsky 1982; Karlsson et al. 1999; Heyman and Wigzell 1984; Wiersma et al. 1989; Enriquez-Rincon and Klaus 1984; Greenbury and Moore 1968; Whited Collisson et al. 1979, 1984) have been found. In anti-RhD prophylaxis, evidence points towards non-epitope-specific suppression since IgG specific for the Kell blood group antigen can inhibit responses against both Kell and Duffy antigens (Woodrow et al. 1975). Moreover, it has been calculated that the dose of anti-RhD which efficiently inhibits RhD immunization covers only around 20 % of all RhD epitopes (Pollack 1984). The idea that epitope specificity proves Fc independence, whereas non-epitope specificity proves Fc dependence, has been challenged. IgG specific for a hapten (NIP) administered together with SRBC-NIP suppressed only the NIP-specific response when the hapten was present at low density but suppressed both NIP- and SRBC-specific responses when the hapten was present at high density (Getahun and Heyman 2009). Moreover, hapten-specific F(ab')$_2$ fragments could suppress responses to erythrocyte determinants when injected together with haptenated erythrocytes (Karlsson et al. 1999). The best explanation for these findings is that IgG (or F(ab')$_2$ fragments) binding to a high density epitope sterically hinders recognition also of neighboring epitopes. Therefore, epitope-specific suppression will be seen during low epitope-density conditions and non-epitope-specific suppression during high epitope-density conditions, and both may be explained by epitope masking (Heyman 1999).

9.2.3 *9.2.3 F(ab')$_2$ Fragments and IgG-Mediated Suppression*

To determine whether the Fc part is required it has been tested whether F(ab')$_2$ fragments are suppressive. Such studies have given discrepant results, some showing that F(ab')$_2$ fragments are efficient suppressors (Karlsson et al. 1999;

Cerottini et al. 1969; Tao and Uhr 1966) and others that F(ab')$_2$ fragments lose their suppressive capacity (Brüggemann and Rajewsky 1982; Nicholas 1969; Chan and Sinclair 1973; Enriquez-Rincon and Klaus 1984). Although this approach may seem straightforward, it has some inherent difficulties. F(ab')$_2$ fragments have a shorter half-life in vivo than intact IgG (Spiegelberg and Weigle 1965), owing to lack of binding to the protective neonatal Fc receptor, FcRn (Junghans and Anderson 1996). This may lead to inability to suppress, owing to rapid elimination of IgG from the circulation rather than to absence of the Fc part. On the other hand, F(ab')$_2$ fragments may be contaminated with intact IgG, leading to false Fc-independent suppression.

9.2.4 Fc Receptors and IgG-Mediated Suppression

There are five types of Fc receptors for IgG (reviewed in Nimmerjahn and Ravetch 2006; Kuo et al. 2010). FcRn binds all IgG subclasses and is important for transport of IgG from mother to fetus as well as in protecting IgG from proteolytic degradation. The ligand-binding alpha chain is associated with beta-2-microglobulin, required for expression of FcRn. FcγRI, FcγRIII, and FcγRIV are all activating Fc receptors with a plethora of effector functions. Their individual alpha chains are associated with a common FcRγ chain which signals by an immunoreceptor tyrosine-based activation motif (ITAM) in its cytoplasmic domain. Deletion of the FcRγ chain abolishes the function and expression of all activating FcγRs. FcγRIIB consists only of a ligand-binding alpha chain with an immunoreceptor tyrosine-based inhibitory motif (ITIM) in its cytoplasmic domain. When the ITIM is brought in close proximity to ITAMs in other immune receptors (e.g., BCR, FcγRI, FcγRIII, FcγRIV, FcεRI), ITAM signaling is inhibited, leading to negative regulation of many cellular functions (reviewed in Daeron and Lesourne 2006). As mentioned, investigations of F(ab')$_2$-mediated suppression as well as epitope specificity of suppression have given discrepant results, possibly owing to inherent limitations in their usefulness in determining Fc dependence of suppression. In an alternative approach, IgG-mediated suppression was investigated in mice lacking Fc receptors for IgG. It was found that IgG administered prior to antigen suppresses primary antibody responses to SRBC equally well in wild-type mice and in mice lacking FcγRIIB (FcγRIIB$^{-/-}$), FcγRI+III+IV (FcRγ$^{-/-}$), FcγRIIB+I+III+IV (FcγRIIB x FcRγ$^{-/-}$ double knockouts), or FcRn (β2-microglobulin$^{-/-}$) (Karlsson et al. 1999, 2001). In addition, suppression of secondary responses and suppression by IgG administered several days after the antigen were equally efficient in FcγRIIB$^{-/-}$ as in wild-type mice (Karlsson et al. 2001). Noteworthy is that IgG-mediated suppression in vitro was dependent on FcγRIIB, illustrating the differences between in vitro and in vivo models (Karlsson et al. 2001).

9.2.5 Complement and IgG-Mediated Suppression

Complement activation by mouse IgG (Heyman et al. 1988a) or human anti-RhD (Freedman et al. 1980) is not required for the suppressive effect.

9.2.6 Antigen Localization and IgG-Mediated Suppression

SRBC administered iv to mice are rapidly eliminated from the circulation and localization of this antigen in vivo has been infrequently studied. Suppressive doses of IgG, passively administered before ^{51}Cr-labeled erythrocytes, do not change the total amount of SRBC in the spleen (Dennert 1971). Histologic examination of splenic sections showed that the number of biotinylated SRBC in the marginal zone was markedly reduced in mice given SRBC-specific IgG together with the SRBC (Getahun and Heyman 2009). A large (2.5 MDa) virus-like particle, Qβ-VLP, consisting of multimers of a coat protein from the bacteriophage Qβ, was prevented from being trapped in the follicles by specific IgG administered prior to the antigen (Link et al. 2012).

9.2.7 T Cells and IgG-Mediated Suppression

Antibody responses to thymus-dependent antigens require the presence of $CD4^+$ T helper cells. While B cell activation is easy to follow by measuring serum titers of specific antibody, in vivo activation of specific T cells is difficult to assess. One way to detect specific $CD4^+$ T cells in vivo is to use the DO11.10 mouse strain which carries transgenic T cell receptor alpha and beta chains specific for an ovalbumin (OVA) peptide presented in the MHC class II molecule I-A^d (Murphy et al. 1990). OVA-specific T cells can be measured by staining the transgenic T cell receptor with a clonotypic antibody. Using this system, it was found that while IgG efficiently suppressed the antibody response, T cell proliferation was only marginally reduced (Getahun and Heyman 2009). Other, more indirect, ways to assess T helper cell induction support the conclusion that IgG does not inhibit priming of T helper cells in spite of a strong inhibition of antibody responses (Karlsson et al. 1999; Brinc et al. 2007).

9.2.8 The Mechanism Behind IgG-Mediated Suppression

In spite of intensive efforts to understand the mechanism behind IgG-mediated feedback suppression, it is still not clear how it operates. Three major hypotheses have been proposed: (i) inhibition of B cell activation via FcγRIIB, (ii) epitope masking, and (iii) antigen clearance/deviation. Inhibition of B cell activation by co-cross-linking of the inhibitory FcγRIIB and the BCR (which contains Ig-alpha/Ig-beta chains with ITAMs) takes place in vitro (Amigorena et al. 1992; Muta et al. 1994). It has been postulated that IgG-antigen complexes in vivo co-cross-link these receptors, leading to negative regulation of the antigen-specific B cell. The fact that murine IgG3, although it does not bind FcγRIIB, efficiently suppresses antibody responses was hard to reconcile with this hypothesis (Brüggemann and Rajewsky 1982; Heyman and Wigzell 1984). Subsequently, the observation that IgG suppressed erythrocyte responses equally well in FcγRIIB-deficient and wild-type mice excluded central B cell inhibition as the only explanation for IgG-mediated suppression (at least in mice) (Karlsson et al. 1999, 2001). This should not be taken as an indication that the receptor does not play a role in other situations in vivo (Nimmerjahn and Ravetch 2006). However, it seems likely that FcγRIIB modifies, rather than completely prevents, antibody responses. An example of this modifying role in antibody feedback regulation will be discussed in Sect. 9.3.2.3.

A simple explanation for IgG-mediated suppression would be that the IgG antibodies by binding to the antigen hide it from recognition by B cells. Thus, no B cell activation would take place and the lack of response would be perceived as suppression. This is referred to as epitope masking and is difficult to prove experimentally. Compatible with epitope masking are the observations of efficient suppression by F(ab')$_2$ fragments (Karlsson et al. 1999; Cerottini et al. 1969; Tao and Uhr 1966), epitope specificity of suppression (at low epitope density) (Getahun and Heyman 2009), non-epitope specificity (at high epitope density) (Karlsson et al. 1999; Getahun and Heyman 2009), independence of FcγRs (Karlsson et al. 1999, 2001), independence of complement activation (Heyman et al. 1988a; Freedman et al. 1980), and lack of suppression of T helper cell priming (Karlsson et al. 1999; Brinc et al. 2007). The observations that IgG is suppressive also when administered 1–5 days after the antigen (Henry and Jerne 1968; Karlsson et al. 2001; Chan and Sinclair 1973) suggest that IgG interferes with the binding between antigen and B cells and appear more easy to reconcile with epitope masking than with increased clearance/deviation of antigen. Possibly, it reflects the same principle as postulated during affinity maturation in germinal centers where high-affinity B cells are believed to outcompete low-affinity B cells for access to antigen presented on FDCs. Other observations, such as lack of suppression by F(ab')$_2$ fragments (Brüggemann and Rajewsky 1982; Nicholas 1969; Chan and Sinclair 1973; Enriquez-Rincon and Klaus 1984) and non-epitope-specific suppression during low epitope-density conditions (Woodrow et al. 1975; Pollack 1984), argue against epitope masking.

A third possibility to explain IgG-mediated suppression is that IgG-opsonized antigen is eliminated and thereby prevented from interacting with the immune system. This could be caused by increased phagocytosis or by deviation of antigen from areas where immune interactions take place, which in the case of iv-administered antigen primarily are the splenic follicles. In murine models, SRBC is cleared from the circulation within minutes, with or without IgG, and differences in

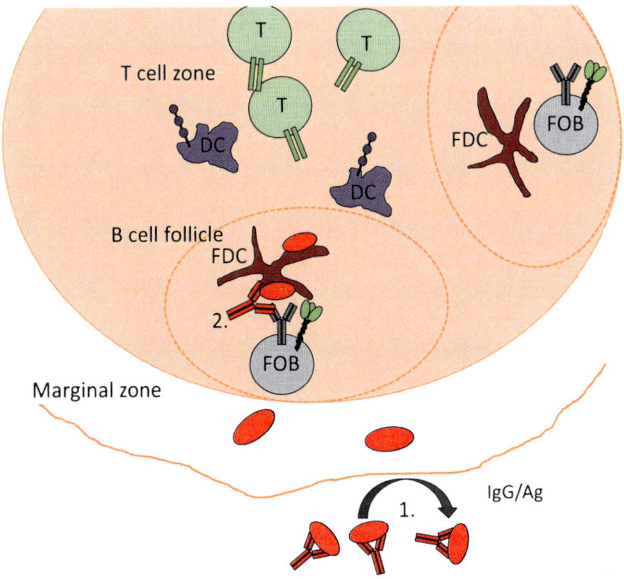

Fig. 9.2 *IgG-mediated suppression*. IgG administered prior to, or up to 5 days after, large particulate antigens like erythrocytes can completely inhibit the antibody response. This ability is used clinically to prevent Rhesus immunization of pregnant RhD⁻ women. IgG can suppress in the absence of IgG Fc receptors and does not require that IgG activates complement. The likely mechanisms of action are that (*1*) IgG bound to the antigen prevents it from reaching the B cell follicle, either by deviating the antigen to other organs, where no induction of immune responses can take place, or by inducing efficient elimination via phagocytosis, and/or (*2*) that IgG by binding to the antigen hides it from B cells (epitope masking), thus effectively competing with B cells for access to the antigen

clearance rates are difficult to measure (Getahun and Heyman 2009; Dennert et al. 1971). IgG decreased the amount of SRBC in the marginal zone (Getahun and Heyman 2009) and prevented follicular localization of large virus-like particles, Qβ-VLP (Link et al. 2012). These observations suggest that IgG causes deviation of antigen from follicles, thus preventing formation of the germinal centers required for class switch recombination and somatic hypermutation. In humans, RhD^+ allogeneic erythrocytes remain in the circulation for several days. Anti-RhD increases clearance of RhD^+ erythrocytes, but whether there is a correlation between clearance and prevention of RhD immunization is not clear (Beliard 2006).

Since IgG-mediated suppression is unperturbed in mice lacking FcγRIIB, it seems unlikely that central B cell inhibition plays a major role (Karlsson et al. 1999, 2001; Heyman et al. 2001). During suppressive conditions, no evidence of clonal deletion or active suppression of B cells is seen, but instead the B cells act as if they have never encountered the antigen (Brinc et al. 2008b). This is what would be expected both after epitope masking and if antigen did not localize to B cell follicles (Fig. 9.2). Experimental support for both mechanisms exist, and mathematical modeling has suggested that both contribute to suppression (Na et al. 2006).

Their relative role probably varies depending on the experimental conditions. When antigen is already deposited in secondary lymphoid organs at the time of IgG administration, epitope masking is likely to play a major role, whereas clearance/deviation may be more important when antigen remains for a long time in the circulation, as would be the case with allogeneic RhD^+ erythrocytes.

9.3 Enhancement of Antibody Responses

All antibody isotypes except IgD have been shown to enhance antibody responses although studies of IgA are rare. Monoclonal IgA was able to enhance induction of memory cells, but had no effect on a primary response (Klaus 1979; Coulie and Van Snick 1985). Below, the enhancing effects of IgM, IgG, and IgE will be discussed separately.

9.3.1 IgM-Mediated Enhancement

9.3.1.1 IgM Enhances Antibody Responses to Erythrocytes, KLH, and Malaria Parasites

An early indication that IgM was able to enhance responses to erythrocytes was found during attempts to understand the mechanism behind RhD prophylaxis (Clarke et al. 1963). When RhD^- male volunteers were given plasma containing IgM anti-RhD together with RhD^+ blood, immunization against the RhD was enhanced and not suppressed (Clarke et al. 1963). Studies in animal models using SRBC have confirmed this observation (Henry and Jerne 1968; Wason 1973; Dennert 1971; Schrader 1973; Heyman et al. 1982; Whited Collisson et al. 1983; Applequist et al. 2000; Heyman et al. 1985). IgM also enhances responses to keyhole limpet hemocyanine (KLH), a large protein with a molecular weight of >1 MDa (Coulie and Van Snick 1985; Youd et al. 2002), and to malaria parasites (Harte et al. 1983). Enhancement of responses to small proteins like OVA has not been reported. IgM can only enhance responses against suboptimal doses of antigen (Henry and Jerne 1968; Powell et al. 1982; Lehner et al. 1983). Not only primary responses (Henry and Jerne 1968; Wason 1973; Dennert 1971; Schrader 1973; Heyman et al. 1982; Whited Collisson et al. 1983; Applequist et al. 2000; Heyman et al. 1985) but also induction of immunological memory (Heyman and Wigzell 1985; Youd et al. 2002) and germinal centers (Ferguson et al. 2004) are augmented. IgM enhances the production of IgM (Henry and Jerne 1968; Wason 1973; Dennert 1971; Schrader 1973; Heyman et al. 1982; Whited Collisson et al. 1983; Applequist et al. 2000), all IgG subclasses (Heyman et al. 1985), and IgE (Strannegård and Belin 1971). In order for enhancement to take place, IgM must be administered within a few hours of the antigen (Wason 1973; Powell et al. 1982; Dennert 1973); IgM administered 1–2 days after SRBC instead suppressed the antibody response (Wason 1973).

9.3.1.2 Complement and IgM-Mediated Enhancement

Monoclonal IgM which, owing to a point mutation in the third constant domain of the mu-heavy chain, is unable to bind C1q and to induce hemolysis (Shulman et al. 1987), cannot enhance antibody responses (Heyman et al. 1988b). In analogy, monomeric IgM, which is unable to activate complement, is unable to enhance antibody responses (Youd et al. 2002) and IgM does not enhance in mice lacking C3 or complement receptors 1 and 2 (CR1/2) (Applequist et al. 2000; Heyman et al. 1988b). Taken together, these observations suggest that IgM-mediated enhancement of antibody responses depends on the ability of IgM to activate complement. However, enhancement does not require that the complement cascade is activated to its final stage where a membrane-attack complex is generated and osmotic lysis of the target cell occurs: IgM enhances well in C5-deficient mice (Heyman et al. 1988b) and IgM enhances responses to KLH, which is a protein and cannot be lysed (Enriquez-Rincon and Klaus 1984; Coulie and Van Snick 1985; Youd et al. 2002). In order to bind C1q, the C1q binding sites of IgM must be exposed. This has been thought to be accomplished by a conformation change in the IgM molecule after it has bound to its antigen. Recent data suggest that the C1q-binding sites will be exposed towards the solution when IgM binds with several of its subunits to an antigen without any obvious conformation change (Czajkowsky and Shao 2009). The requirement for correct exposure of C1q binding sites may explain why only responses against large antigens are enhanced. These antigens would enable IgM to attach with several of its antigen-binding domains and facilitate the orientation/conformation change required for C1q binding.

The connection between a normal antibody response and complement was first made when mice depleted of C3 by treatment with cobra venom factor were shown to have severely impaired antibody responses (Pepys 1976). This finding was confirmed in animals genetically lacking C3 (O´Neil et al. 1988; Fischer et al. 1996), and a similar phenotype was seen in C1q-, C2-, and C4-deficient animals (Fischer et al. 1996; Böttger et al. 1985; Cutler et al. 1998; Rutemark et al. 2011) as well as in C4-deficient humans (Jackson et al. 1979). Complement can be activated by three different pathways: the alternative, the lectin, and the classical pathway. C1q initiates classical complement activation after having bound to IgM or IgG antibodies forming an immune complex with an antigen. Therefore, the crucial role of C1q in primary antibody responses appears paradoxical since very little specific antibody is present in naïve mice to form the immune complexes required to activate the classical pathway. Based on two sets of observations, a role of natural IgM has been postulated to explain this enigma. First, mice lacking secretory IgM ($sIgM^{-/-}$) have lower antibody responses than normal mice, and the responses can be restored with IgM from normal mouse serum (Ehrenstein et al. 1998; Baumgarth et al. 2000). Second, enhancement of antibody responses by specific IgM is complement dependent (Applequist et al. 2000; Youd et al. 2002; Heyman et al. 1988b). These findings suggested that natural IgM binding to antigen, albeit with low affinity, activates C1q and initiates a primary antibody response which would then be further enhanced by early specific IgM. This idea was tested experimentally in mice which produced secretory IgM, but where a mutation had been inserted in the gene encoding, the mu-heavy chain rendering all IgM molecules unable to activate complement. Surprisingly, these mice had normal antibody responses (Rutemark et al. 2011), suggesting that the role of natural IgM in promoting antibody responses is not dependent on complement activation.

9.3.1.3 Fc Receptors and IgM-Mediated Enhancement

As discussed in Chap. 2, Fc receptors for IgM have been described. However, their role in IgM-mediated enhancement of antibody responses has not yet been examined.

9.3.1.4 T Cells and IgM-Mediated Enhancement

IgM does not enhance in T cell-deficient (nude) mice, showing that IgM cannot circumvent the need for T cells in responses to T-dependent antigens (Lehner et al. 1983). Whether IgM upregulates T cell proliferation is not known.

9.3.1.5 Antigen Localization and IgM-Mediated Enhancement

Complement is important for splenic localization of antigens administered iv (Link et al. 2012; Papamichail et al. 1975; Klaus and Humphrey 1977; Davies et al. 1993), and it has been shown that specific IgM enhances localization of antigen to the spleen in parallel with enhancing the antibody response (Dennert 1971; Harte et al. 1983). Detailed studies of splenic sections revealed that natural as well as specific IgM promoted trapping of NP-BSA and Qβ dimers in splenic follicles (Link et al. 2012; Youd et al. 2002; Ferguson et al. 2004), but whether this was paralleled by an enhanced antibody response was not reported. Both antigens are small, and therefore, IgM would not be expected to enhance antibody responses against them. IgM-induced follicular antigen trapping is dependent on CR1/2 and C3, and the immune complexes appear to be transported by CR1/2^{+} marginal zone B cells into follicles (Link et al. 2012; Ferguson et al. 2004).

9.3.1.6 The Mechanism Behind IgM-Mediated Enhancement

Given that enhancement of antibody responses by specific IgM depends on complement, the mechanism behind is likely to be closely related to how complement regulates antibody responses. This is believed to depend on the complement receptors 1 (CD35) and 2 (CD21), primarily expressed on B cells and FDCs. Murine CR1/2 is derived from the same gene (*Cr2*) by alternative splicing, and $Cr2^{-/-}$ mice lack both receptors (Molina et al. 1990; Kurtz et al. 1990). Lack of CR1/2 results in a similar phenotype as lack of C1q, C2, C4, and C3 (reviewed in Roozendaal and Carroll 2007). C1q, C2, and C4 are required for the generation of the split products C3b, iC3b, C3dg, and C3d, which constitute the ligands for CR1/2. Therefore, it seems likely that the impaired antibody responses seen in the absence of C1, C3, C4, and CR1/2 are all caused by lack of CR1/2-mediated effects, either indirectly, owing to failure to cleave C3 and to generate receptor ligands, or directly by the absence of CR1/2.

There are two major, not mutually exclusive, hypotheses for how CR1/2, as well as the early complement factors and IgM, augment antibody responses (Fig. 9.3). The first is based on the in vitro findings that co-cross-linking of the BCR and the co-receptor complex CR2/CD19/CD81 on the B cell surface lowers the threshold for B cell activation by a 100-fold (Carter et al. 1988; Carter and Fearon 1992; Cherukuri et al. 2001). Possibly, antigen-IgM-complement complexes could co-cross-link the two receptor complexes in vivo and cause more efficient B cell signaling, resulting in enhanced antibody responses. The fact that CR1/2 expressed on B cells, and not on FDCs, sometimes play a major role for antibody responses, is compatible with this hypothesis (Ahearn et al. 1996; Croix et al. 1996).

The other proposed model is that complement facilitates trapping of antigen in splenic follicles (Fig. 9.3). Formation of germinal centers in the follicles is required for class switch recombination and affinity maturation of the antibody response, and germinal center formation is impaired in C3- and CR1/2-deficient mice (Papamichail et al. 1975; Ahearn et al. 1996; Mattsson et al. 2011). A likely scenario is that IgM binds to antigen and activates complement, leading to deposition of C3 fragments on the antigen. Marginal zone B cells, expressing high levels of CR1/2, shuttle between the marginal zone and the follicle (Cinamon et al. 2008) and have been shown to deliver antigen-complement-IgM to FDCs (Link et al. 2012; Ferguson et al. 2004). Importantly, CR1/2 expression on marginal zone B cells is sometimes dispensable for a normal antibody response, suggesting that alternative routes for antigen transport exist (Mattsson et al. 2011; Fang et al. 1998; Brockman et al. 2006). When antigen-complement reaches the follicle, FDCs capture the complexes via CR1/2, supported by the findings that CR1/2 expression on FDCs is often crucial for normal antibody responses (Mattsson et al. 2011; Fang et al. 1998; Brockman et al. 2006). Compatible with the antigen trapping model is that IgM increases the total amount of antigen in the spleen (Dennert 1971; Harte et al. 1983), that IgM enhances responses only to suboptimal antigen doses (Henry and Jerne 1968; Powell et al. 1982; Lehner et al. 1983), and that the resulting response mimics that seen with a higher dose of antigen alone (Henry and Jerne 1968; Heyman et al. 1985).

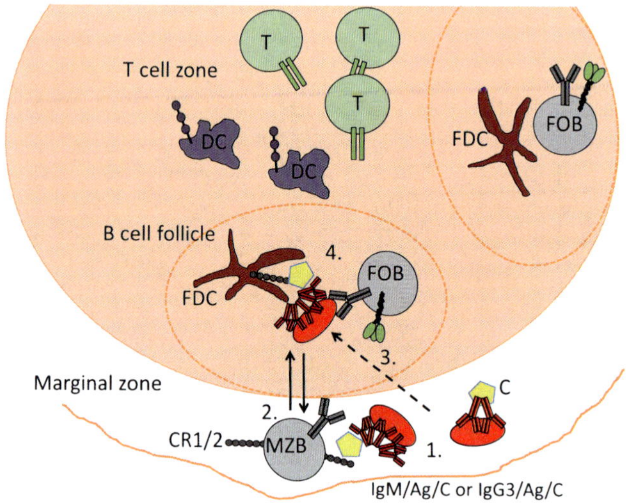

Fig. 9.3 *IgM- and IgG3-mediated enhancement*. IgM administered together with large antigens like erythrocytes or KLH, and IgG3 administered with soluble small proteins, enhance antibody responses. Both pathways require the presence of CR1/2 and C3, and IgM must be able to activate complement in order to be enhancing. The likely mechanisms of action are that (*1*) IgM or IgG3 bind to antigen and activate complement leading to deposition of C3 fragments (which are the ligands for CR1/2) on the Ag-Ab complex, (*2*) Ag-Ab-C complexes are transported into follicles by CR1/2⁺ marginal zone B cells (MZB) or (*3*) Ag-Ab-C complexes reach the follicles by unknown pathways, and (*4*) Ag-Ab-C complexes are trapped on CR1/2⁺ FDCs which present the antigen in an efficient way to follicular B cells (FOB)

9.3.2 IgG-Mediated Enhancement

Low doses of IgG, or IgG administered in preformed complexes (Enriquez-Rincon and Klaus 1984; Dennert 1971), have occasionally been shown to enhance responses to SRBC, although the dominating effect of IgG administered with this antigen is suppressive. Another example of the dual effect of IgG antibodies is that hapten-specific monoclonal IgG suppresses responses to haptenated SRBC but enhances responses to haptenated soluble proteins (Wiersma et al. 1989; Enriquez-Rincon and Klaus 1984).

9.3.2.1 IgG1, IgG2a, IgG2b, and IgG3 Enhance Antibody Responses to Protein Antigens

It was early recognized that complexes of a protein antigen and antisera induced stronger antibody responses than antigen administered alone (von Behring and Wernicke 1892; Stoner and Terres 1963). When monoclonal antibodies became available, it was shown that all murine subclasses can enhance primary antibody responses to proteins such as KLH, OVA, BSA, and tetanus toxoid (Wiersma et al. 1989;

Enriquez-Rincon and Klaus 1984; Coulie and Van Snick 1985; Wernersson et al. 1999; Wiersma 1991, 1992; Diaz de Ståhl et al. 2003; Getahun et al. 2004). Also secondary responses and induction of memory are enhanced by IgG (Klaus 1979; Coulie and Van Snick 1985; Diaz de Ståhl and Heyman 2001) and somatic mutation in germinal centers increases (Nie et al. 1997). IgG-mediated enhancement is potent, and the antibody responses can be enhanced by several 100-fold (Coulie and Van Snick 1985; Wernersson et al. 1999). The affinity of IgG antibodies does not correlate with their enhancing effect (Wiersma et al. 1989; Coulie and Van Snick 1985).

9.3.2.2 Complement and IgG-Mediated Enhancement

IgG3-mediated enhancement is impaired in mice lacking C3 or CR1/2 (Diaz de Ståhl et al. 2003), whereas enhancement is normal in mice lacking the Fc receptor for IgG3 (FcγRI) (Diaz de Ståhl et al. 2003; Hjelm et al. 2005). These observations suggest that IgG3 operates via the complement system. Whether enhancement mediated by IgG1, IgG2a, and IgG2b also requires complement has been a subject of discussion. Early studies showed that C3 depletion of mice abolished trapping of IgG-antigen complexes in B cell follicles and IgG-mediated enhancement of B cell priming (Enriquez-Rincon and Klaus 1984; Papamichail et al. 1975; Klaus 1978). Later, the ability of a panel of monoclonal IgG antibodies to activate complement was found to correlate with their ability to enhance antibody responses (Wiersma et al. 1989; Coulie and Van Snick 1985). Whereas these findings suggested that IgG-mediated enhancement was dependent on complement activation, other studies argued against this. Monoclonal IgG2a antibodies, which owing to a point mutation had lost their ability to activate complement, enhanced antibody responses as efficiently as the wild-type IgG2a (Wiersma et al. 1990). Murine IgG1 is generally a poor complement activator, although some monoclonal IgG1 clones can activate complement (Wiersma et al. 1989). Studies using two different monoclonal IgG1 antibodies, one which could and one which could not activate complement, showed that both were efficient enhancers (Wiersma et al. 1990). Moreover, IgG2a can enhance responses in mice lacking CR1/2 (Applequist et al. 2000), whereas IgG1 and IgG2a cannot enhance antibody responses in $FcR\gamma^{-/-}$ mice (which have a normal complement system but lack all activating IgG Fc receptors) (Wernersson et al. 1999; Getahun et al. 2004).

Thus, it appears that in order to enhance antibody responses, IgM and IgG3 depend more on complement activation, while IgG1, IgG2a, and IgG2b depend more on Fc-receptor-mediated functions. This may possibly be explained by the various strategies the antibody isotypes use to activate complement. Whereas one IgM molecule is sufficient to bind C1q and initiate the classical pathway, two or more IgG molecules must cooperate in order to bind C1q. This is usually achieved by several IgG molecules binding to neighboring epitopes on an antigen. However, unlike the other IgG subclasses, murine IgG3 has the capacity to facilitate binding of other IgG3 molecules via so-called Fc-mediated cooperativity. This means that one IgG3 molecule binding to the surface of, e.g., group A streptococci, can attract additional IgG3 molecules via its Fc part (Cooper et al. 1991; Greenspan and Cooper 1992). This self-aggregation is also evidenced by the tendency of IgG3 to precipitate in the cold (Abdelmoula et al. 1989). Since one IgM or IgG3 molecule (which attracts additional IgG3 molecules) suffices to initiate the complement cascade, it seems likely that these isotypes are more prone to use the complement system in vivo than are IgG2a and IgG2b. Complement activation by IgG2a and IgG2b will require high concentrations of specific IgG and also sufficient epitope density to allow binding of two IgG molecules closely together. These two requirements are more difficult to achieve than binding of single IgM or IgG3 molecules to one epitope.

9.3.2.3 Fc Receptors and IgG-Mediated Enhancement

The Fc receptor for IgG3 has been elusive but is now thought to be FcγRI (Gavin et al. 1998; Barnes et al. 2002). IgG3-mediated enhancement is normal in $FcR\gamma^{-/-}$ mice (lacking FcγRI, III, and IV) as well as in $Fc\gamma RI^{-/-}$ mice (Diaz de Ståhl et al. 2003; Hjelm et al. 2005). In contrast, enhancement by IgG1, IgG2a, and IgG2b was impaired in $FcR\gamma^{-/-}$ mice (Wernersson et al. 1999; Getahun et al. 2004). This implies that IgG3 does not use Fc receptors for mediating enhancement, whereas IgG1, IgG2a, and IgG2b do. As discussed above, this correlates inversely with the use of complement by the various isotypes. The role of the negatively regulating FcγRIIB in antibody responses to complexes of IgG and soluble antigens has been studied. In contrast to the lack of importance of this receptor in IgG-mediated suppression of responses to erythrocytes (Karlsson et al. 1999, 2001), it played a significant role in IgG1-, IgG2a-, and IgG2b-mediated enhancement (Wernersson et al. 1999; Getahun et al. 2004). This was evidenced by an "enhanced enhancement" in $Fc\gamma RIIB^{-/-}$ mice: although the antibody response in wild-type mice was significantly enhanced by these IgG subclasses, the antibody response was dramatically higher in $Fc\gamma RIIB^{-/-}$ mice, probably owing to lack of negative regulation (Wernersson et al. 1999; Getahun et al. 2004). These findings support the idea mentioned earlier (Sect. 9.2.8) that FcγRIIB is a modulating rather than a completely inhibiting receptor. In contrast, IgE- and IgG3-mediated enhancement was normal in $Fc\gamma RIIB^{-/-}$ mice (Wernersson et al. 1999; Diaz de Ståhl et al. 2003). IgE acts via the low-affinity receptor for IgE, CD23, (Gustavsson et al. 1994; Getahun et al. 2005), and IgG3 most likely via CR1/2 (Diaz de Ståhl et al. 2003). Neither of these receptors contain ITAMs and therefore are not expected to be affected by FcγRIIB. The ability of FcγRIIB to downregulate enhancement by IgG1, IgG2a, and IgG2b supports the observations that this enhancement is mediated by a receptor containing the common FcRγ chain (Wernersson et al. 1999). It was previously mentioned that the enhancing capacity of panels of monoclonal IgG antibodies correlated with their ability to activate complement (Wiersma et al. 1989; Coulie and Van Snick 1985). The poorly enhancing antibodies generally belonged to the IgG1 subclass, which does not activate complement very well. IgG1 binds better to inhibitory FcγRs than do IgG2a and IgG2b (Nimmerjahn and Ravetch 2005). Therefore, the observed correlation between IgG1 and lack of enhancement (Wiersma et al. 1989; Coulie and Van Snick 1985) may alternatively be explained by IgG1 having a higher tendency to inhibit enhancement than the other subclasses rather than to its lower ability to activate complement.

9.3.2.4 T Cells and IgG-Mediated Enhancement

Antigens administered with polyclonal IgG and monoclonal IgG2a induce higher proliferation of specific T cells in vivo than antigens administered alone (Getahun et al. 2004; Hamano et al. 2000; Terres et al. 1974; de Jong et al. 2006). The enhancement is dependent on activating IgG Fc receptors (Getahun et al. 2004; Hamano et al. 2000) and on the presence of $CD11c^{+}$ cells (de Jong et al. 2006). In contrast, monoclonal IgG3 was a poor inducer of $CD4^{+}$ T cell proliferation both in vitro and in vivo (Hjelm et al. 2005).

9.3.2.5 Antigen Localization and IgG-Mediated Enhancement

Several reports show that specific IgG or hyperimmune sera increases localization of protein antigens in splenic follicles, and in some cases, this was accompanied by an enhanced induction of B memory cells (Link et al. 2012; Klaus 1978, 1979).

9.3.2.6 The Mechanism Behind IgG-Mediated Enhancement

Regulation of antibody responses by IgG is an extremely complex affair. Whereas all IgG subclasses suppress responses to particulate antigens, they are also able to enhance responses to soluble antigens, and one and the same hapten-specific IgG antibody can have dual effects (Wiersma et al. 1989; Enriquez-Rincon and Klaus 1984). There are several Fc receptors for IgG, expressed on different immune cells. Often more than one FcγR is expressed on the same cell type, and the various IgG subclasses bind to more than one FcγR. There are both activating and inhibitory FcγRs and the relative affinity of a given IgG antibody or immune serum for activating and inhibitory receptors will affect their immunoregulatory function (Nimmerjahn and Ravetch 2005). To further add to the complexity, inhibitory ITAMs have been described (Blank et al. 2009). Most IgG subclasses activate complement and may influence the immune response via this route. These factors make it difficult to precisely elucidate how IgG enhances antibody responses, and most likely several mechanisms are operating. This would in particular be true under physiological conditions when a mixture of IgG subclasses are produced upon immunization.

IgG3-mediated enhancement largely depends on complement (Diaz de Ståhl et al. 2003; Hjelm et al. 2005). As discussed above in relation to IgM-mediated enhancement, involvement of complement may, e.g., result in increased antigen capture by FDCs, increased transportation of immune complexes into follicles by marginal zone B cells, or increased B cell signaling owing to co-cross-linking of BCR and CR2/CD19/CD81. To date, there is little data allowing differentiation between these possibilities as explanations, but in analogy with IgM-mediated enhancement, follicular trapping is likely to play an important role (Fig. 9.3).

IgG1-, IgG2a-, and IgG2b-mediated enhancement has been shown to require activating FcγRs (Wernersson et al. 1999; Getahun et al. 2004; Diaz de Ståhl and Heyman 2001; Wiersma et al. 1990), although contribution of complement in some situations cannot be ruled out. Trapping on FDCs is unlikely since the most abundantly expressed IgG Fc receptor on these cells is FcγRIIB, and IgG1-, IgG2a-, and IgG2b-mediated enhancement was even more pronounced in FcγRIIB$^{-/-}$ than in wild-type mice (Wernersson et al. 1999). Moreover, enhancement by IgG2a requires that FcγRs be expressed on a bone marrow-derived cell, whereas expression on FDCs is dispensable (Diaz de Ståhl and Heyman 2001). A more plausible mechanism is that FcγR$^+$ antigen presenting cells take up and process IgG-antigen complexes more efficiently than antigen alone. This would lead to enhanced antigen presentation to CD4$^+$ T cells and subsequently to enhanced antibody production. In line with this are studies showing that IgG enhances proliferation of CD4$^+$ T cells in vivo (Getahun et al. 2004; Hamano et al. 2000; Terres et al. 1974; de Jong et al. 2006) and that bone marrow-derived cells (Wernersson et al. 1999), more precisely CD11c$^+$ cells (de Jong et al. 2006), are required (Fig. 9.4).

9.3.3 IgE-Mediated Enhancement

9.3.3.1 IgE Enhances Antibody Responses to Small Proteins

In the beginning of the 1990s, it was hypothesized that IgE would be able to potentiate immune responses by forming IgE immune complexes which would bind to the low-affinity receptor for IgE, CD23 (Mudde et al. 1990). An important basis for this idea was that human and mouse B cells had been shown to take up IgE-antigen complexes via CD23 and efficiently present the antigen to CD4$^+$ T cells in vitro (Kehry and Yamashita 1989; Pirron et al. 1990). When the ability of IgE to regulate antibody responses in vivo was tested experimentally, it was found that small quantities of IgE, administered in close temporal relationship with antigen, indeed was a potent enhancer of the antibody response to this antigen (Heyman et al. 1993). Administration of 10–100 micrograms of monoclonal TNP-specific IgE induced a more than 100-fold stronger IgG anti-BSA response than BSA-TNP administered alone (Heyman et al. 1993). Responses to proteins such as BSA, OVA, and tetanus toxoid could be enhanced, whereas no effect was seen against SRBC (Gustavsson et al. 1994; Hjelm et al. 2008). Primary and secondary antibody responses as well as induction of germinal centers were enhanced, and there was no obvious skewing of the Th1/Th2 balance (Gustavsson et al. 1994; Hjelm et al. 2008). Although

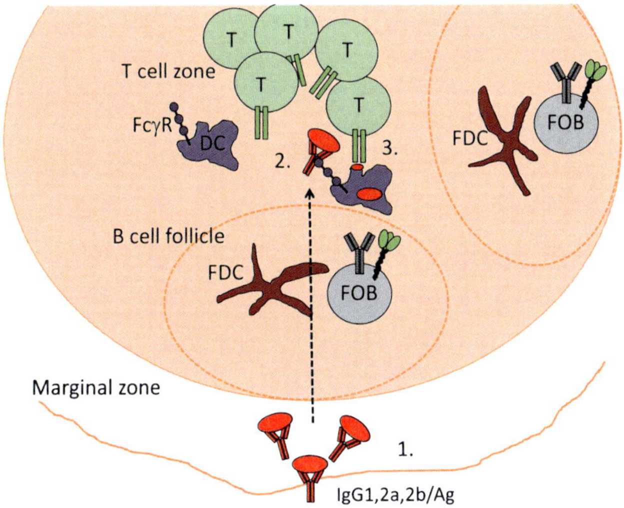

Fig. 9.4 *IgG1-, IgG2a-, and IgG2b-mediated enhancement.* IgG1, IgG2a, and IgG2b enhance antibody responses to soluble proteins like OVA, BSA, KLH, and tetanus toxoid. Enhancement requires the presence of activating FcγRs and affects also CD4⁺ antigen-specific T cells. The likely mechanism is that (*1*) IgG-Ag complexes reach the marginal zone after being injected iv and (*2*) via unknown routes reach CD11c⁺ dendritic cells which bind IgG-Ag complexes to their FcγRs, endocytose them, and (*3*) present antigenic peptides to CD4⁺ T cells. These become activated, proliferate, and subsequently help specific B cells to produce antibodies

enhancement of the IgG response was most marked, also specific IgM, IgA, and IgE production was upregulated (Gustavsson et al. 1994; Westman et al. 1997).

9.3.3.2 Fc Receptors and IgE-Mediated Enhancement

The most well-known IgE receptor is the high-affinity receptor, FcεRI, which is primarily expressed on mast cells and basophils. However, murine IgE binds with low affinity to CD23, FcγRIIB, FcγRIII, and FcγRIV (Conrad 1990; Takizawa et al. 1992; Mancardi et al. 2008). IgE-mediated enhancement of antibody responses was completely abrogated when CD23 was blocked by a monoclonal antibody (Heyman et al. 1993) and in CD23-knockout mice (Getahun et al. 2005; Hjelm et al. 2008; Fujiwara et al. 1994; Gustavsson et al. 2000). Moreover, antigen targeted to CD23 by covalent coupling to anti-CD23 antibodies induced enhanced antibody responses (Squire et al. 1994). Enhancement was normal mice lacking FcγRIII, FcγRIV, and FcεRI (FcRγ$^{-/-}$) (Wernersson et al. 1999) or FcγRIIB (Getahun et al. 2004). Taken together, these observations show that IgE-mediated enhancement is exclusively dependent on CD23. This is the only Fc receptor that does not belong to the super-immunoglobulin family (reviewed in Conrad 1990; Gould and Sutton 2008). In mice, CD23 is mainly expressed on B cells and FDCs and exists in at least two isoforms, CD23a and CD23b. CD23a is constitutively expressed on B cells and

FDCs, whereas CD23b requires IL-4 for its expression. IgE enhances equally well in IL-4-deficient as in wild-type mice, suggesting that only the a isoform is involved in IgE-mediated enhancement (Hjulström et al. 1995). Studies in bone marrow chimeric mice showed that CD23+ B cells, but not CD23+ FDCs, were required (Getahun et al. 2005; Gustavsson et al. 2000).

9.3.3.3 Complement and IgE-Mediated Enhancement

IgE does not activate complement, but human CD23 and human CR2 are a receptor-ligand pair (Aubry et al. 1992). However, IgE-mediated enhancement is unperturbed in mice lacking CR1/2 or C3 (Applequist et al. 2000; Diaz de Ståhl et al. 2003).

9.3.3.4 T Cells and IgE-Mediated Enhancement

The DO11.10 experimental system described above was used to test whether IgE could enhance proliferation of T helper cells. Wild-type mice, transferred with transgenic OVA-specific CD4+ T cells, were immunized with TNP-specific IgE and OVA-TNP or with OVA-TNP alone. IgE markedly enhanced proliferation and activation of OVA-specific CD4+ T cells (Getahun et al. 2005; Hjelm et al. 2008).

9.3.3.5 Antigen Localization and IgE-Mediated Enhancement

The fate of IgE-complexed antigen has been followed in vivo. Five minutes after iv injection, a majority of CD23+ B cells in blood had bound the antigen, whereas few B cells had bound uncomplexed antigen (Hjelm et al. 2008). After 30 min, IgE-complexed antigen was found on follicular B cells in the splenic follicles (Hjelm et al. 2008).

9.3.3.6 The Mechanism Behind IgE-Mediated Enhancement

In human and mouse in vitro systems, IgE-antigen complexes are efficiently endocytosed, processed, and presented to T cells by B cells which have taken up the immune complexes via CD23 (Kehry and Yamashita 1989; Pirron et al. 1990; van der Heijden et al. 1993; Bheekha Escura et al. 1995; Carlsson et al. 2007). The massive proliferation of antigen-specific CD4+ T cells induced by IgE antigen in vivo is also dependent on CD23+ B cells (Getahun et al. 2005). Therefore, it seemed logical that enhanced B-cell-mediated antigen presentation would explain IgE-mediated upregulation of T and B cell responses in vivo. However, the fact that CD23+ B cells were required for the transportation of IgE antigen into follicles (Hjelm et al. 2008) opened for the possibility that transportation, and not presentation, was the effector function exerted by the CD23+ B cells. Recent work indeed shows that the cells

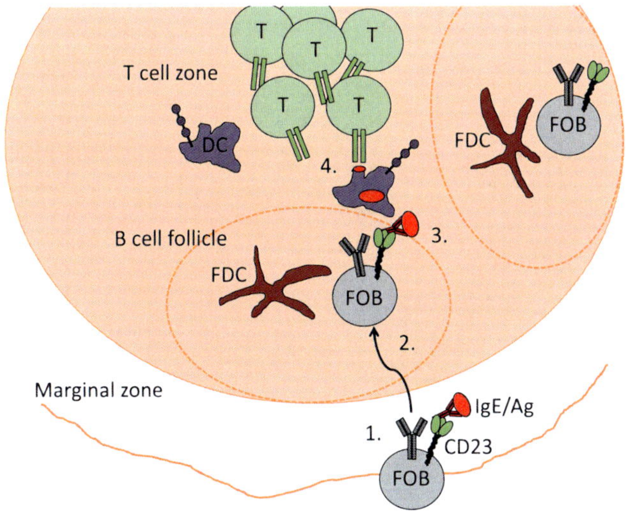

Fig. 9.5 *IgE-mediated enhancement*. IgE enhances antibody responses to soluble proteins like OVA, BSA, and tetanus toxoid. Enhancement requires the presence of the low-affinity receptor for IgE, FcεRII (CD23), and affects also CD4⁺ antigen-specific T cells. The likely mechanism is that (*1*) IgE-Ag complexes are captured by CD23⁺ peripheral B cells which (*2*) transport them to follicles and, at the T-B zone border, (*3*) deliver IgE-Ag to CD11c⁺ dendritic cells which (*4*) endocytose the antigen and present antigenic peptides to CD4⁺ T cells. These become activated, proliferate, and subsequently help specific B cells to produce antibodies

presenting IgE antigen to CD4⁺ T cells in vivo are not CD23⁺ B cells but CD11c⁺ cells, supposedly dendritic cells (Henningsson et al. 2011). Taken together, available data suggest that IgE-antigen complexes injected iv rapidly bind to recirculating CD23⁺ B cells which transport the antigen to splenic follicles. The follicular B cells then deliver their IgE-antigen cargo to dendritic cells, probably at the border between the B and T cell zone. Dendritic cells would then take up the antigen in a CD23-independent manner, since murine dendritic cells do not express CD23, process, and present it to CD4⁺ T cells which will efficiently help specific B cells to produce antibodies (Fig. 9.5).

9.4 Concluding Remarks

9.4.1 Biological and Clinical Role of IgG-Mediated Regulation

Not only passively administered but also endogenously produced IgG is suppressive (Heyman and Wigzell 1985; Bystryn et al. 1970). This, together with the observations that IgG can inhibit ongoing antibody responses (Henry and Jerne 1968; Karlsson et al. 2001; Chan and Sinclair 1973), suggests that IgG-mediated suppression plays a physiological role in downregulating antibody responses. The most important clinical role is in the successful prevention of RhD immunization and hemolytic disease of the newborn. The challenge here is to define the underlying mechanism so that efficient monoclonal IgG anti-D antibodies can be generated and used in therapy. The enhancing potential of IgG, seen with small antigens, is likely to play a major role for efficient secondary responses.

9.4.2 *Biological and Clinical Role of IgM-Mediated Enhancement*

Observations that natural IgM plays a role to initiate antibody responses suggest that IgM is important to jump-start primary antibody responses. Specific IgM produced early in a primary response will presumably further feedback enhance the response.

9.4.3 *The Biological and Clinical Role of IgE-Mediated Enhancement*

The biologic role of IgE-mediated enhancement is not clear. Normally, IgE levels in serum are very low and may not reach the concentrations required for induction of significant enhancement. Ten but not 1 μg specific IgE/mouse enhance antibody responses (Heyman et al. 1993). Assuming that a mouse has 2 ml blood, this corresponds to 5 μg/ml of specific IgE in serum. For comparison, the concentration of total IgE in a healthy adult is approximately 0.3 μg/ml, suggesting that the levels of specific IgE will be too low for an efficient enhancing effect. Hypothetically, higher concentrations of IgE may be produced locally in healthy individuals. Specific IgE increases in virus infections, and this may play a role in potentiating early antibody responses (Welliver et al. 1982). Allergic diseases are also accompanied by high levels of specific IgE, and it has been suggested that preexisting IgE in sensitized patients may bind to the allergen and start a vicious circle, enhancing the unwanted response to the allergen. This idea has received some experimental support from human in vitro studies (van der Heijden et al. 1993; van Neerven et al. 1999, 2004). Clinical trials where CD23 is blocked by monoclonal antibodies have reported a decrease in IgE concentrations, but no clinical effect on asthma (Rosenwasser et al. 2003).

In conclusion, antibodies are potent physiological regulators of specific immune responses. Elucidation of their mechanisms of action will increase our understanding of basic immunology and hopefully lead to new therapeutic applications.

References

Abdelmoula M, Spertini F, Shibata T, Gyotoku Y, Luzuy S, Lambert PH et al (1989) IgG3 is the major source of cryoglobulins in mice. J Immunol 143:526–532

Ahearn JM, Fischer MB, Croix D, Goerg S, Ma M, Xia J et al (1996) Disruption of the Cr2 locus results in a reduction in B-1a cells and in an impaired B cell response to T-dependent antigen. Immunity 4:251–262

Amigorena S, Bonnerot C, Drake JR, Choquet D, Hunziker W, Guillet J-G et al (1992) Cytoplasmic domain heterogeneity and functions of IgG Fc receptors in B lymphocytes. Science 256:1808–1812

Applequist SE, Dahlström J, Jiang N, Molina H, Heyman B (2000) Antibody production in mice deficient for complement receptors 1 and 2 can be induced by IgG/Ag and IgE/Ag, but not IgM/Ag complexes. J Immunol 165:2398–2403

Aubry J-P, Pochon S, Graber P, Jansen KU, Bonnefoy J-Y (1992) CD21 is a ligand for CD23 and regulates IgE production. Nature 358:505–507

Barnes N, Gavin AL, Tan PS, Mottram P, Koentgen F, Hogarth PM (2002) FcγRI-deficient mice show multiple alterations to inflammatory and immune responses. Immunity 16(3):379–389

Baumgarth N, Herman OC, Jager GC, Brown LE, Herzenberg LA, Chen J (2000) B-1 and B-2 cell-derived immunoglobulin M antibodies are nonredundant components of the protective response to influenza virus infection. J Exp Med 192:271–280

Beliard R (2006) Monoclonal anti-D antibodies to prevent alloimmunization: lessons from clinical trials. Transfus Clin Biol 13(1–2):58–64

Bheekha Escura R, Wasserbauer E, Hammerschmid F, Pearce A, Kidd P, Mudde GC (1995) Regulation and targeting of T-cell immune responses by IgE and IgG antibodies. Immunology 86:343–350

Blank U, Launay P, Benhamou M, Monteiro RC (2009) Inhibitory ITAMs as novel regulators of immunity. Immunol Rev 232(1):59–71

Böttger EC, Hoffmann T, Hadding U, Bitter-Suermann D (1985) Influence of genetically inherited complement deficiencies on humoral immune response in guinea pigs. J Immunol 135:4100–4107

Bowman JM (1988) The prevention of Rh immunization. Transfus Med Rev 2:129–150

Brinc D, Lazarus AH (2009) Mechanisms of anti-D action in the prevention of hemolytic disease of the fetus and newborn. Hematology Am Soc Hematol Educ Program 185–191

Brinc D, Le-Tien H, Crow AR, Freedman J, Lazarus AH (2007) IgG-mediated immunosuppression is not dependent on erythrocyte clearance or immunological evasion: implications for the mechanism of action of anti-D in the prevention of haemolytic disease of the newborn? Br J Haematol 139(2):275–279

Brinc D, Le-Tien H, Crow AR, Siragam V, Freedman J, Lazarus AH (2008a) Transfusion of IgG-opsonized foreign red blood cells mediates reduction of antigen-specific B cell priming in a murine model. J Immunol 181(2):948–953, July 15

Brinc D, Le-Tien H, Crow AR, Siragam V, Freedman J, Lazarus AH (2008b) Immunoglobulin G-mediated regulation of the murine immune response to transfused red blood cells occurs in the absence of active immune suppression: implications for the mechanism of action of anti-D in the prevention of haemolytic disease of the fetus and newborn? Immunology 124(1):141–146

Brockman MA, Verschoor A, Zhu J, Carroll MC, Knipe DM (2006) Optimal long-term humoral responses to replication-defective herpes simplex virus require CD21/CD35 complement receptor expression on stromal cells. J Virol 80(14):7111–7117

Brody NI, Walker JG, Siskind GW (1967) Studies on the control of antibody synthesis. Interaction of antigenic competition and suppression of antibody formation by passive antibody on the immune response. J Exp Med 126:81–91

Brüggemann M, Rajewsky K (1982) Regulation of the antibody response against hapten-coupled erythrocytes by monoclonal anti-hapten antibodies of various isotypes. Cell Immunol 71:365–373
Bystryn J-C, Graf MW, Uhr JW (1970) Regulation of antibody formation by serum antibody. II. Removal of specific antibody by means of exchange transfusion. J Exp Med 132:1279–1287
Carlsson F, Hjelm F, Conrad DH, Heyman B (2007) IgE enhances specific antibody and T cell responses in mice overexpressing CD23. Scand J Immunol 66:261–270
Carter RH, Fearon DT (1992) CD19: lowering the threshold for antigen receptor stimulation of B lymphocytes. Science 256:105–107
Carter RH, Spycher MO, Ng YC, Hoffmann R, Fearon DT (1988) Synergistic interaction between complement receptor type 2 and membrane IgM on B-lymphocytes. J Immunol 141:457–463
Cerottini JC, McConahey PJ, Dixon FJ (1969) The immunosuppressive effect of passively administered antibody IgG fragments. J Immunol 102:1008–1015
Chan PL, StC Sinclair NR (1973) Regulation of the immune response. VI. Inability of F(ab')2 antibody to terminate established immune responses and its ability to interfere with IgG antibody-mediated immunosuppression. Immunology 24:289–310
Cherukuri A, Cheng PC, Sohn HW, Pierce SK (2001) The CD19/CD21 complex functions to prolong B cell antigen receptor signaling from lipid rafts. Immunity 14(2):169–179
Cinamon G, Zachariah MA, Lam OM, Foss FW Jr, Cyster JG (2008) Follicular shuttling of marginal zone B cells facilitates antigen transport. Nat Immunol 9(1):54–62
Clarke CA, Donohoe WTA, Woodrow JC, Finn R, Krevans JR, Kulke W et al (1963) Further experimental studies on the prevention of Rh haemolytic disease. Br Med J 1:979–984
Conrad DH (1990) FcεRII/CD23: the low affinity receptor for IgE. Annu Rev Immunol 8:623–645
Cooper LJ, Schimenti JC, Glass DD, Greenspan NS (1991) H chain C domains influence the strength of binding of IgG for streptococcal group A carbohydrate. J Immunol 146(8):2659–2663
Coulie P, Van Snick J (1985) Enhancement of IgG anti-carrier responses by IgG2-anti-hapten antibodies in mice. Eur J Immunol 15:793–798
Croix DA, Ahearn JM, Rosengard AM, Han S, Kelsoe G, Ma M et al (1996) Antibody response to a T-dependent antigen requires B cell expression of complement receptors. J Exp Med 183:1857–1864
Cutler AJ, Botto M, van Essen D, Rivi R, Davies KA, Gray D et al (1998) T cell-dependent immune response in C1q-deficient mice: defective interferon γ production by antigen-specific T cells. J Exp Med 187:1789–1797
Czajkowsky DM, Shao Z (2009) The human IgM pentamer is a mushroom-shaped molecule with a flexural bias. Proc Natl Acad Sci USA 106(35):14960–14965
Daeron M, Lesourne R (2006) Negative signaling in Fc receptor complexes. Adv Immunol 89:39–86
Davies KA, Erlendsson K, Beynon HLC, Peters AM, Steinsson K, Valdimarsson H et al (1993) Splenic uptake of immune complexes in man is complement-dependent. J Immunol 151:3866–3873
de Jong JM, Schuurhuis DH, Ioan-Facsinay A, Welling MM, Camps MG, van der Voort EI et al (2006) Dendritic cells, but not macrophages or B cells, activate major histocompatibility complex class II-restricted CD4+ T cells upon immune-complex uptake in vivo. Immunology 119(4):499–506
Dennert G (1971) The mechanism of antibody-induced stimulation and inhibition of the immune response. J Immunol 106:951–955
Dennert G (1973) Effects of IgM on the in vivo and in vitro immune response. Proc Soc Exp Biol Med 143:889–893
Dennert G, Pohlit H, Rajewsky K (1971) Co-operative antibody: a concentrating device. In: Mäkelä O, Cross A, Kosunen TU (eds) Cell interactions and receptor antibodies in immune responses. Academic, London, pp 3–7

Diaz de Ståhl T, Heyman B (2001) IgG2a-mediated enhancement of antibody responses is dependent of FcRγ+ bone marrow-derived cells. Scand J Immunol 54:1–7

Diaz de Ståhl T, Dahlström J, Carroll MC, Heyman B (2003) A role for complement in feedback-enhancement of antibody responses by IgG3. J Exp Med 197:1183–1190

Dresser DW (1990) Feedback by early and late primary antisera on the primary and secondary adoptive immune responses of mice to burro erythrocytes. Cell Immunol 127:405–419

Ehrenstein MR, O'Keefe TL, Davies SL, Neuberger MS (1998) Targeted gene disruption reveals a role for natural secretory IgM in the maturation of the primary immune response. Proc Natl Acad Sci USA 95(17):10089–10093

Enriques-Rincon F, Klaus GGB (1984) Differing effects of monoclonal anti-hapten antibodies on humoral responses to soluble or particulate antigens. Immunology 52:129–136

Enriquez-Rincon F, Klaus GGB (1984) Follicular trapping of hapten-erythrocyte-antibody complexes in mouse spleen. Immunology 52:107–116

Fang Y, Xu C, Fu Y-X, Holers VM, Molina H (1998) Expression of complement receptors 1 and 2 on follicular dendritic cells is necessary for the generation of a strong antigen-specific IgG response. J Immunol 160:5273–5279

Ferguson AR, Youd ME, Corley RB (2004) Marginal zone B cells transport and deposit IgM-containing immune complexes onto follicular dendritic cells. Int Immunol 16(10):1411–1422

Fischer MB, Ma M, Goerg S, Zhou X, Xia J, Finco O et al (1996) Regulation of the B cell response to T-dependent Ags by classical pathway complement. J Immunol 157:549–556

Freedman J, Massey A, Chaplin H, Monroe MC (1980) Assessment of complement binding by anti-D and anti-M antibodies employing labelled antiglobulin antibodies. Br J Haematol 45(2):309–318

Fujiwara H, Kikutani H, Suematsu S, Naka T, Yoshida K, Yoshida K et al (1994) The absence of IgE antibody-mediated augmentation of immune responses in CD23-deficient mice. Proc Natl Acad Sci USA 91:6835–6839

Gavin AL, Barnes N, Dijstelbloem HM, Hogarth PM (1998) Identification of the mouse IgG3 receptor: implications for antibody effector function at the interface between innate and adaptive immunity. J Immunol 160:20–23

Getahun A, Heyman B (2009) Studies on the mechanism by which antigen-specific IgG suppresses primary antibody responses: evidence for epitope masking and decreased localization of antigen in the spleen. Scand J Immunol 70(3):277–287

Getahun A, Dahlström J, Wernersson S, Heyman B (2004) IgG2a-mediated enhancement of Ab- and T-cell responses and its relation to inhibitory and activating FcγRs. J Immunol 172:5269–5276

Getahun A, Hjelm F, Heyman B (2005) IgE enhances antibody and T cell responses in vivo via CD23+ B Cells. J Immunol 175(3):1473–1482

Gould HJ, Sutton BJ (2008) IgE in allergy and asthma today. Nat Rev Immunol 8(3):205–217

Greenbury CL, Moore DH (1968) Non-specific antibody-induced suppression of the immune response. Nature 219:526–527

Greenspan NS, Cooper LJN (1992) Intermolecular cooperativity: a clue to why mice have IgG3? Immunol Today 13:164–168

Grundbacher FJ (1992) Behring's discovery of diphtheria and tetanus antitoxins. Immunol Today 13(5):188–190

Gustavsson S, Hjulström S, Liu T, Heyman B (1994) CD23/IgE-mediated regulation of the specific antibody response in vivo. J Immunol 152:4793–4800

Gustavsson S, Wernersson S, Heyman B (2000) Restoration of the antibody response to IgE/antigen complexes in CD23-deficient mice by CD23+ spleen or bone marrow cells. J Immunol 164:3990–3995

Hamano Y, Arase H, Saisho H, Saito T (2000) Immune complex and Fc receptor-mediated augmentation of antigen presentation for in vivo Th cell responses. J Immunol 164(12):6113–6119

Harte PG, Cooke A, Playfair JHL (1983) Specific monoclonal IgM is a potent adjuvant in murine malaria vaccination. Nature 302:256–258

Henningsson F, Ding Z, Dahlin JS, Linkevicius M, Carlsson F, Grönvik KO et al (2011) IgE-mediated enhancement of CD4 T cell responses in mice requires antigen presentation by CD11c cells and not by B cells. PLoS One 6(7):e21760

Henry C, Jerne N (1968) Competition of 19S and 7S antigen receptors in the regulation of the primary immune response. J Exp Med 128:133–152

Heyman B (1999) Antibody feedback suppression: towards a unifying concept? Immunol Lett 68:41–45

Heyman B (2000) Regulation of antibody responses via antibodies, complement, and Fc receptors. Annu Rev Immunol 18:709–737

Heyman B (2003) Feedback regulation by IgG antibodies. Immunol Lett 88:157–161

Heyman B, Wigzell H (1984) Immunoregulation by monoclonal sheep erythrocyte specific IgG antibodies. Suppression is correlated to level of antigen binding and not to isotype. J Immunol 132:1136–1143

Heyman B, Wigzell H (1985) IgM enhances and IgG suppresses immunological memory in mice. Scand J Immunol 21:255–266

Heyman B, Andrighetto S, Wigzell H (1982) Antigen dependent IgM-mediated enhancement of the sheep erythrocyte response in mice. Evidence for induction of B cells with specificities other than that of the injected antibodies. J Exp Med 155:994–1009

Heyman B, Hobbs M, Weigle WO (1985) IgM-mediated enhancement of in vivo anti-sheep erythrocyte antibody responses: isotype analysis of the enhanced responses. Cell Immunol 92:134–141

Heyman B, Wiersma E, Nose M (1988a) Complement activation is not required for IgG-mediated suppression of the antibody response. Eur J Immunol 18:1739–1743

Heyman B, Pilström L, Shulman MJ (1988b) Complement activation is required for IgM-mediated enhancement of the antibody response. J Exp Med 167:1999–2004

Heyman B, Liu T, Gustavsson S (1993) In vivo enhancement of the specific antibody response via the low affinity receptor for IgE. Eur J Immunol 23:1739–1742

Heyman B, Dahlström J, Diaz de Ståhl T, Getahun A, Wernersson S, Karlsson MCI (2001) No evidence for a role of FcγRIIB in suppression of in vivo antibody responses to erythrocytes by passively administered IgG. Scand J Immunol 53:331–334

Hjelm F, Carlsson F, Verbeek S, Heyman B (2005) IgG3-mediated enhancement of the antibody response is normal in Fc gammaRI-deficient mice. Scand J Immunol 62(5):453–461

Hjelm F, Carlsson F, Getahun A, Heyman B (2006) Antibody-mediated regulation of the immune response. Scand J Immunol 64(3):177–184

Hjelm F, Karlsson MCI, Heyman B (2008) A novel B-cell mediated transport of IgE-immune complexes to the follicle of the spleen. J Immunol 180:6604–6610

Hjulström S, Landin A, Jansson L, Holmdahl R, Heyman B (1995) No role of interleukin-4 in CD23/IgE-mediated enhancement of the murine antibody response in vivo. Eur J Immunol 25:1469–1472

Jackson CG, Ochs HD, Wedgwood RJ (1979) Immune response of a patient with deficiency of the fourth component of complement and systemic lupus erythematosus. N Engl J Med 300:1124–1129

Junghans RP, Anderson CL (1996) The protection receptor for IgG catabolism is the beta-2-microglobulin-containing neonatal intestinal transport receptor. Proc Natl Acad Sci USA 93:5512–5516

Karlsson MCI, Wernersson S, Diaz de Ståhl T, Gustavsson S, Heyman B (1999) Efficient IgG-mediated suppression of primary antibody responses in Fc-gamma receptor-deficient mice. Proc Natl Acad Sci USA 96:2244–2249

Karlsson MCI, Getahun A, Heyman B (2001) FcγRIIB in IgG-mediated suppression of antibody responses: different impact in vivo and in vitro. J Immunol 167:5558–5564

Kehry MR, Yamashita LC (1989) Low-affinity IgE receptor (CD23) function on mouse B cells: role in IgE-dependent antigen focusing. Proc Natl Acad Sci USA 86:7556–7560
Klaus GGB (1978) The generation of memory cells. II. Generation of B memory cells with preformed antigen-antibody complexes. Immunology 34:643–652
Klaus GGB (1979) Generation of memory cells. III. Antibody class requirements for the generation of B-memory cells by antigen-antibody complexes. Immunology 37:345–351
Klaus GGB, Humphrey JH (1977) The generation of memory cells. I. The role of C3 in the generation of B memory cells. Immunology 33:31
Krieger N, Pesce A, Michael JG (1983) Immunoregulation of the anti-bovine serum albumin response by polyclonal and monoclonal antibodies. Cell Immunol 80:279–287
Kumpel BM, Elson CJ (2001) Mechanism of anti-D-mediated immune suppression - a paradox awaiting resolution? Trends Immunol 22:26–31
Kuo TT, Baker K, Yoshida M, Qiao SW, Aveson VG, Lencer WI et al (2010) Neonatal Fc receptor: from immunity to therapeutics. J Clin Immunol 30(6):777–789
Kurtz CB, O'Toole E, Christensen SM, Weis JH (1990) The murine complement receptor gene family IV: alternative splicing of Cr2 gene transcripts predicts two distinct gene products that share homologous domains with both human CR2 and CR1. J Immunol 144:3581–3591
Lehner P, Hutchings P, Lydyard PM, Cooke A (1983) Regulation of the immune response by antibody II. IgM-mediated enhancement: dependency on antigen dose, T-cell requirement and lack of evidence for an idiotype-related mechanism. Immunology 50:503–509
Link A, Zabel F, Schnetzler Y, Titz A, Brombacher F, Bachmann MF (2012) Innate immunity mediates follicular transport of particulate but not soluble protein antigen. J Immunol 188:3724–3733
Mancardi DA, Iannascoli B, Hoos S, England P, Daeron M, Bruhns P (2008) FcgammaRIV is a mouse IgE receptor that resembles macrophage FcepsilonRI in humans and promotes IgE-induced lung inflammation. J Clin Invest 118(11):3738–3750
Mattsson J, Yrlid U, Stensson A, Schon K, Karlsson MC, Ravetch JV et al (2011) Complement activation and complement receptors on follicular dendritic cells are critical for the function of a targeted adjuvant. J Immunol 187(7):3641–3652
Molina H, Kinoshita T, Inoue K, Carel J-C, Holers VM (1990) A molecular and immunochemical characterization of mouse CR2: evidence for a single gene model of mouse complement receptors 1 and 2. J Immunol 145:2974–2983
Möller G (1985) Antibody-mediated suppression of the immune response is determinant specific. Eur J Immunol 15:409–412
Möller G, Wigzell H (1965) Antibody synthesis at the cellular level. Antibody-induced suppression of 19S and 7S antibody responses. J Exp Med 121:969–989
Mudde GC, Hansel TT, van Reijsen FC, Osterhoff BF, Bruijnzeel-Koomen CAFM (1990) IgE: an immunoglobulin specialized in antigen capture? Immunol Today 11:440–443
Murphy KM, Heimberger AB, Loh DY (1990) Induction by antigen of intrathymic apoptosis of CD4+CD8+TCRlo thymocytes in vivo. Science 250:1720–1723
Muta T, Kurosaki T, Misulovin Z, Sanchez M, Nussenzweig MC, Ravetch JV (1994) A 13-amino-acid motif in the cytoplasmic domain of FcγRIIB modulates B-cell receptor signalling. Nature 368:70–73
Na D, Kim D, Lee D (2006) Mathematical modeling of humoral immune response suppression by passively administered antibodies in mice. J Theor Biol 241(4):830–851
Nicholas R StC. Sinclair (1969) Regulation of the immune response. I. Reduction in ability of specific antibody to inhibit longlasting IgG immunological priming after removal of the Fc fragment. J Exp Med 129:1183–1201
Nie X, Basu S, Cerny J (1997) Immunization with immune complex alters the repertoire of antigen-reactive B cells in the germinal centers. Eur J Immunol 27(12):3517–3525
Nimmerjahn F, Ravetch JV (2005) Divergent immunoglobulin g subclass activity through selective Fc receptor binding. Science 310(5753):1510–1512, December 2

Nimmerjahn F, Ravetch JV (2006) Fcgamma receptors: old friends and new family members. Immunity 24(1):19–28
Nimmerjahn F, Ravetch JV (2010) Antibody-mediated modulation of immune responses. Immunol Rev 236:265–275
O'Neil KM, Ochs SR, Heller SR, Cork LC, Morris JM, Winkelstein JA (1988) Role of C3 in humoral immunity. Defective antibody production in C3-deficient dogs. J Immunol 140:1939–1945
Papamichail M, Gutierrez C, Embling P, Johnson P, Holborow EJ, Pepys MB (1975) Complement dependence of localisation of aggregated IgG in germinal centers. Scand J Immunol 4:343–347
Pepys MB (1976) Role of complement in the induction of immunological responses. Transplant Rev 32:93–120
Pirron U, Schlunck T, Prinz JC, Rieber EP (1990) IgE-dependent antigen focusing by human B lymphocytes is mediated by low-affinity receptor for IgE. Eur J Immunol 20:1547–1551
Pollack W (1984) In haemolytic disease of the newborn (Garratty, G., ed). Am Assoc Blood Banks 53–66
Powell R, Hutchings P, Cooke A, Lydyard PM (1982) Antibody mediated regulation of immune responses I. Enhancement of specific antibody responses through IgM antibodies. Immunol Lett 4:253–258
Quintana IZ, Silveira AV, Möller G (1987) Regulation of the antibody response to sheep erythrocytes by monoclonal IgG antibodies. Eur J Immunol 17:1343–1349
Roozendaal R, Carroll MC (2007) Complement receptors CD21 and CD35 in humoral immunity. Immunol Rev 219:157–166
Rosenwasser LJ, Busse WW, Lizambri RG, Olejnik TA, Totoritis MC (2003) Allergic asthma and an anti-CD23 mAb (IDEC-152): results of a phase I, single-dose, dose-escalating clinical trial. J Allergy Clin Immunol 112(3):563–570
Rutemark C, Alicot E, Bergman A, Ma M, Getahun A, Ellmerich S et al (2011) Requirement for complement in antibody responses is not explained by the classic pathway activator IgM. Proc Natl Acad Sci USA 108(43):E934–E942
Safford JJW, Tokuda S (1971) Antibody-mediated suppression of the immune response: effect on the development of immunologic memory. J Immunol 107:1213–1225
Schrader JW (1973) Regulation of the immune response by IgM antibody: a paradoxical suppression of the in vitro primary immune response to sheep erythrocytes by IgM. Aust J Exp Biol Med Sci 51:333–346
Shulman MJ, Collins C, Pennell N, Hozumi N (1987) Complement activation by IgM: evidence for the importance of the third constant domain of the μ heavy chain. Eur J Immunol 17:549–554
Spiegelberg HL, Weigle WO (1965) The catabolism of homologous and heterologous 7s gamma globulin fragments. J Exp Med 121:323–338
Squire CM, Studer EJ, Lees A, Finkelman FD, Conrad DH (1994) Antigen presentation is enhanced by targeting antigen to the Fc epsilon RII by antigen-anti-Fc epsilon RII conjugates. J Immunol 152(9):4388–4396
Stoner RD, Terres G (1963) Enhanced antitoxin responses in irradiated mice elicited by complexes of tetanus toxoid and specific antibody. J Immunol 91:761–770
Strannegård Ö, Belin L (1970) Suppression of reagin synthesis in rabbits by passively administered antibody. Immunology 20:775–785
Strannegård Ö, Belin L (1971) Enhancement of reagin formation in rabbits by passively administered 19S antibody. Immunology 20:427–431
Takizawa F, Adamczewski M, Kinet J-P (1992) Identification of the low affinity receptor for immunoglobulin E on mouse mast cells and macrophages as FcγRII and FcγRIII. J Exp Med 176:469–476
Tao TW, Uhr JW (1966) Capacity of pepsin-digested antibody to inhibit antibody formation. Nature 212:208–209

Terres G, Habicht GS, Stoner RD (1974) Carrier-specific enhancement of the immune response using antigen-antibody complexes. J Immunol 112:804–811
Uhr JW, Möller G (1968) Regulatory effect of antibody on the immune response. Adv Immunol 8:81–127
Urbaniak SJ, Greiss MA (2000) RhD haemolytic disease of the fetus and the newborn. Blood Rev 14(1):44–61
van der Heijden FL, van Neerven RJJ, van Katwijk M, Bos JD, Kapsenberg ML (1993) Serum-IgE-facilitated allergen presentation in atopic disease. J Immunol 150:1–8
van Neerven RJJ, Wikborg T, Lund G, Jacobsen B, Brich-Nielsen Å, Arnved J et al (1999) Blocking antibodies induced by specific allergy vaccination prevent the activation of CD4+ T cells by inhibiting serum-IgE-facilitated allergen presentation. J Immunol 163:2944–2952
van Neerven RJJ, Arvidsson M, Ipsen H, Sparholt SH, Rak S, Würtzen PA (2004) A double-blind, placebo-controlled birch allergy vaccination study: inhibition of CD23-mediated serum-immunoglobulin E-facilitated allergen presentation. Clin Exp Allergy 34:420–428
von Behring E, Wernicke E (1892) Über Immunisierung und Heilung von Versuchstieren bei der Diphterie. Z Hyg Infektionskrankheit 12:10–44
Walker JG, Siskind GW (1968) Studies on the control of antibody synthesis. Effect of antibody affinity upon its ability to suppress antibody formation. Immunology 14:21–28
Wason WM (1973) Regulation of the immune response with antigen specific IgM antibody: a dual role. J Immunol 110:1245–1252
Welliver RC, Wong DT, Middleton E Jr, Sun M, McCarthy N, Ogra PL (1982) Role of parainfluenza virus-specific IgE in pathogenesis of croup and wheezing subsequent to infection. J Pediatr 101(6):889–896
Wernersson S, Karlsson M, Dahlström J, Mattsson R, Verbeek JS, Heyman B (1999) IgG-mediated enhancement of Ab responses is low in FcRγ chain deficient mice and increased in FcγRII deficient mice. J Immunol 163:618–622
Westman S, Gustavsson S, Heyman B (1997) Early expansion of secondary B cells after primary immunization with antigen complexed with IgE. Scand J Immunol 46:10–15
Whited-Collisson E, Andersson B, Lamon EW (1979) Modulation of hapten-specific responses with anticarrier antibody: I. differential effects of IgM and IgG anticarrier on primary direct and indirect hapten-specific plaque-forming cells. Proc Soc Exp Biol Med 162:194–198
Whited Collisson E, Andersson B, Rönnholm M, Lamon E (1983) Potentiation of antibody responses by specific IgM: specificity and thymus dependency. Cell Immunol 79:44–55
Whited Collisson E, Andersson B, Lamon EW (1984) Avidities of hapten-specific antibodies when the responses are modulated by anti-carrier antibodies. Immunology 53:443–449
Wiersma EJ (1991) Effects of IgG antibodies and complement on the immune response. Thesis Uppsala University,p 26
Wiersma EJ (1992) Enhancement of the antibody response to protein antigens by specific IgG under different experimental conditions. Scand J Immunol 36:193–200
Wiersma EJ, Coulie PG, Heyman B (1989) Dual immunoregulatory effects of monoclonal IgG-antibodies: suppression and enhancement of the antibody response. Scand J Immunol 29:439–448
Wiersma EJ, Nose M, Heyman B (1990) Evidence of IgG-mediated enhancement of the antibody response in mice without classical pathway complement activation. Eur J Immunol 20:2585–2589
Woodrow JC, Clarke CA, Donohow WT, Finn R, McConnell RB, Sheppard PM et al (1975) Mechanism of Rh prophylaxis: an experimental study on specificity of immunosuppression. Br Med J 2:57–59
Youd ME, Ferguson AR, Corley RB (2002) Synergistic roles of IgM and complement in antigen trapping and follicular localization. Eur J Immunol 32:2328–2337

Chapter 10
Engineered Antibody Derivatives in Preclinical and Clinical Development

Christian Kellner and Matthias Peipp

Abstract Monoclonal antibodies are established treatment options for cancer and autoimmune diseases, but despite obvious clinical success, response rates are still unsatisfactory. Thus, a variety of approaches were pursued to improve antibody therapy resulting in the development of next generation monoclonal antibodies or antibody-derived therapeutic proteins which are in various stages of preclinical or clinical development. These include modified intact antibody molecules with modulated effector functions, antibody conjugates developed to efficiently deliver cytotoxic compounds to the tumor, bispecific antibodies redirecting cytotoxicity by various effector cell populations, as well as antibody derivatives and antibody fragments solely containing selected parts of an antibody with various potential applications. In this chapter, general strategies enhancing the efficacy of intact antibodies and their derivatives are described, and selected examples of molecules that already have entered clinical trials are introduced.

Keywords Fc engineering • Immunotoxin • Antibody-drug conjugate (ADC) • Bispecific antibody • Antibody-directed enzyme prodrug therapy (ADEPT) • Antibody therapy • Antibody derivatives

10.1 General Considerations

Today, monoclonal antibodies are widely used in the clinic and are established treatment options for cancer as well as autoimmune diseases (Chan and Carter 2010; Weiner et al. 2010; Oldham and Dillman 2008). Accordingly, "block buster"

C. Kellner, Ph.D • M. Peipp Ph.D (✉)
Division of Stem Cell Transplantation and Immunotherapy, 2nd Medical Department, Christian-Albrechts-University Kiel, Schittenhelmstr. 12, 24105 Kiel, Germany
e-mail: c.kellner@med2.uni-kiel.de; m.peipp@med2.uni-kiel.de

F. Nimmerjahn (ed.), *Molecular and Cellular Mechanisms of Antibody Activity*,
DOI 10.1007/978-1-4614-7107-3_10,

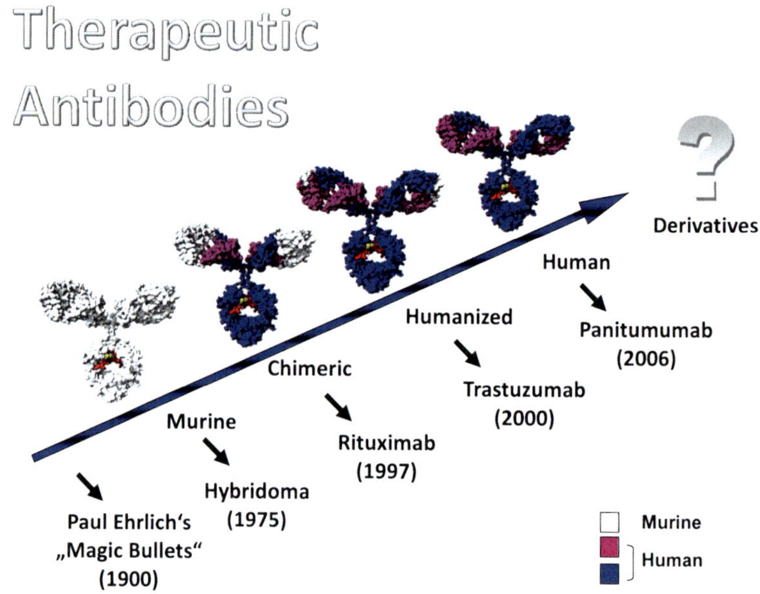

Fig. 10.1 *Engineering antibodies and antibody derivatives for clinical application*. Immunogenicity of murine antibodies initially used in clinical trials represented a major issue. The development of novel technologies such as chimerization, humanization, as well as technologies allowing the generation of fully human antibodies were key to the clinical success. Now, a next wave of modified antibodies and antibody derivatives will have to prove clinical benefit. The IgG model used for illustration is based on the pd-file (Clark 1997) provided by Dr. Mike Clark on the official website (http://www.path.cam.ac.uk/~mrc7/)

antibodies as represented by rituximab and trastuzumab have dramatically changed treatment strategies of different types of cancer such as lymphomas and breast cancer. Despite obvious clinical success, still not all patients optimally benefit from antibody therapy. Novel insights into the mechanisms of action and new engineering technologies inspired the design of second and third generation monoclonal antibodies (Carter 2006; Presta 2008). These technologies include humanization, generation of fully human antibodies, affinity maturation, and optimization of the antibody's Fc-portion (Figs. 10.1 and 10.2). Such modifications offered the opportunity to precisely improve both direct and indirect effector functions including blocking and signaling properties, induction of ADCC, phagocytosis, and CDC (Fig. 10.3). Engineered monoclonal antibodies just have entered clinical trials and in the long term will hopefully demonstrate improved performance. Besides engineering of "naked" monoclonal antibodies, a variety of antibody conjugates have been developed to deliver cytotoxic compounds to the tumor site (Allen 2002). These include radionuclides, bacterial- or plant-derived toxins, as well as chemotherapeutic agents.

In addition to modifying whole antibody molecules, molecular engineering allowed the design and evaluation of antibody derivatives and antibody fragments that make use of selected parts of an antibody like the Fc domain or the antibodies' Fv regions (Holliger and Hudson 2005; Beck and Reichert 2011). In this chapter,

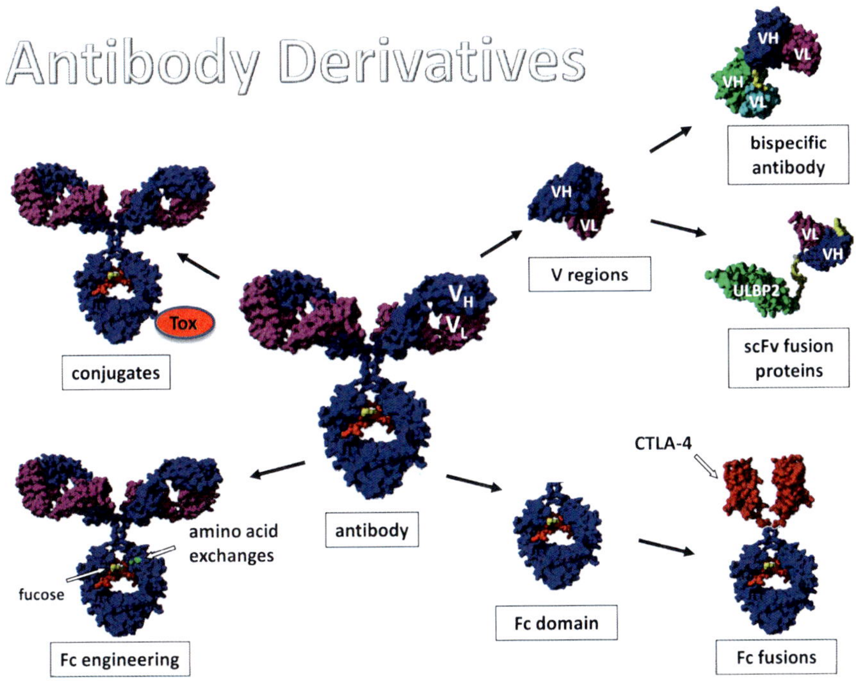

Fig. 10.2 *Design of engineered antibodies and antibody derivatives*. Antibody derivatives make use of selected parts of an antibody resulting in molecules with novel biological activity and rationally designed mechanisms of action. *Tox*=cytotoxic component/substance, *blue*=heavy-chain domains; *pink*=light-chain domains. (Note: model for CTLA-4-Fc fusion protein does not represent a calculated model; it should be recognized as a schematic representation.) The IgG model is based on the pd-file (Clark 1997) provided by Dr. Mike Clark on the official website (http://www.path.cam.ac.uk/~mrc7/)

general engineering strategies of full antibodies and antibody derivatives are described (Fig. 10.2; Tables 10.1–10.4), and selected examples for molecules that already have entered clinical trials are given (Table 10.5).

10.2 Fc Engineering of IgG1 Antibodies to Enhance Cytotoxicity

Animal models suggested that the cytotoxic potential of monoclonal antibodies depends on enhanced binding affinity of their Fc domains to activating FcγR relative to the inhibitory FcγRIIb isoform (Clynes et al. 2000; Nimmerjahn and Ravetch 2005). A clear contribution of complement-mediated killing was found in some animal models (Di Gaetano et al. 2003; Golay et al. 2006), while in others, these data could not be confirmed (Nimmerjahn and Ravetch 2005). Moreover, the situation may be even more complex, because some observations implied that additional factors such as tumor burden, tumor localization, or the tumor

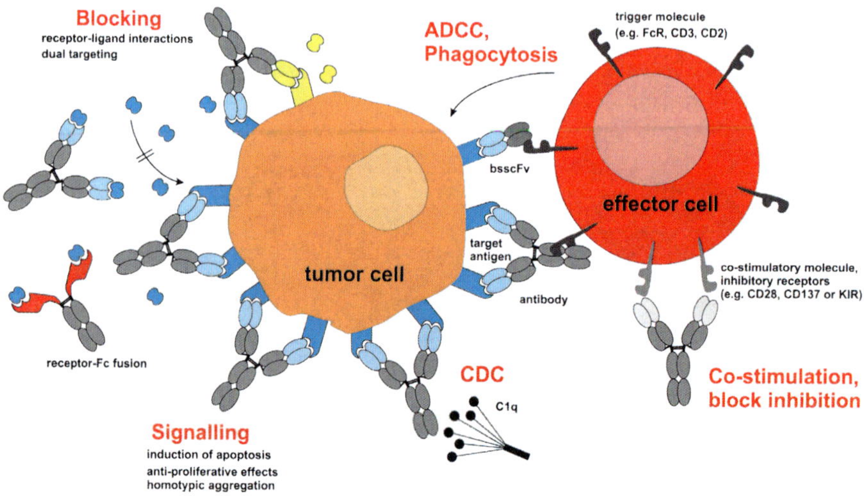

Fig. 10.3 *Mechanisms of action triggered by monoclonal antibodies and antibody-based derivatives*. Antibodies and antibody-based derivatives are able to trigger various effector mechanisms to elicit therapeutic activity

microenvironment may differentially determine the contribution of different mechanisms of tumor cell elimination, as evidenced in mouse models using CD20 antibodies (Gong et al. 2005; Boross et al. 2011). Data from clinical trials analyzing Fcγ receptor (FcR) polymorphisms from patients receiving antibody therapy revealed that homozygous expression of FcγRIIIa-V158 and FcγRIIa-H131 correlated with improved response rates and in some cases with a better overall survival (Weng and Levy 2003; Musolino et al. 2008). Together these data pointed to an important role for Fc-mediated effector mechanisms in antibody therapy. Based on these findings several strategies have been developed to enhance or diminish Fc interactions of full-length IgG antibodies with components of the human immune system, such as FcR or complement factors (Carter 2006; Presta 2008). Especially the IgG_1 antibody isotype representing the most widely used antibody isotype in antibody therapy was subject of diverse engineering strategies.

10.2.1 Enhancing Effector Cell-Mediated Cytotoxicity (ADCC) and Phagocytosis

While distinct structural requirements for the effector functions of IgG antibodies were apparent from early studies (Jefferis and Lund 2002), analyses of the co-crystal structure of the IgG Fc with FcγRIII suggested that direct interaction sites were located in the protein moiety of IgG, while only minor contacts between sugar residues of the Fc and FcγRIII were observed (Sondermann et al. 2000). Since the activation of leukocyte FcR was strictly dependent on Fc glycosylation, it was supposed that the oligosaccharide moiety exerted its influence indirectly by modulation of the

Table 10.1 Selected engineered Fc variants with enhanced ADCC activity

Variant	FcγRIIIa binding	FcγRIIb binding	IIIa/IIb profile	ADCC induction	Fold reduction in EC50 value	Complement activation	Reference
Wild type	↑	↑	1	↑	–	+	
S298A-E333A-K334A	↑↑	↓	10	↑↑↑	10–100	No data available	(Shields et al. 2001)
S239D-I332E	↑↑↑	↑↑	4	↑↑↑	10–100	+	(Lazar et al. 2006)
S239D-I332E-A330L	↑↑↑↑	↑↑↑	9	↑↑↑	10–100	–	(Lazar et al. 2006)
F243L-R292P-Y300L-V305I-P396L	↑↑↑	↑(↑)	7	↑↑↑	10–100	No data available	(Stavenhagen et al. 2007)
AME-D (specific mutations not disclosed)	↑↑↑	N.a.	N.a.	↑↑↑	5–10	No data available	(Bowles et al. 2006)

IIIa/IIb fold FcγRIIIa binding/FcγRIIb binding, *EC50* effective concentration 50 %

↑ = enhanced activity/binding compared to wt; ↓ = reduced activity/binding compared to wt

Reprinted with permission (Peipp et al. 2011)

Table 10.2 Selected engineered IgG1-Fc variants with enhanced CDC activity

Variant	ID	C1q binding (KD in nM)	C1q fold binding	CDC (fold) potency	ADCC (fold) potency	Reference
Wild type		48	1	1	1	
K326W			3	2	0	(Idusogie et al. 2001)
K326W-E333S			5	2	0	(Idusogie et al. 2001)
S267E-H268F-S324T	EFT	1.0	47	6.9	0.045	(Bernett et al. 2010)
S267E-H268F-S324T-G236A-I332E	EFT+AE			23	1.2	(Bernett et al. 2010)
IgG1-IgG3 chimera	113 F		7.1	3.7	1	(Bernett et al. 2010; Natsume et al. 2008)

Reprinted with permission (Peipp et al. 2011)

Table 10.3 Selected engineered IgG1-Fc variants with enhanced FcRn binding

Variant	FcRn-binding K_D (nM)	Method K_D determination	FcRn binding (fold)	Enhancement of half-life in primates	ADCC induction	Complement activation	Reference
Wild type	2460	Biacore	1	1	+	+	
M252Y-S254T-T256E (YTE)	230	Biacore	11	3–4			(Oganesyan et al. 2009; Dall'Acqua et al. 2002, 2006)
T250Q-M428L		ELISA	28	2.5	+	+	(Datta-Mannan et al. 2007a, b; Hinton et al. 2006)
N434A	570	Biacore	4	2	+	+	(Yeung et al. 2009)
H433K-N434F	660	Biacore	4				
T307Q-N434A	240[a]	Biacore	10[a]	2.2			(Yeung et al. 2010)
T307A-E380A-N434A		ELISA	11				(Petkova et al. 2006)
M428L-N434S	218	Biacore	11	3–4			(Zalevsky et al. 2010)

[a]Affinity for primate FcRn (affinities often roughly identical to that obtained for hFcRn)

(Reprinted with permission (Peipp et al. 2011))

Table 10.4 Miscellaneous Fc variants with altered biological function

Variant	Species	Isotype	Expression system	Glycosylation status	FcR binding	Reference
N297Q/A	Human	IgG_1	Mammalian	agly.	Residual FcγRI binding	(Sazinsky et al. 2008)
E382V-M428I	Human	IgG_1	Bacterial	agly.	Selective binding to FcγRI	(Jung et al. 2010)
			Mammalian	gly.	Binding similar to wild type	
S298G-T299A	Human		Mammalian	agly.	Selective binding to FcγRI, FcγRIIa, FcγRIIb; no binding to FcγRIIIa or C1q	(Sazinsky et al. 2008)
D265A	Murine	IgG_1	Mammalian		No binding to FcγRIIb, FcγRIII	(Baudino et al. 2008)
		$IgG_{2a/b}$	Mammalian	gly.	No binding to FcγRs	
G236R-L328R	Human	IgG_1	Mammalian		No FcγR interactions	(Horton et al. 2011)
S267E-L328F	Human	IgG_1	Mammalian		400-fold enh. FcγRIIb binding	(Horton et al. 2011)
G236A	Human	IgG_1	Mammalian		Enhanced FcγRIIa binding	(Richards et al. 2008)

agly aglycosylated, *gly* glycosylated

Table 10.5 Selected examples for engineered antibodies and antibody-based derivatives in clinical testing

Antibody/derivative	Description/proposed mechanism	Indication	Status	Reference
Fc-engineered antibodies				
Obinutuzumab (GA101)	Type II CD20 antibody, low-fucosylated IgG_1; apoptosis, non-apoptotic programmed cell death, enhanced ADCC	B-cell lymphoma	Phase I and II	(Salles et al. 2012)
GA201	EGFR antibody, low-fucosylated IgG_1; receptor blockade, enhanced ADCC	Advanced solid tumors	Phase I	(Paz-Ares et al. 2011)
Mogamulizumab (KW-0761)	Anti-CCR4 antibody, non-fucosylated IgG_1; enhanced ADCC	Adult T-cell leukemia-lymphoma	Phase II (approved in Japan 2012)	(Ishida et al. 2012)
XmAb®2513	CD30 antibody, protein-engineered; enhanced ADCC	Hodgkin's Lymphoma	Phase I	NCT00606645
AME-133v (LY2469298)	CD20 antibody, protein-engineered; enhanced ADCC	Follicular lymphoma	Phase I	(Forero-Torres et al. 2012)
PRO131921	CD20 antibody, protein-engineered; enhanced ADCC	NHL, CLL	Phase I/II	(Alduaij and Illidge 2011)
Immunotoxins and antibody-drug conjugates				
Gemtuzumab ozogamicin	Humanized IgG_4 conjugated to calicheamycin; cleavage of DNA	AML	Withdrawn	(Castaigne et al. 2012; Estey 2012)
Trastuzumab emtansine (T- DM1)	Humanized IgG_1 conjugated to DM1; inhibition of microtubule assembly, ADCC, receptor downregulation	Breast cancer	Approved 2013	(Burris et al. 2011)
Brentuximab vedotin	Chimeric CD30 antibody conjugated to MMAE via cleavable peptide linker, inhibition of tubule formation	Hodgkin lymphoma, systemic anaplastic large cell lymphoma	Approved 2011	(Younes et al. 2010)
BL22	CD22-specific dsFv fused to a truncated version of Pseudomonas exotoxin A; inhibition of protein synthesis, induction of apoptosis	Hairy cell leukemia	Phase II	(Kreitman et al. 2009)

(continued)

Table 10.5 (continued)

Antibody/derivative	Description/proposed mechanism	Indication	Status	Reference
Moxetumomab Pasudotox (CAT-8015 or HA22)	Affinity maturated version of BL22; inhibition of protein synthesis, induction of apoptosis	Hairy cell leukemia	Phase I	(Kreitman et al. 2012)
^{90}Y-labeled ibritumomab tiuxetan (Zevalin)	^{90}Y-labeled murine IgG_1 antibody	Relapsed indolent NHL	Approved 2002	
^{131}I-labeled tositumomab (Bexxar)	^{131}I-labeled murine IgG_{2a}/lambda antibody	Follicular, non-Hodgkin lymphoma	Approved 2003	
Bispecific antibodies				
Blinatumomab	bsscFv; [CD19 × CD3]; T-cell recruitment	NHL; ALL	Phase I and II	(Bargou et al. 2008; Topp et al. 2011)
Catumaxomab	Triomab; [EpCAM × CD3]; T-cell recruitment	Malignant ascites	Approved 2009	(Seimetz et al. 2010)
AFM13	Tandab; [CD30 × CD16a]; NK-cell recruitment	Hodgkin lymphoma	Phase I	NCT01221571
MM-111	bsscFv fused to HSA; [Her2 × Her3]; Blocking,	Solid tumors	Phase I	NCT00911898
MEHD7945A	Two-in-one IgG Blocking, Fc-mediated effects	Epithelial tumors		NCT01207323
CVX-241	CrossMab; [Ang2 × VEGF]; Blocking	Solid tumors	Phase I	NCT01004822
Immunocytokines				
Hu14.18-IL2	Humanized antibody against GD2 fused to IL-2	Neuroblastoma	Phase II	(Shusterman et al. 2010)
Fc fusion proteins				
Etanercept	ECD of TNF-receptor 2 fused to Fc; Blocking	Rheumatoid arthritis	Approved 1998	
Abatacept	ECD of CTLA-4 fused to Fc	Rheumatoid arthritis	Approved 2005	
Alefacept	First extracellular domain of LFA-3 fused to IgG1-Fc; inhibition of T-cell activation	Plaque psoriasis	Approved 2003 Withdrawn 2011	
Romiplostim (AMG-531)	IgG1-Fc fusion with a peptide that binds to the thrombopoietin receptor; agonist	Chronic immune thrombocytopenic purpura (ITP)	Approved 2008	

Sgp130-Fc	ECD of gp130 fused to Fc; inhibition of IL-6 trans-signaling	Inflammatory diseases	Preclinical; phase I planned for 2012	(Waetzig and Rose-John 2012)
Pretargeting agents				
B9E9FP	Streptavidin fused to CD20 scFv	NHL	Phase I	(Forero et al. 2004)
Rituximab-SA	Streptavidin conjugated to CD20 IgG_1	NHL	Phase I/II	(Weiden et al. 2000)
ADEPT				
A5CP	Anti-CEA F(ab')2 conjugated to CPG2	Solid tumors	Phase I	(Napier et al. 2000)
MFECP1	Anti-CEA scFv fused to CPG2	Solid tumors	Phase I	(Mayer et al. 2006)

Fc conformation (Wright and Morrison 1997). More refined analysis revealed that also interactions between the FcR-bound carbohydrate structures and the Fc-bound oligosaccharide are critical for the strength of the Fc-FcR interaction (Ferrara et al. 2006, 2011). Based on these findings, two approaches became evident to manipulate Fc function: altering the glycosylation profile (glyco-engineering) (Jefferis 2009) or exchange of select amino acids (protein-engineering) to enhance binding to FcR.

10.2.1.1 Glyco-engineered Antibodies

The N-linked biantennary complex-type oligosaccharide attached to N297 is buried between the two CH_2 domains, forming extensive contacts with the protein backbone within CH_2 (Jefferis 2009). Especially non-fucosylated antibodies demonstrated enhanced ADCC activity (Shinkawa et al. 2003). Consequently, several strategies have been pursued to reduce core fucosylation (van Berkel et al. 2010; Umana et al. 1999; Davies et al. 2001; Yamane-Ohnuki et al. 2004). The resulting increased ADCC activity could be explained by a higher binding affinity to FcγRIIIa on NK cells (Shields et al. 2002). The high affinity between glyco-engineered antibodies and FcγRIIIa is partially mediated by interactions formed between the carbohydrate attached at N162 of the FcR and regions of the Fc part that are only accessible when non-fucosylated (Ferrara et al. 2006). As FcγRIIIa and FcγRIIIb are the only human Fcγ receptors which are glycosylated at this position, the proposed interactions might in part explain the observation that glyco-engineered antibodies have selectively increased FcγRIII-binding affinity, while affinities to other Fc receptors remained unchanged (Ferrara et al. 2006). Recently, also direct interaction of the Fc-bound and FcR-attached oligosaccharide have been suggested as an additional mechanism how low-fucosylated antibodies gain the observed affinity enhancement (Ferrara et al. 2011). Functionally, non-fucosylated forms of rituximab were more potent in inducing ADCC than fucosylated rituximab, and this difference was independent of the FcγRIIIa-158-V/F polymorphism. In contrast to V-carriers showing a 10-fold greater activity than F-carriers for fucosylated rituximab, there was no significant difference observed when non-fucosylated rituximab was used (Niwa et al. 2004). Consequently, the use of low-fucose antibodies may improve the therapeutic effects of CD20-directed therapy independent of the FcγRIIIa allotype. Recently, PMN opsonized with low-fucose antibodies were reported to demonstrate stronger phagocytic activity of tumor cells (Shibata-Koyama et al. 2009). However, low-fucose content was correlated with diminished neutrophil-mediated ADCC, demonstrating that glyco-engineering may not universally increase ADCC activity (Peipp et al. 2008). Glyco-engineered antibodies have now entered clinical trials (Ishida et al. 2012; Paz-Ares et al. 2011; Salles et al. 2012), but it is too early to draw definitive conclusions whether these antibodies demonstrate superior activity. Interpretation in some cases may not allow the conclusion that a superior performance is indeed due to glyco-engineering, because as in the case of GA101 (glyco-engineered CD20 antibody) and KW-0761 (glyco-engineered anti-CCR4 antibody), no clinical experience with corresponding non-glyco-engineered variants of these antibodies are available.

10.2.1.2 Protein-Engineered Antibodies

Using various techniques, including "alanine scanning," yeast display or computational design algorithms and high-throughput screening, human IgG_1 Fc variants with higher affinity for FcγRIIIa and enhanced ADCC activity were isolated. While most of the identified mutations are located in the binding interface of Fc and FcR, some Fc variants with enhanced ADCC activity were identified with mutated amino acids in the CH_3 domain of the antibody (Stavenhagen et al. 2007). These data suggest that some mutants increase Fc receptor binding indirectly, e.g., by altering the structural properties of the Fc domain or by influencing the Fc glycosylation. Lazar and colleagues (Lazar et al. 2006) identified a number of variants with significant enhancements in binding affinity to both human FcγRIIIa-V158 and FcγRIIIa-F158 alleles along with an improved FcγRIIIa/IIb ratio and enhanced ADCC potential. The substitution G236A selectively improved binding to FcγRIIa resulting in enhanced phagocytosis by macrophages (Richards et al. 2008). Interestingly, variant S239D/I332E elicited CDC comparable to wild-type IgG_1, while the addition of a A330L mutation reduced CDC activity. The set of S239D/I332E and S239D/I332E/A330L variants provides the option for enhancing ADCC with or without complement fixation. Superior B-cell depletion by Fc-optimized CD20 antibodies was demonstrated in cynomolgus monkeys (Lazar et al. 2006). Interestingly, the S239D/I332E variant proved to be essential for B-cell depletion in cynomolgus monkeys using a CD19-directed antibody (Horton et al. 2008) (Zalevsky et al. 2009). Thus, also target antigens such as CD19, which formerly proved ineffective for therapy with conventional antibodies, may be reconsidered for Fc-engineered antibodies. Antibodies with protein-engineered Fc variants (e.g., directed against CD30, ClinicalTrials.gov Identifier: NCT00606645; AME-133v (LY2469298), (Forero-Torres et al. 2012); PRO131921, (Alduaij and Illidge 2011)) have already entered the first phases of clinical testing, but similarly to glyco-engineered antibodies, it is too early to draw definitive conclusions whether this strategy delivers the expected benefits.

10.2.2 Altered Complement Activation (CDC)

Since the contribution of complement to antibodies' therapeutic efficacy is discussed controversial (Di Gaetano et al. 2003; Wang et al. 2009), and complement activation may contribute to antibodies' side effects in patients, approaches to either enhance or to diminish complement activation have been proposed. Mutational analyses of human IgG_1 revealed that the C1q-binding region is centered around D270, K322, P329, and P331 in the CH_2 domain. More detailed analysis showed that amino acids K326 and E333 profoundly influenced complement activation (Idusogie et al. 2001). A K326W substitution provided an increase in C1q binding (3-fold) and complement-mediated killing (2-fold). The K326W/E333S double mutant demonstrated additive increases in C1q binding (5-fold), but no further increase in CDC activity. Interestingly, both variants were completely inactive in triggering ADCC.

Recently, 38 antibody variants were screened for enhanced CDC activity (Bernett et al. 2010). Variant "EFT" demonstrated a 47-fold higher binding affinity and 7-fold increased cytotoxic activity, but this variant exerted dramatically reduced ADCC activity. To restore ADCC activity, the "EFT" variant was combined with amino acid substitutions that demonstrated enhanced ADCC activity. This combination indeed restored ADCC activity to that of wild-type IgG. Tested in the background of a CD19 antibody that was not competent in triggering complement lysis, this variant triggered potent CDC. Together, these data demonstrated that antibodies can be converted by Fc engineering to variants that potently trigger CDC. As a drawback, these results illustrated that simultaneously enhancing CDC and ADCC activity represents a major challenge. To date, no CDC-optimized antibodies have entered clinical trials.

10.3 Alternative Isotypes to Recruit a Different Spectrum of Effector Functions

As an alternative to engineering IgG_1 antibodies, other antibody isotypes may be considered for therapeutic use in the future. Different antibody isotypes show unique FcR- and complement-binding profiles, thereby triggering a distinct spectrum of effector functions (reviewed in Jefferis 2012). For example, human IgG_2 antibodies have been demonstrated to potently trigger ADCC by recruiting myeloid effector cells, while lacking cytotoxic capacity with NK cells (Schneider-Merck et al. 2010). IgG_3 antibodies may also represent an interesting alternative to IgG_1 antibodies as they potently trigger ADCC and mediate CDC especially at low antigen densities (Garred et al. 1989). Besides IgG antibodies, IgA antibodies show interesting features in triggering antitumor cytotoxicity. This isotype potently triggers ADCC mediated by neutrophils, monocytes, and macrophages (Lohse et al. 2011). This is due to its high potential in triggering FcαRI (CD89), making it an ideal candidate to recruit myeloid effector cells against tumor cells. In general, considering antibody isotypes other than IgG_1 may allow the recruitment of different effector functions. In combination with IgG antibodies, this may result in synergistic effects leading to more efficient elimination of tumor cells.

10.4 Diminishing/Eliminating Selected Cytotoxic Effector Functions

For certain antibody-based therapeutic strategies, the triggering of cytotoxic effector functions via FcR or complement engagement is not wanted. For example, immunoconjugates or immunotoxins (see below) may trigger off target damage to healthy tissue when taken up by FcR-expressing cells. In these situations, often antibody fragments or IgG_4 antibodies are used to reduce or even abolish both FcR-binding and complement activation abilities. As an alternative, engineered IgG-Fc variants have been tested. Several variants of the murine and human IgG_1 and IgG_2 Fc domains have been reported and show different characteristics in terms of residual FcR and complement binding. Several mutants show altered/reduced FcR and complement binding by exchanging amino acids that are most likely directly involved in protein-protein interaction (D265A, murine IgG; G236R-L328R, human IgG_1) (Baudino et al. 2008). In contrast, the N297Q/A variants prevent Fc glycosylation, resulting in a proposed conformational change described as closed conformation of the Fc domain resulting in highly diminished binding to basically all FcγR and C1q. In the background of CD3 antibodies, these strategies partially eliminated the unwanted mitogenic activity of such molecules (Renders and Valerius 2003).

In two independent approaches, aglycosylated variants with restored binding activity to selected FcγR have been described (Sazinsky et al. 2008; Jung et al. 2010). These strategies allowed the generation of human IgG_1 variants with very unique FcR-binding profiles and biologic functions. While in the first approach, amino acid exchanges in the glycosylation motif around amino acid position N297 were identified that led to aglycosylated variants produced in mammalian cells, in the second approach, variants were identified with a novel bacterial-based screening system. Mutants directly involving the Fc glycosylation motif allow the production of aglycosylated antibodies in various expression hosts, such as mammalian cells, yeast, and bacteria. In contrast mutants isolated in the second approach have to be produced in bacteria, because the glycosylation site in these variants was not altered.

10.5 Fc Engineering to Suppress B Cell Functions

Besides engineering the Fc domain to enhance or prevent activation of cytotoxic effector functions such as ADCC or CDC, also strategies to shut down B-cell functions have been described. To achieve this, CD19 as part of the B-cell co receptor complex was targeted via a modified CD19 antibody harboring an Fc domain with enhanced FcγRIIb-binding capacity, thereby cross-linking CD19 with FcγRIIb on B cells, resulting in enhanced ITIM phosphorylation of FcγRIIb and leading to suppressed BCR-induced calcium mobilization and proliferation of human B cells from healthy volunteers and systemic lupus erythematosus patients (Horton et al. 2011). In a similar approach, coengagement of FcγRIIb and IgE B-cell receptors suppressed IgE production by blocking IgE-positive B-cell activation and differentiation to IgE-secreting plasma cells (Chu et al. 2012). This novel approach may represent an effective therapy for allergic asthma but has not entered clinical trial yet.

10.6 Molecules with Modified Pharmacokinetics

10.6.1 Engineering of Therapeutic Antibodies

Clinical responses in rituximab-treated patients were correlated with favorable pharmacokinetics (Berinstein et al. 1998). Therefore, modulating the serum retention half-life of therapeutic antibodies represents an attractive approach, which may reduce the amount of required antibodies and may allow prolonging treatment intervals. The prolonged retention half-life of IgG requires interaction with the neonatal FcR (FcRn) (Ghetie and Ward 2000). FcRn binding prevents lysosomal degradation of IgG and triggers sorting to the cell surface, where IgG is released. Thereby, FcRn actively controls serum half-life of endogenous and therapeutic antibodies.

Analysis of the crystal structure of FcRn with Fc has mapped their interaction site to the interface between CH_2 and CH_3 in IgG molecules (Burmeister et al. 1994). Considering the potential clinical relevance of this interaction, engineering approaches to alter binding characteristics of IgG to FcRn were obvious. From early mutational analysis and by considering the biology of FcRn, it became evident that introduced mutations should not interfere with the pH-dependency of FcRn binding. Various Fc variants have been described which had enhanced FcRn-binding affinity at pH6 while maintaining low affinity at pH7.4. Different variants prolonged serum half-life in nonhuman primate models by two- to fourfold, but the mechanism by which the gain in affinity was achieved is not fully understood and may well vary between the different mutations. The crystal structure of the "YTE" mutant Fc was compared to wt IgG_1-Fc, and very few structural changes were observed. These data suggested that the improvement in binding is due to direct, local amino acid interactions (Oganesyan et al. 2009). Importantly, antibodies carrying a M428L-N434S Fc variant demonstrated half-life extensions up to fivefold in human FcRn transgenic mice and threefold in cynomolgus monkeys and demonstrated significantly improved antitumor activity in xenografted mouse models (Zalevsky et al. 2010). To date, such variants have not been tested in a clinical setting.

10.6.2 Enhancing Pharmacokinetic Properties of Therapeutic Proteins and Antibody Derivatives

Antibody Fc domains may not only be used to mediate effector functions but were also introduced into therapeutic proteins as dimerization domain and/or to prolong the blood retention times of many protein domains in vivo (Beck and Reichert 2011). Etanercept and abatacept, Fc fusion proteins of IgG-Fc and the extracellular domains of TNF receptor or CTLA-4, represent prototypic molecules that are already clinically approved for the treatment of rheumatoid arthritis. Similarly designed biological agents, such as gp130-Fc, are in the various stages of preclinical and clinical development and demonstrate interesting features (Waetzig and Rose-John 2012).

Besides fusing Fc domains, additional promising strategies have been developed to enhance the in vivo retention times of antibody derivatives. These include fusion of albumin allowing recycling via FcRn, fusion of albumin- or IgG-binding domains, and introduction of glycosylation sites (Müller et al. 2007; Unverdorben et al. 2012; Hutt et al. 2012; Stork et al. 2008; Stork et al. 2007).

10.7 Bispecific Antibodies

Bispecific antibodies combine two binding sites with different specificities in one single protein. Bispecific antibodies have been produced in a variety of formats including full-length IgG antibodies, IgG-like molecules and antibody fragments (Fig. 10.4). Traditionally, bispecific antibodies were generated by hybrid hybridoma (quadroma) technology or by chemical conjugation. However, these formats were associated with severe shortcomings which prevented extensive clinical evaluation. Quadroma antibodies produced severe side effects, which may be due to Fc-mediated systemic effector cell activation resulting in cytokine release syndromes. Moreover, human anti-mouse antibody responses were induced by immunogenic murine sequences contained in quadroma antibodies (Segal et al. 1999). Additionally, these first generation bispecific antibodies were associated with production problems.

These limitations were addressed by the development of novel bispecific antibody formats. Production of bispecific IgG antibodies was facilitated by procedures forcing the formation of bispecific heterodimers, for example, by the generation of mouse IgG_{2a} and rat IgG_{2b} hybrid antibodies resulting in bispecific and trifunctional triomabs (Zeidler et al. 1999) or by introducing knob-into-hole mutations into CH_3 domains (Merchant et al. 1998). Two-in-one antibodies represent another elegant way to generate bispecific molecules. These are IgG molecules containing two identical Fab-binding moieties which were engineered to bind two different targets (Bostrom et al. 2009; Schaefer et al. 2011a). Another approach is represented by the Ig crossover technology. In combination with the knob into holes strategy which assures heterodimeric association of heavy chains, correct pairing of light and cognate heavy chains is promoted by exchanging the heavy-chain and light-chain domains within the Fab of one half of the bispecific IgG molecule (Schaefer et al. 2011b). However, various other emerging technologies for production of bispecific antibodies or antibody-like molecules have been explored, including dual-variable-domain immunoglobulins (DVD-Ig), the kappa-lambda body technology, the strand-exchanged engineered domain (SEED) platform, and the DuoBody platform (Schuurman et al. 2008; Davis et al. 2010; Wu et al. 2007; Dhimolea and Reichert 2012). In their simplest format, however, bispecific antibodies consist only of antibody V-regions. Such bispecific scFv (bsscFvs) are produced as monomeric single-chain molecules such as tandem scFvs or as heterodimeric proteins such as diabodies

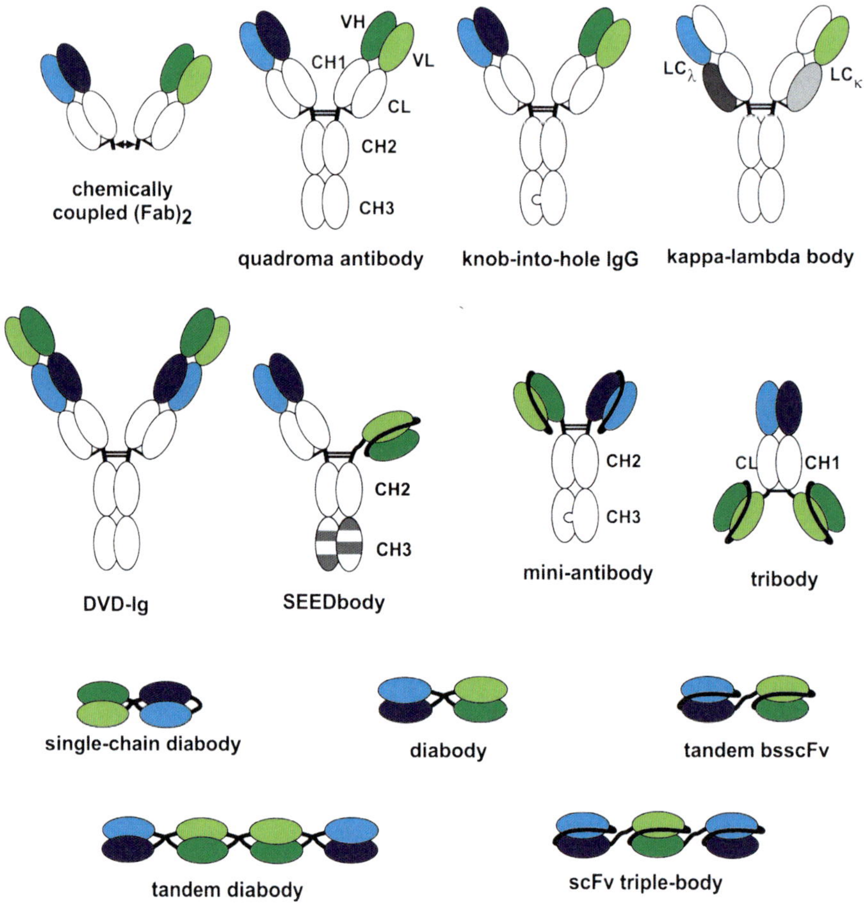

Fig. 10.4 *Examples for bispecific antibody formats*. Bispecific antibodies combine two binding sites with different specificities in one single protein and have been designed in a huge variety of formats. All of which display distinct advantages and pitfalls. Selected formats are displayed. For more detailed information refer to (Muller and Kontermann 2010). *Blue* v-domains = specificity 1. *Green* v-domains = specificity 2

or dual-affinity re-targeting proteins (Johnson et al. 2010). Due to their small size of approx. 50 kDa and the lack of an Fc, bsscFvs are anticipated to have a high tissue penetration potential, avoid uptake by various FcR, and are supposed to be less immunogenic. Further modifications of the formats generated molecules with three or more binding sites to achieve bi- or multivalent-targeting formats (Kellner et al. 2008; Shahied et al. 2004; Kipriyanov et al. 1999). However, such small-sized molecules are rapidly cleared from the blood with plasma retention half-lives of only a few hours or less. Thus, methods were established to improve pharmacokinetic properties, for example, by including protein domains such as human serum albumin (HSA) binding to FcRn (Müller et al. 2007; Kontermann 2009).

One application of bispecific antibodies is recruiting cytotoxic effector cells. Typically, one binding site of a bispecific molecule is specific for a surface structure expressed by the target cell, whereas the second site engages a stimulatory surface receptor on the effector cells, such as T cells, NK cells, monocytes, macrophages, and granulocytes. Each effector cell population expresses different sets of activating receptors which may serve as valuable trigger molecules for bispecific antibodies. These include activating FcγR [FcγRIIIa (CD16a) on NK cells and macrophages, FcγRIIa (CD32a) on granulocytes and macrophages, and FcγRIa (CD64) on macrophages and activated granulocytes], the FcαR I (CD89; macrophages and granulocytes), and CD3 (T cells). Bispecific antibodies may circumvent at least some shortcomings associated with monoclonal antibodies whose therapeutic efficacy may be compromised by insufficient interactions between the Fc portion and the FcR in vivo. For example, bispecific antibodies can be designed to avoid binding to non-activating or inhibitory FcR such as FcγRIIIb and FcγRIIb. The use of antigen-combining sites from antibodies recognizing epitopes distinct from the Fc-binding site offers the opportunity to generate bispecific antibodies which may even trigger FcR when occupied by endogenous immunoglobulin. Apart from that, it is assumed that bispecific antibodies are not hampered by components of the complement system, and their efficacy should not be influenced by FcR-allelic variants. Moreover, bispecific antibodies may bind FcR with comparably high affinities, which at least in part may account for the observed enhanced cytotoxic potential exerted by individual bispecific antibodies in comparison with the corresponding parental monoclonal antibodies (Shahied et al. 2004; Glorius et al. 2013). Finally, bispecific antibodies offer the opportunity to recruit T cells, which lack expression of FcR and therefore are not available as effector cells for monoclonal antibodies (Staerz et al. 1985).

Promising in vitro results and data obtained in animal models encouraged the clinical testing of individual bispecific antibodies. For example, catumaxomab, a bispecific [EpCAM × CD3] triomab produced promising results and was approved by the EMEA in 2009 for the treatment of malignant ascites (Linke et al. 2010; Seimetz et al. 2010). More recently, blinatumomab, a [CD19 × CD3] tandem bsscFv belonging to the class of "Bispecific T-cell engager" (BiTE), has demonstrated efficacy in phase I and phase II clinical trials in patients with non-Hodgkin lymphomas and B-precursor acute lymphoblastic leukemia (Bargou et al. 2008; Topp et al. 2011; Baeuerle and Reinhardt 2009). Whereas clinical benefits were observed with bispecific antibodies redirecting T cells, no convincing effects have been described for bispecific antibodies recruiting NK or myeloid effector cells by engagement of FcR. However, clinical experience is limited to first generation bispecific antibodies, which may not represent ideal formats. However, the tandem diabody (tandab) AFM13 has entered a phase I clinical trial in patients with Hodgkin lymphoma (NCT01221571). AFM13 contains two binding sites for CD30 and the trigger molecule FcγRIIIa and avoids binding to the non-activating FcγRIIIb isoform expressed by granulocytes. It is expected that the selectivity for FcγRIIIa may translate into improved clinical performance.

Other potential applications of bispecific antibodies include blockade of receptor-ligand interactions and induction of antiproliferative or pro-apoptotic effects. Being able to bind two different epitopes on the same or on a different antigen bispecific antibodies may have an improved potential to neutralize one single cytokine (or the corresponding receptor) or to simultaneously neutralize two different cytokines or receptors. For example, MM-111 consisting of two scFv with specificities for either Her2 or Her3 fused to modified HSA was designed to block heterodimerization of Her2 and Her3 receptors and to inhibit Her2/Her3 signaling and tumor growth (McDonagh et al. 2012). MM-111 is currently tested in phase I/II study in patients with solid tumors (NCT00911898). MEHD7945A, a EGFR- and Her3-specific two-in-one antibody, efficiently blocked receptor signaling and moreover mediated ADCC via its Fc portion and had a higher efficacy in tumor mouse models than monospecific antibodies (Schaefer et al. 2011a). MEHD7945A is currently evaluated in a phase I clinical study (NCT01207323). For direct induction of apoptosis through dual targeting of two different surface receptors multivalent formats such as the dock and lock (DNL) system, which allow the generation of bispecific hexavalent antibodies consisting of an IgG molecule associated to 4 Fab fragments appear especially attractive (Rossi et al. 2006, 2009).

10.8 Enhancing Immune Responses with Immunocytokines and Immuno ligands

Antibodies were also used to target proteins with immunomodulatory functions to aberrant cells to stimulate the body´s own effector mechanisms. For example, antibodies or antibody fragments were linked to a variety of immunoregulatory cytokines including interleukin-2 (IL-2), interleukin-12, granulocyte-macrophage colony-stimulating factor (GM-CSF), and TNF, all of which being capable to enhance antitumor immune responses (Pasche and Neri 2012; Kontermann 2012). Immunocytokines based on full IgG are firstly able to elicit Fc-mediated effector functions and secondly provide a costimulating signal augmenting both adaptive and innate immune responses. Moreover, by antibody-mediated delivery, on the one hand local concentrations of the cytokine in the tumor microenvironment may be achieved which are high enough to induce immune responses, and on the other hand toxic side effects may be reduced which were associated with systemic treatment (Lode et al. 1998a). For example, even when untargeted IL-2 was given at high doses, which were already associated with severe side effects, at the tumor-side still suboptimal concentrations were achieved (Lode et al. 1998b).

Most experiences are available with immunocytokines containing IL-2. IL-2 has, for example, been fused to antibodies targeting the disialoganglioside GD2 (hu14.18-IL2), EpCAM (EMD 273066), or the EGFR. These immunocytokines mediated elimination of established tumors and eradicated metastasis in melanoma, colon carcinoma, and neuroblastoma mouse models (Becker et al. 1996a; Pancook et al. 1996; Xiang et al. 1997). Importantly, the immunocytokines had a greater efficacy than a combination of the antibody plus IL-2 (Becker et al. 1996a). The antitumoral effects included stimulation of T- and NK-cell activities (Lode et al. 1998b;

Becker et al. 1996b). Targeted IL-2 may also enhance ADCC induced by the linked antibody moiety. Moreover, antibody-targeted IL-2 has been shown to enhance vaccination-induced immune responses and the immunogenicity of soluble antigens was increased by antibody-IL2 fusion proteins directed against the antigen (Niethammer et al. 2001; Dela Cruz et al. 2005). These encouraging data led to clinical trials performed with immunocytokines containing IL-2 and targeting EpCAM (EMD 273066; huKS-IL-2) or GD2 (hu14.18-IL-2). EMD 273066 was tested in a clinical phase I study with patients with androgen-independent prostate cancer (Ko et al. 2004). In clinical phase I trials with adult patients with malignant melanoma or pediatric patients with neuroblastoma, the hu14.18-IL-2 fusion protein exerted biological activity and demonstrated clinical tolerability (King et al. 2004; Osenga et al. 2006). In a subsequent phase II in a patient cohort with recurrent or refractory neuroblastoma complete responses were obtained only in patients with minimal residual disease, while patients with bulky disease did not respond to treatment (Shusterman et al. 2010). Apart from IL-2 fusion proteins based on full antibodies recently several scFv- and diabody-based fusion proteins entered clinical trials and demonstrated encouraging results (reviewed in Pasche and Neri 2012), but further studies are warranted to further proof this interesting strategy.

In another approach, antibodies or antibody fragments were linked to ligands binding to activating or co-stimulatory receptors on effector cells. NK-cell cytotoxicity was elicited against tumor cells by recombinant fusion proteins between tumor-directed scFv and ligands for NKG2D, an stimulatory receptor expressed by NK cells and some T cells (Kellner et al. 2012; Stamova et al. 2011; von Strandmann et al. 2006). Ligands for NKG2D are normally absent or only poorly expressed by healthy tissues but become upregulated upon cellular stress such as malignant transformation or viral infections and function as alert signals for NK cells. For example, the NKG2D-specific ligand ULBP2 was fused to a CD20 scFv. The recombinant fusion protein sensitized lymphoma cells for NK-cell-mediated lysis and synergistically augmented ADCC by monoclonal antibodies, suggesting that co-ligation of FcγRIII and a different activating receptor and the activation of signaling pathways may provide an attractive strategy to further improve NK-cell-mediated responses (Kellner et al. 2012). Similar approaches to enhance NK-cell-mediated ADCC have been described by triggering other stimulatory receptors or by blocking critical inhibitory receptors, such as CD137 or KIR, respectively (Figure 3; Kohrt et al. 2011, 2012; Binyamin et al. 2008).

10.9 Direct Arming of Antibodies with Cell-Death Inducing Agents/Molecules

Antibody-drug conjugates (ADC) and recombinant immunotoxins represent promising agents in tumor therapy. These agents are bifunctional molecules that are composed of a targeting moiety (monoclonal antibody, antibody fragment, ligand) and a cytotoxic compound that is either chemically cross-linked or genetically fused

(Carter and Senter 2008; Pastan et al. 2006; Schirrmann et al. 2009). The targeting moiety is used to deliver cytotoxic compounds to distinct antigen-positive cells. In the past, a variety of substances have been tested for molecule design. These include chemotherapeutic agents, radioisotopes, as well as toxins derived from bacteria and plants (Wu and Senter 2005). In many cases, the toxic component only displays cytotoxicity when internalized. Accordingly, for the development of immunoconjugates and genetic toxin fusions, target antigens with a high internalization capacity have been selected. These include CD7, CD22, CD30, or CD33 on leukemias and lymphomas (Pastan et al. 2006; Peipp et al. 2002; Schwemmlein et al. 2006; Sutherland et al. 2006) and EGFR or HER2 on solid tumors (Carter and Senter 2008; Pastan et al. 2006; Wu and Senter 2005; Wels et al. 1992, 1995).

Numerous molecules have been clinically tested, but only a small number have been clinically approved (Table 10.1). Gemtuzumab ozogamicin represents a prototypic ADC conjugate composed of calicheamycin chemically linked to a CD33 IgG_4 antibody (Linenberger 2005; Sievers et al. 2001) (note: GO has been withdrawn from the market). ^{90}Y-labeled ibritumomab tiuxetan (Zevalin) and ^{131}I-labeled tositumomab (Bexxar) are approved CD20 radioimmunoconjugates which have demonstrated improved response rates and progression-free survival in lymphoma patients (Witzig et al. 2002; Davis et al. 2004).

Recently, novel ADC demonstrated impressive clinical results. These conjugates make use of novel linker technologies, allowing stable covalent linkage of the cytotoxic moiety and release from the antibody when reaching specific intracellular compartments. Trastuzumab-DM1 (derivative of maytansine 1) showed promising results in phase I and phase II clinical trials in patients with HER2-positive metastatic breast cancer (Krop et al. 2010; Lewis Phillips et al. 2008; Burris et al. 2011). Treatment with brentuximab vedotin, a chimeric CD30 antibody linked to the antimitotic agent monomethyl auristatin E (MMAE) resulted in tumor regression for most patients with relapsed or refractory CD30-positive Hodgkin lymphomas in a phase I study (Younes et al. 2010). Mechanistically, the ADC is rapidly internalized after CD30 binding and is transported to lysosomes. In this compartment, the peptide linker is selectively cleaved (Sutherland et al. 2006; Doronina et al. 2003). MMAE is then released into the cell, where it binds tubulin and promotes cell cycle arrest and apoptosis.

A CD22-directed immunotoxin (BL22), representing a group of single-chain immunotoxins with scFv fragments fused to a truncated version of *Pseudomonas aeruginosa* exotoxin A, showed high response rates in phase II clinical trials in patients with hairy cell leukemia, achieving up to 47 % complete remissions (Kreitman et al. 2001, 2009). Mechanistically, Pseudomonas exotoxin A-based immunotoxins, after receptor-mediated endocytosis, are routed through the trans-golgi network to the endoplasmic reticulum and escape to the cytoplasm via distinct protein domains. Although immunogenicity was a concern for using such molecules, especially in studies with hematological tumors neutralizing antibodies have only been found infrequently. Novel engineering approaches eliminating potentially immunogenic B-cell epitopes may further reduce the immunogenic potential, thereby expanding its applicability (Pastan et al. 2011). Irrespective of the molecule format, selection of the targeting antibody is critical. Epitope specificity may severely compromise surface redistribution, internalization, and the intracellular routing of the receptor and routing of the ADC/receptor complex. Therefore, rational selection of promising antibody candidates is critical for conjugate design (Kellner et al. 2011).

Novel approaches aim to harness human proteins as effector molecules inducing apoptosis in the target cell which are expected to exhibit a lower immunogenicity than the plant or bacteria-derived toxins. These include antibody-targeted human RNases (De Lorenzo et al. 2004; Menzel et al. 2008) and fusion proteins between scFv fragments and pro-apoptotic proteins such as BID (Wang et al. 2010) and granzyme B (Stahnke et al. 2008). Similar to immunotoxins, these molecules are delivered by the antibody component to the tumor cells and induce apoptosis in the targets cells upon internalization.

In a different approach, scFv fragments were fused to soluble ligands for death receptors belonging to TNF/TGF superfamily, e.g., TRAIL (Stieglmaier et al. 2008; Bremer et al. 2006) and FasL (Samel et al. 2003). Upon binding to the tumor cells and aggregation, the ligands are able to cross-link the cognate death receptor and induce apoptosis. One advantage of this approach which does not need internalization of the construct is that it also facilitates killing of bystander tumor cells not expressing the target antigen.

10.10 Pretargeting Radionuclides and Prodrugs

The technique of pretargeting offers the opportunity to target cytotoxic agents such as radionuclides, chemotherapeutic agents, toxins, or prodrugs to the tumor site (Fig. 10.5) (Carter 2001). In contrast to direct arming approaches, chemical conjugation procedures or genetic fusion approaches are avoided. These strategies require the application of at least two separate molecules and aim to improve tumor/blood ratios – thereby reducing systemic toxicities associated with conventional radiotherapy or chemotherapeutic agents (Sharkey and Goldenberg 2011). Moreover, pretargeting separates the processes of tumor targeting and delivery of the effector molecule (Reilly 2006). Pretargeting may, for example, provide certain benefits in radioimmunotherapy of solid tumors, because antibodies directly armed with radionuclides are often associated with severe toxicities towards healthy tissues including the bone marrow due to unfavorable tumor/blood ratios. Indeed, pretargeted radioimmunotherapy had less toxicity and was similar efficacious to deliver radioactivity to the tumor as conventional radioimmunotherapy in mouse xenograft tumor models (Axworthy et al. 2000).

Typically, in a first step, the unlabeled pretargeting agent (a monoclonal antibody or an antibody derivative) is allowed to localize at the tumor site (Fig. 10.5). In a second step, excess molecules are removed from circulation by administration of a clearing agent. In the third step, radioactivity is delivered to the tumor by

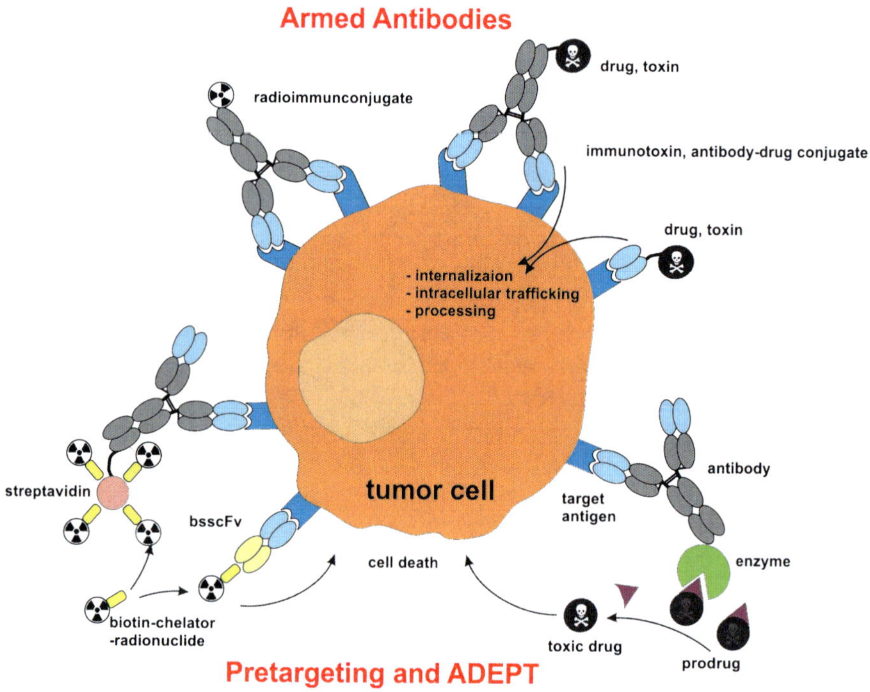

Fig. 10.5 *Armed antibodies and pretargeting strategies. Armed antibodies*: ADC are bifunctional molecules that are composed of a targeting moiety and a cytotoxic compound. The targeting moiety is used to deliver cytotoxic compounds such as chemotherapeutic agents, radioisotopes, as well as protein toxins to distinct antigen-positive cells. *Pretargeting*: In a first step, the unlabeled pretargeting agent (a monoclonal antibody or an antibody derivative) is allowed to localize at the tumor site. In a second step, excess molecules are removed. In a third step, radioactivity is delivered to the tumor by administration of radiolabeled molecules reactive to the antibody. *ADEPT*: in the first step, the infused antibody-enzyme fusion protein is allowed to localize to the tumor. In the second step, a prodrug is administered after clearance of enzyme from the circulation. The enzyme cleaves the prodrug to generate the active drug locally at the tumor site

administration of radiolabeled molecules reactive to the antibody. These are ideally low molecular weight molecules which exhibit a strong tumor penetration potential but rapidly cleared from the blood. Common pretargeting strategies use bispecific antibodies with a first specificity to a tumor-associated antigen and a second specificity binding the radionuclide carrier. Other approaches harnessed the tight interactions between biotin and avidin or streptavidin with a binding constant of 10^{-15} M. Moreover, avidin or streptavidin is able to bind up to four radiolabeled biotin molecules, the reagents were not toxic and were predisposed to modification by radioactive labeling or chemical conjugation (Sharkey and Goldenberg 2011). For example, for pretargeting, a conjugate (or a fusion) between a monoclonal antibody or an antibody fragment and streptavidin was used, and radioactivity was delivered by radionuclide-chelate/biotin molecules.

With pretargeted radioimmunotherapy, promising data were obtained in animal models. For example, a CD20 scFv was genetically fused to streptavidin. In xenograft mouse models, the fusion protein efficiently localized at the tumor site and enabled specific targeting of a subsequently administered radionuclide-labeled biotin molecule (Schultz et al. 2000). However, clinical experience is still limited, and most clinical studies aimed to identify optimal pretargeting conditions and to determine the maximal tolerated doses (Sharkey and Goldenberg 2011). One phase II trial was performed with NRLU-10-SA, an EpCAM-specific IgG_1 conjugated to streptavidin, and ^{90}Y-DOTA-biotin (110 mCi/m2) in patients with colorectal cancer: However, no objective responses were reported (Knox et al. 2000). In a phase I/II study, a rituximab-streptavidin conjugate was tested in combination with a DOTA-biotin ligand labeled with ^{111}In in NHL patients (Weiden et al. 2000). Objective responses were observed in four of seven patients, and toxicities were generally mild. B9E9FP, a CD20 scFv genetically fused to streptavidin, in combination with 11In/90Y-DOTA-biotin demonstrated in a phase I trial in patients with NHL safety and efficacy with objective responses in three out of 15 patients (Forero et al. 2004).

The antibody-directed enzyme prodrug therapy (ADEPT) aims to locally restrict the activity of drugs to the tumor site in using antibody-enzyme conjugates and non-cytotoxic prodrugs (Fig. 10.5) (Begent et al. 2010). A tumor-reactive antibody component is used as a targeting device to deliver a nonhuman enzyme to the tumor. After clearance of the antibody-enzyme conjugate from the blood stream, an inactive prodrug, a substrate for the pretargeted enzyme, is applied. As a result, the prodrug is converted into an active cytotoxic agent by the enzyme predominantly at the tumor site. Since its first formulation in 1974, a variety of different antibody formats, enzymes, and prodrugs have been investigated to optimize the concept. For example, a genetic fusion protein consisting of a CEA-directed scFv and carboxy-peptidase G2 in combination with a bis-iodo phenol (BIP) mustard prodrug was evaluated in a phase I clinical study in patients with colorectal or other CEA-expressing tumors (Mayer et al. 2006).

10.11 Conclusions

Antibodies are widely used in clinical applications and represent standard treatment options in various settings. Limitations in therapeutic outcome of patients treated with monoclonal antibodies and a more detailed understanding of biological processes involved in the pathophysiology of various malignancies have inspired the development of second and third generation antibodies which were rationally selected to trigger distinct effector functions. In addition, antibody engineering technologies have been developed that allow selectively enhancing certain effector functions.

As outlined above, besides engineering monoclonal antibodies, a variety of antibody-based derivatives with novel rationally designed mechanisms of action, such as bispecific antibodies and immunoconjugates have been developed and demonstrated convincing clinical benefit. These technologies together with a more profound mechanistic understanding of the complex molecular and cellular networks and cascades involved in the development and progression of various diseases may allow the design of fit-for-purpose biological agents with customized effector functions and pharmacokinetic characteristics that are optimally suited in a distinct clinical situation.

Acknowledgements We apologize to all investigators whose important work in the field could not be cited because of space limitations.

References

Alduaij W, Illidge TM (2011) The future of anti-CD20 monoclonal antibodies: are we making progress? Blood 117(11):2993–3001

Allen TM (2002) Ligand-targeted therapeutics in anticancer therapy. Nat Rev Cancer 2(10): 750–763

Axworthy DB, Reno JM, Hylarides MD, Mallett RW, Theodore LJ, Gustavson LM et al (2000) Cure of human carcinoma xenografts by a single dose of pretargeted yttrium-90 with negligible toxicity. Proc Natl Acad Sci USA 97(4):1802–1807

Baeuerle PA, Reinhardt C (2009) Bispecific T-cell engaging antibodies for cancer therapy. Cancer Res 69(12):4941–4944

Bargou R, Leo E, Zugmaier G, Klinger M, Goebeler M, Knop S et al (2008) Tumor regression in cancer patients by very low doses of a T cell-engaging antibody. Science 321(5891):974–977

Baudino L, Shinohara Y, Nimmerjahn F, Furukawa J, Nakata M, Martinez-Soria E et al (2008) Crucial role of aspartic acid at position 265 in the CH2 domain for murine IgG2a and IgG2b Fc-associated effector functions. J Immunol 181(9):6664–6669

Beck A, Reichert JM (2011) Therapeutic Fc-fusion proteins and peptides as successful alternatives to antibodies. mAbs 3(5):415–416

Becker JC, Pancook JD, Gillies SD, Mendelsohn J, Reisfeld RA (1996a) Eradication of human hepatic and pulmonary melanoma metastases in SCID mice by antibody-interleukin 2 fusion proteins. Proc Natl Acad Sci USA 93(7):2702–2707

Becker JC, Pancook JD, Gillies SD, Furukawa K, Reisfeld RA (1996b) T cell-mediated eradication of murine metastatic melanoma induced by targeted interleukin 2 therapy. J Exp Med 183(5):2361–2366

Begent R, Sharma S, Chester K (2010) Antibody-Dependent Enzyme Prodrug Therapy (ADEPT). In: Kontermann R, Dübel, S (eds.): Antibody Engineering, Volume 1, 2nd Edition, Springer-Verlag, Berlin Heidelberg

Berinstein NL, Grillo-Lopez AJ, White CA, Bence-Bruckler I, Maloney D, Czuczman M et al (1998) Association of serum Rituximab (IDEC-C2B8) concentration and anti-tumor response in the treatment of recurrent low-grade or follicular non-Hodgkin's lymphoma. Ann Oncol 9(9):995–1001

Bernett MJ, Karki S, Moore GL, Leung IW, Chen H, Pong E et al (2010) Engineering fully human monoclonal antibodies from murine variable regions. J Mol Biol 396(5):1474–1490

Binyamin L, Alpaugh RK, Hughes TL, Lutz CT, Campbell KS, Weiner LM (2008) Blocking NK cell inhibitory self-recognition promotes antibody-dependent cellular cytotoxicity in a model of anti-lymphoma therapy. J Immunol 180(9):6392–6401

Boross P, Jansen JH, de Haij S, Beurskens FJ, van der Poel CE, Bevaart L et al (2011) The in vivo mechanism of action of CD20 monoclonal antibodies depends on local tumor burden. Haematologica 96(12):1822–1830

Bostrom J, Yu SF, Kan D, Appleton BA, Lee CV, Billeci K et al (2009) Variants of the antibody herceptin that interact with HER2 and VEGF at the antigen binding site. Science 323(5921):1610–1614
Bowles JA, Wang SY, Link BK, Allan B, Beuerlein G, Campbell MA et al (2006) Anti-CD20 monoclonal antibody with enhanced affinity for CD16 activates NK cells at lower concentrations and more effectively than rituximab. Blood 108(8):2648–2654
Bremer E, van Dam G, Kroesen BJ, de Leij L, Helfrich W (2006) Targeted induction of apoptosis for cancer therapy: current progress and prospects. Trends Mol Med 12(8):382–393
Burmeister WP, Huber AH, Bjorkman PJ (1994) Crystal structure of the complex of rat neonatal Fc receptor with Fc. Nature 372(6504):379–383
Burris HA 3rd, Rugo HS, Vukelja SJ, Vogel CL, Borson RA, Limentani S et al (2011) Phase II study of the antibody drug conjugate trastuzumab-DM1 for the treatment of human epidermal growth factor receptor 2 (HER2)-positive breast cancer after prior HER2-directed therapy. J Clin Oncol 29(4):398–405
Carter P (2001) Improving the efficacy of antibody-based cancer therapies. Nat Rev Cancer 1(2):118–129
Carter PJ (2006) Potent antibody therapeutics by design. Nat Rev Immunol 6(5):343–357
Carter PJ, Senter PD (2008) Antibody-drug conjugates for cancer therapy. Cancer J 14(3): 154–169
Castaigne S, Pautas C, Terre C, Raffoux E, Bordessoule D, Bastie JN et al (2012) Effect of gemtuzumab ozogamicin on survival of adult patients with de-novo acute myeloid leukaemia (ALFA-0701): a randomised, open-label, phase 3 study. Lancet 379(9825):1508–1516
Chan AC, Carter PJ (2010) Therapeutic antibodies for autoimmunity and inflammation. Nat Rev Immunol 10(5):301–316
Chu SY, Horton HM, Pong E, Leung IW, Chen H, Nguyen DH et al (2012) Reduction of total IgE by targeted coengagement of IgE B-cell receptor and FcgammaRIIb with Fc-engineered antibody. J Allergy Clin Immunol 129(4):1102–1115
Clark MR (1997) IgG effector mechanisms. Chem Immunol 65:88–110
Clynes RA, Towers TL, Presta LG, Ravetch JV (2000) Inhibitory Fc receptors modulate in vivo cytoxicity against tumor targets. Nat Med 6(4):443–446
Dall'Acqua WF, Woods RM, Ward ES, Palaszynski SR, Patel NK, Brewah YA et al (2002) Increasing the affinity of a human IgG1 for the neonatal Fc receptor: biological consequences. J Immunol 169(9):5171–5180
Dall'Acqua WF, Kiener PA, Wu H (2006) Properties of human IgG1s engineered for enhanced binding to the neonatal Fc receptor (FcRn). J Biol Chem 281(33):23514–23524
Datta-Mannan A, Witcher DR, Tang Y, Watkins J, Wroblewski VJ (2007a) Monoclonal antibody clearance. Impact of modulating the interaction of IgG with the neonatal Fc receptor. J Biol Chem 282(3):1709–1717
Datta-Mannan A, Witcher DR, Tang Y, Watkins J, Jiang W, Wroblewski VJ (2007b) Humanized IgG1 variants with differential binding properties to the neonatal Fc receptor: relationship to pharmacokinetics in mice and primates. Drug Metab Dispos 35(1):86–94
Davies J, Jiang L, Pan LZ, LaBarre MJ, Anderson D, Reff M (2001) Expression of GnTIII in a recombinant anti-CD20 CHO production cell line: Expression of antibodies with altered glycoforms leads to an increase in ADCC through higher affinity for FC gamma RIII. Biotechnol Bioeng 74(4):288–294
Davis TA, Kaminski MS, Leonard JP, Hsu FJ, Wilkinson M, Zelenetz A et al (2004) The radioisotope contributes significantly to the activity of radioimmunotherapy. Clin Cancer Res 10(23):7792–7798
Davis JH, Aperlo C, Li Y, Kurosawa E, Lan Y, Lo KM et al (2010) SEEDbodies: fusion proteins based on strand-exchange engineered domain (SEED) CH3 heterodimers in an Fc analogue platform for asymmetric binders or immunofusions and bispecific antibodies. Protein Eng Des Sel 23(4):195–202
De Lorenzo C, Arciello A, Cozzolino R, Palmer DB, Laccetti P, Piccoli R et al (2004) A fully human antitumor immunoRNase selective for ErbB-2-positive carcinomas. Cancer Res 64(14):4870–4874

Dela Cruz JS, Morrison SL, Penichet ML (2005) Insights into the mechanism of anti-tumor immunity in mice vaccinated with the human HER2/neu extracellular domain plus anti-HER2/neu IgG3-(IL-2) or anti-HER2/neu IgG3-(GM-CSF) fusion protein. Vaccine 23(39):4793–4803

Dhimolea E, Reichert JM (2012) World bispecific antibody summit, 27–28 Sept 2011, Boston. *mAbs* 4(1):4–13

Di Gaetano N, Cittera E, Nota R, Vecchi A, Grieco V, Scanziani E et al (2003) Complement activation determines the therapeutic activity of rituximab in vivo. J Immunol 171(3):1581–1587

Doronina SO, Toki BE, Torgov MY, Mendelsohn BA, Cerveny CG, Chace DF et al (2003) Development of potent monoclonal antibody auristatin conjugates for cancer therapy. Nat Biotechnol 21(7):778–784

Estey E (2012) Treatment of AML: resurrection for gemtuzumab ozogamicin? Lancet 379(9825):1468–1469

Ferrara C, Stuart F, Sondermann P, Brunker P, Umana P (2006) The carbohydrate at FcgammaRIIIa Asn-162. An element required for high affinity binding to non-fucosylated IgG glycoforms. J Biol Chem 281(8):5032–5036

Ferrara C, Grau S, Jager C, Sondermann P, Brunker P, Waldhauer I et al (2011) Unique carbohydrate-carbohydrate interactions are required for high affinity binding between FcgammaRIII and antibodies lacking core fucose. Proc Natl Acad Sci USA 108(31): 12669–12674

Forero A, Weiden PL, Vose JM, Knox SJ, LoBuglio AF, Hankins J et al (2004) Phase 1 trial of a novel anti-CD20 fusion protein in pretargeted radioimmunotherapy for B-cell non-Hodgkin lymphoma. Blood 104(1):227–236

Forero-Torres A, de Vos S, Pohlman BL, Pashkevich M, Cronier DM, Dang NH et al (2012) Results of a phase 1 study of AME-133v (LY2469298), an Fc-engineered humanized monoclonal anti-CD20 antibody, in FcgammaRIIIa-genotyped patients with previously treated follicular lymphoma. Clin Cancer Res 18(5):1395–1403

Garred P, Michaelsen TE, Aase A (1989) The IgG subclass pattern of complement activation depends on epitope density and antibody and complement concentration. Scand J Immunol 30(3):379–382

Ghetie V, Ward ES (2000) Multiple roles for the major histocompatibility complex class I- related receptor FcRn. Annu Rev Immunol 18:739–766

Glorius P, Baerenwaldt A, Kellner C, Staudinger M, Parren PWHI, van de Winkel JGJ et al (2013) The novel tribody [$(CD20)_2$xCD16] efficiently triggers effector cell-mediated lysis of malignant B cells. Leukemia 27(1):190–201

Golay J, Cittera E, Di Gaetano N, Manganini M, Mosca M, Nebuloni M et al (2006) The role of complement in the therapeutic activity of rituximab in a murine B lymphoma model homing in lymph nodes. Haematologica 91(2):176–183

Gong Q, Ou Q, Ye S, Lee WP, Cornelius J, Diehl L et al (2005) Importance of cellular microenvironment and circulatory dynamics in B cell immunotherapy. J Immunol 174(2):817–826

Hinton PR, Xiong JM, Johlfs MG, Tang MT, Keller S, Tsurushita N (2006) An engineered human IgG1 antibody with longer serum half-life. J Immunol 176(1):346–356

Holliger P, Hudson PJ (2005) Engineered antibody fragments and the rise of single domains. Nat Biotechnol 23(9):1126–1136

Horton HM, Bernett MJ, Pong E, Peipp M, Karki S, Chu SY et al (2008) Potent in vitro and in vivo activity of an Fc-engineered anti-CD19 monoclonal antibody against lymphoma and leukemia. Cancer Res 68(19):8049–8057

Horton HM, Chu SY, Ortiz EC, Pong E, Cemerski S, Leung IW et al (2011) Antibody-mediated coengagement of FcgammaRIIb and B cell receptor complex suppresses humoral immunity in systemic lupus erythematosus. J Immunol 186(7):4223–4233

Hutt M, Farber-Schwarz A, Unverdorben F, Richter F, Kontermann RE (2012) Plasma half-life extension of small recombinant antibodies by fusion to immunoglobulin-binding domains. J Biol Chem 287(7):4462–4469

Idusogie EE, Wong PY, Presta LG, Gazzano-Santoro H, Totpal K, Ultsch M et al (2001) Engineered antibodies with increased activity to recruit complement. J Immunol 166(4):2571–2575

Ishida T, Joh T, Uike N, Yamamoto K, Utsunomiya A, Yoshida S et al (2012) Defucosylated anti-CCR4 monoclonal antibody (KW-0761) for relapsed adult T-cell leukemia-lymphoma: a multicenter phase II study. J Clin Oncol 30(8):837–842

Jefferis R (2009) Glycosylation as a strategy to improve antibody-based therapeutics. Nat Rev Drug Discov 8(3):226–234

Jefferis R (2012) Isotype and glycoform selection for antibody therapeutics. Arch Biochem Biophys 526(2):159–166

Jefferis R, Lund J (2002) Interaction sites on human IgG-Fc for FcgammaR: current models. Immunol Lett 82(1–2):57–65

Johnson S, Burke S, Huang L, Gorlatov S, Li H, Wang W et al (2010) Effector cell recruitment with novel Fv-based dual-affinity re-targeting protein leads to potent tumor cytolysis and in vivo B-cell depletion. J Mol Biol 399(3):436–449

Jung ST, Reddy ST, Kang TH, Borrok MJ, Sandlie I, Tucker PW et al (2010) Aglycosylated IgG variants expressed in bacteria that selectively bind FcgammaRI potentiate tumor cell killing by monocyte-dendritic cells. Proc Natl Acad Sci USA 107(2):604–609

Kellner C, Bruenke J, Stieglmaier J, Schwemmlein M, Schwenkert M, Singer H et al (2008) A novel CD19-directed recombinant bispecific antibody derivative with enhanced immune effector functions for human leukemic cells. J Immunother 31(9):871–884

Kellner C, Bleeker WK, Lammerts van Bueren JJ, Staudinger M, Klausz K, Derer S et al (2011) Human kappa light chain targeted Pseudomonas exotoxin A–identifying human antibodies and Fab fragments with favorable characteristics for antibody-drug conjugate development. J Immunol Methods 371(1–2):122–133

Kellner C, Hallack D, Glorius P, Staudinger M, Mohseni Nodehi S, de Weers M et al (2012) Fusion proteins between ligands for NKG2D and CD20-directed single-chain variable fragments sensitize lymphoma cells for natural killer cell-mediated lysis and enhance antibody-dependent cellular cytotoxicity. Leukemia 26(4):830–834

King DM, Albertini MR, Schalch H, Hank JA, Gan J, Surfus J et al (2004) Phase I clinical trial of the immunocytokine EMD 273063 in melanoma patients. J Clin Oncol 22(22):4463–4473

Kipriyanov SM, Moldenhauer G, Schuhmacher J, Cochlovius B, Von der Lieth CW, Matys ER et al (1999) Bispecific tandem diabody for tumor therapy with improved antigen binding and pharmacokinetics. J Mol Biol 293(1):41–56

Knox SJ, Goris ML, Tempero M, Weiden PL, Gentner L, Breitz H et al (2000) Phase II trial of yttrium-90-DOTA-biotin pretargeted by NR-LU-10 antibody/streptavidin in patients with metastatic colon cancer. Clin Cancer Res 6(2):406–414

Ko YJ, Bubley GJ, Weber R, Redfern C, Gold DP, Finke L et al (2004) Safety, pharmacokinetics, and biological pharmacodynamics of the immunocytokine EMD 273066 (huKS-IL2): results of a phase I trial in patients with prostate cancer. J Immunother 27(3):232–239

Kohrt HE, Houot R, Goldstein MJ, Weiskopf K, Alizadeh AA, Brody J et al (2011) CD137 stimulation enhances the antilymphoma activity of anti-CD20 antibodies. Blood 117(8):2423–2432

Kohrt HE, Houot R, Weiskopf K, Goldstein MJ, Scheeren F, Czerwinski D et al (2012) Stimulation of natural killer cells with a CD137-specific antibody enhances trastuzumab efficacy in xenotransplant models of breast cancer. J Clin Invest 122(3):1066–1075

Kontermann RE (2009) Strategies to extend plasma half-lives of recombinant antibodies. BioDrugs 23(2):93–109

Kontermann RE (2012) Antibody-cytokine fusion proteins. Arch Biochem Biophys 526(2):194–205

Kreitman RJ, Wilson WH, Bergeron K, Raggio M, Stetler-Stevenson M, FitzGerald DJ et al (2001) Efficacy of the anti-CD22 recombinant immunotoxin BL22 in chemotherapy-resistant hairy-cell leukemia. N Engl J Med 345(4):241–247

Kreitman RJ, Stetler-Stevenson M, Margulies I, Noel P, Fitzgerald DJ, Wilson WH et al (2009) Phase II trial of recombinant immunotoxin RFB4(dsFv)-PE38 (BL22) in patients with hairy cell leukemia. J Clin Oncol 27(18):2983–2990

Kreitman RJ, Tallman MS, Robak T, Coutre S, Wilson WH, Stetler-Stevenson M et al (2012) Phase I trial of anti-CD22 recombinant immunotoxin moxetumomab pasudotox (CAT-8015 or HA22) in patients with hairy cell leukemia. J Clin Oncol 30(15):1822–1828
Krop IE, Beeram M, Modi S, Jones SF, Holden SN, Yu W et al (2010) Phase I study of trastuzumab-DM1, an HER2 antibody-drug conjugate, given every 3 weeks to patients with HER2-positive metastatic breast cancer. J Clin Oncol 28(16):2698–2704
Lazar GA, Dang W, Karki S, Vafa O, Peng JS, Hyun L et al (2006) Engineered antibody Fc variants with enhanced effector function. Proc Natl Acad Sci USA 103(11):4005–4010
Lewis Phillips GD, Li G, Dugger DL, Crocker LM, Parsons KL, Mai E et al (2008) Targeting HER2-positive breast cancer with trastuzumab-DM1, an antibody-cytotoxic drug conjugate. Cancer Res 68(22):9280–9290
Linenberger ML (2005) CD33-directed therapy with gemtuzumab ozogamicin in acute myeloid leukemia: progress in understanding cytotoxicity and potential mechanisms of drug resistance. Leukemia 19(2):176–182
Linke R, Klein A, Seimetz D (2010) Catumaxomab: clinical development and future directions. mAbs 2(2):129–136
Lode HN, Xiang R, Becker JC, Gillies SD, Reisfeld RA (1998a) Immunocytokines: a promising approach to cancer immunotherapy. Pharmacol Ther 80(3):277–292
Lode HN, Xiang R, Dreier T, Varki NM, Gillies SD, Reisfeld RA (1998b) Natural killer cell-mediated eradication of neuroblastoma metastases to bone marrow by targeted interleukin-2 therapy. Blood 91(5):1706–1715
Lohse S, Derer S, Beyer T, Klausz K, Peipp M, Leusen JH et al (2011) Recombinant dimeric IgA antibodies against the epidermal growth factor receptor mediate effective tumor cell killing. J Immunol 186(6):3770–3778
Mayer A, Francis RJ, Sharma SK, Tolner B, Springer CJ, Martin J et al (2006) A phase I study of single administration of antibody-directed enzyme prodrug therapy with the recombinant anti-carcinoembryonic antigen antibody-enzyme fusion protein MFECP1 and a bis-iodo phenol mustard prodrug. Clin Cancer Res 12(21):6509–6516
McDonagh CF, Huhalov A, Harms BD, Adams S, Paragas V, Oyama S et al (2012) Antitumor activity of a novel bispecific antibody that targets the ErbB2/ErbB3 oncogenic unit and inhibits heregulin-induced activation of ErbB3. Mol Cancer Ther 11(3):582–593
Menzel C, Schirrmann T, Konthur Z, Jostock T, Dubel S (2008) Human antibody RNase fusion protein targeting CD30+ lymphomas. Blood 111(7):3830–3837
Merchant AM, Zhu Z, Yuan JQ, Goddard A, Adams CW, Presta LG et al (1998) An efficient route to human bispecific IgG. Nat Biotechnol 16(7):677–681
Muller D, Kontermann RE (2010) Bispecific antibodies for cancer immunotherapy: current perspectives. BioDrugs 24(2):89–98
Müller D, Karle A, Meissburger B, Hofig I, Stork R, Kontermann RE (2007) Improved pharmacokinetics of recombinant bispecific antibody molecules by fusion to human serum albumin. J Biol Chem 282(17):12650–12660
Musolino A, Naldi N, Bortesi B, Pezzuolo D, Capelletti M, Missale G et al (2008) Immunoglobulin G fragment C receptor polymorphisms and clinical efficacy of trastuzumab-based therapy in patients with HER-2/neu-positive metastatic breast cancer. J Clin Oncol 26(11):1789–1796
Napier MP, Sharma SK, Springer CJ, Bagshawe KD, Green AJ, Martin J et al (2000) Antibody-directed enzyme prodrug therapy: efficacy and mechanism of action in colorectal carcinoma. Clin Cancer Res 6(3):765–772
Natsume A, In M, Takamura H, Nakagawa T, Shimizu Y, Kitajima K et al (2008) Engineered antibodies of IgG1/IgG3 mixed isotype with enhanced cytotoxic activities. Cancer Res 68(10):3863–3872
Niethammer AG, Xiang R, Ruehlmann JM, Lode HN, Dolman CS, Gillies SD et al (2001) Targeted interleukin 2 therapy enhances protective immunity induced by an autologous oral DNA vaccine against murine melanoma. Cancer Res 61(16):6178–6184
Nimmerjahn F, Ravetch JV (2005) Divergent immunoglobulin g subclass activity through selective Fc receptor binding. Science 310(5753):1510–1512

Niwa R, Hatanaka S, Shoji-Hosaka E, Sakurada M, Kobayashi Y, Uehara A et al (2004) Enhancement of the antibody-dependent cellular cytotoxicity of low-fucose IgG1 Is independent of FcgammaRIIIa functional polymorphism. Clin Cancer Res 10(18 Pt 1):6248–6255

Oganesyan V, Damschroder MM, Woods RM, Cook KE, Wu H, Dall'acqua WF (2009) Structural characterization of a human Fc fragment engineered for extended serum half-life. Mol Immunol 46(8–9):1750–1755

Oldham RK, Dillman RO (2008) Monoclonal antibodies in cancer therapy: 25 years of progress. J Clin Oncol 26(11):1774–1777

Osenga KL, Hank JA, Albertini MR, Gan J, Sternberg AG, Eickhoff J et al (2006) A phase I clinical trial of the hu14.18-IL2 (EMD 273063) as a treatment for children with refractory or recurrent neuroblastoma and melanoma: a study of the Children's Oncology Group. Clin Cancer Res 12(6):1750–1759

Pancook JD, Becker JC, Gillies SD, Reisfeld RA (1996) Eradication of established hepatic human neuroblastoma metastases in mice with severe combined immunodeficiency by antibody-targeted interleukin-2. Cancer Immunol Immunother 42(2):88–92

Pasche N, Neri D (2012) Immunocytokines: a novel class of potent armed antibodies. Drug Discov Today 17(11–12):583–590

Pastan I, Hassan R, Fitzgerald DJ, Kreitman RJ (2006) Immunotoxin therapy of cancer. Nat Rev Cancer 6(7):559–565

Pastan I, Onda M, Weldon J, Fitzgerald D, Kreitman R (2011) Immunotoxins with decreased immunogenicity and improved activity. Leuk Lymphoma 52(Suppl 2):87–90

Paz-Ares LG, Gomez-Roca C, Delord JP, Cervantes A, Markman B, Corral J et al (2011) Phase I pharmacokinetic and pharmacodynamic dose-escalation study of RG7160 (GA201), the first glycoengineered monoclonal antibody against the epidermal growth factor receptor, in patients with advanced solid tumors. J Clin Oncol 29(28):3783–3790

Peipp M, Kupers H, Saul D, Schlierf B, Greil J, Zunino SJ et al (2002) A recombinant CD7-specific single-chain immunotoxin is a potent inducer of apoptosis in acute leukemic T cells. Cancer Res 62(10):2848–2855

Peipp M, Lammerts van Bueren JJ, Schneider-Merck T, Bleeker WW, Dechant M, Beyer T et al (2008) Antibody fucosylation differentially impacts cytotoxicity mediated by NK and PMN effector cells. Blood 112(6):2390–2399

Peipp M, van de Winkel JG, Valerius T (2011) Molecular engineering to improve antibodies' anti-lymphoma activity. Best Pract Res Clin Haematol 24(2):217–229

Petkova SB, Akilesh S, Sproule TJ, Christianson GJ, Al Khabbaz H, Brown AC et al (2006) Enhanced half-life of genetically engineered human IgG1 antibodies in a humanized FcRn mouse model: potential application in humorally mediated autoimmune disease. Int Immunol 18(12):1759–1769

Presta LG (2008) Molecular engineering and design of therapeutic antibodies. Curr Opin Immunol 20(4):460–470

Reilly RM (2006) Radioimmunotherapy of solid tumors: the promise of pretargeting strategies using bispecific antibodies and radiolabeled haptens. J Nucl Med 47(2):196–199

Renders L, Valerius T (2003) Engineered CD3 antibodies for immunosuppression. Clin Exp Immunol 133(3):307–309

Richards JO, Karki S, Lazar GA, Chen H, Dang W, Desjarlais JR (2008) Optimization of antibody binding to FcgammaRIIa enhances macrophage phagocytosis of tumor cells. Mol Cancer Ther 7(8):2517–2527

Rossi EA, Goldenberg DM, Cardillo TM, McBride WJ, Sharkey RM, Chang CH (2006) Stably tethered multifunctional structures of defined composition made by the dock and lock method for use in cancer targeting. Proc Natl Acad Sci USA 103(18):6841–6846

Rossi EA, Goldenberg DM, Cardillo TM, Stein R, Chang CH (2009) Hexavalent bispecific antibodies represent a new class of anticancer therapeutics: 1. Properties of anti-CD20/CD22 antibodies in lymphoma. Blood 113(24):6161–6171

Salles G, Morschhauser F, Lamy T, Milpied NJ, Thieblemont C, Tilly H, et al (2012) Phase 1 study results of the type II glycoengineered humanized anti-CD20 monoclonal antibody obinutuzumab (GA101) in B-cell lymphoma patients. Blood 119(22):5126–5132

Samel D, Muller D, Gerspach J, Assohou-Luty C, Sass G, Tiegs G et al (2003) Generation of a FasL-based proapoptotic fusion protein devoid of systemic toxicity due to cell-surface antigen-restricted Activation. J Biol Chem 278(34):32077–32082

Sazinsky SL, Ott RG, Silver NW, Tidor B, Ravetch JV, Wittrup KD (2008) Aglycosylated immunoglobulin G1 variants productively engage activating Fc receptors. Proc Natl Acad Sci USA 105(51):20167–20172

Schaefer G, Haber L, Crocker LM, Shia S, Shao L, Dowbenko D et al (2011a) A two-in-one antibody against HER3 and EGFR has superior inhibitory activity compared with monospecific antibodies. Cancer Cell 20(4):472–486

Schaefer W, Regula JT, Bahner M, Schanzer J, Croasdale R, Durr H et al (2011b) Immunoglobulin domain crossover as a generic approach for the production of bispecific IgG antibodies. Proc Natl Acad Sci USA 108(27):11187–11192

Schirrmann T, Krauss J, Arndt MA, Rybak SM, Dubel S (2009) Targeted therapeutic RNases (ImmunoRNases). Expert Opin Biol Ther 9(1):79–95

Schneider-Merck T, Lammerts van Bueren JJ, Berger S, Rossen K, van Berkel PH, Derer S et al (2010) Human IgG2 antibodies against epidermal growth factor receptor effectively trigger antibody-dependent cellular cytotoxicity but, in contrast to IgG1, only by cells of myeloid lineage. J Immunol 184(1):512–520

Schultz J, Lin Y, Sanderson J, Zuo Y, Stone D, Mallett R et al (2000) A tetravalent single-chain antibody-streptavidin fusion protein for pretargeted lymphoma therapy. Cancer Res 60(23): 6663–6669

Schuurman J, Vink T, van de Winkel J, Labrijn AF, Aalberse R, van der Kolfschoten M et al (2008) Inventors; bispecific antibodies and production thereof (WO2008119353)

Schwemmlein M, Peipp M, Barbin K, Saul D, Stockmeyer B, Repp R et al (2006) A CD33-specific single-chain immunotoxin mediates potent apoptosis of cultured human myeloid leukaemia cells. Br J Haematol 133(2):141–151

Segal DM, Weiner GJ, Weiner LM (1999) Bispecific antibodies in cancer therapy. Curr Opin Immunol 11(5):558–562

Seimetz D, Lindhofer H, Bokemeyer C (2010) Development and approval of the trifunctional antibody catumaxomab (anti-EpCAM x anti-CD3) as a targeted cancer immunotherapy. Cancer Treat Rev 36(6):458–467

Shahied LS, Tang Y, Alpaugh RK, Somer R, Greenspon D, Weiner LM (2004) Bispecific minibodies targeting HER2/neu and CD16 exhibit improved tumor lysis when placed in a divalent tumor antigen binding format. J Biol Chem 279(52):53907–53914

Sharkey RM, Goldenberg DM (2011) Cancer radioimmunotherapy. Immunotherapy 3(3):349–370

Shibata-Koyama M, Iida S, Misaka H, Mori K, Yano K, Shitara K et al (2009) Nonfucosylated rituximab potentiates human neutrophil phagocytosis through its high binding for FcgammaRIIIb and MHC class II expression on the phagocytotic neutrophils. Exp Hematol 37(3):309–321

Shields RL, Namenuk AK, Hong K, Meng YG, Rae J, Briggs J et al (2001) High resolution mapping of the binding site on human IgG1 for Fc gamma RI, Fc gamma RII, Fc gamma RIII, and FcRn and design of IgG1 variants with improved binding to the Fc gamma R. J Biol Chem 276(9):6591–6604

Shields RL, Lai J, Keck R, O'Connell LY, Hong K, Meng YG et al (2002) Lack of fucose on human IgG1 N-linked oligosaccharide improves binding to human Fcgamma RIII and antibody-dependent cellular toxicity. J Biol Chem 277(30):26733–26740

Shinkawa T, Nakamura K, Yamane N, Shoji-Hosaka E, Kanda Y, Sakurada M et al (2003) The absence of fucose but not the presence of galactose or bisecting N-acetylglucosamine of human IgG1 complex-type oligosaccharides shows the critical role of enhancing antibody-dependent cellular cytotoxicity. J Biol Chem 278(5):3466–3473

Shusterman S, London WB, Gillies SD, Hank JA, Voss SD, Seeger RC et al (2010) Antitumor activity of hu14.18-IL2 in patients with relapsed/refractory neuroblastoma: a Children's Oncology Group (COG) phase II study. J Clin Oncol 28(33):4969–4975

Sievers EL, Larson RA, Stadtmauer EA, Estey E, Lowenberg B, Dombret H et al (2001) Efficacy and safety of gemtuzumab ozogamicin in patients with CD33-positive acute myeloid leukemia in first relapse. J Clin Oncol 19(13):3244–3254

Sondermann P, Huber R, Oosthuizen V, Jacob U (2000) The 3.2-A crystal structure of the human IgG1 Fc fragment-Fc gammaRIII complex. Nature 406(6793):267–273

Staerz UD, Kanagawa O, Bevan MJ (1985) Hybrid antibodies can target sites for attack by T cells. Nature 314(6012):628–631

Stahnke B, Thepen T, Stocker M, Rosinke R, Jost E, Fischer R et al (2008) Granzyme B-H22(scFv), a human immunotoxin targeting CD64 in acute myeloid leukemia of monocytic subtypes. Mol Cancer Ther 7(9):2924–2932

Stamova S, Cartellieri M, Feldmann A, Bippes CC, Bartsch H, Wehner R et al (2011) Simultaneous engagement of the activatory receptors NKG2D and CD3 for retargeting of effector cells to CD33-positive malignant cells. Leukemia 25(6):1053–1056

Stavenhagen JB, Gorlatov S, Tuaillon N, Rankin CT, Li H, Burke S et al (2007) Fc optimization of therapeutic antibodies enhances their ability to kill tumor cells in vitro and controls tumor expansion in vivo via low-affinity activating Fcgamma receptors. Cancer Res 67(18):8882–8890

Stieglmaier J, Bremer E, Kellner C, Liebig TM, ten Cate B, Peipp M et al (2008) Selective induction of apoptosis in leukemic B-lymphoid cells by a CD19-specific TRAIL fusion protein. Cancer Immunol Immunother 57(2):233–246

Stork R, Muller D, Kontermann RE (2007) A novel tri-functional antibody fusion protein with improved pharmacokinetic properties generated by fusing a bispecific single-chain diabody with an albumin-binding domain from streptococcal protein G. Protein Eng Des Sel 20(11):569–576

Stork R, Zettlitz KA, Muller D, Rether M, Hanisch FG, Kontermann RE (2008) N-glycosylation as novel strategy to improve pharmacokinetic properties of bispecific single-chain diabodies. J Biol Chem 283(12):7804–7812

Sutherland MS, Sanderson RJ, Gordon KA, Andreyka J, Cerveny CG, Yu C et al (2006) Lysosomal trafficking and cysteine protease metabolism confer target-specific cytotoxicity by peptide-linked anti-CD30-auristatin conjugates. J Biol Chem 281(15):10540–10547

Topp MS, Kufer P, Gokbuget N, Goebeler M, Klinger M, Neumann S et al (2011) Targeted therapy with the T-cell-engaging antibody blinatumomab of chemotherapy-refractory minimal residual disease in B-lineage acute lymphoblastic leukemia patients results in high response rate and prolonged leukemia-free survival. J Clin Oncol 29(18):2493–2498

Umana P, Jean-Mairet J, Moudry R, Amstutz H, Bailey JE (1999) Engineered glycoforms of an antineuroblastoma IgG1 with optimized antibody-dependent cellular cytotoxic activity. Nat Biotechnol 17(2):176–180

Unverdorben F, Farber-Schwarz A, Richter F, Hutt M, Kontermann RE (2012) Half-life extension of a single-chain diabody by fusion to domain B of staphylococcal protein A. Protein Eng Des Sel 25(2):81–88

van Berkel PH, Gerritsen J, van Voskuilen E, Perdok G, Vink T, van de Winkel JG et al (2010) Rapid production of recombinant human IgG With improved ADCC effector function in a transient expression system. Biotechnol Bioeng 105(2):350–357

von Strandmann EP, Hansen HP, Reiners KS, Schnell R, Borchmann P, Merkert S et al (2006) A novel bispecific protein (ULBP2-BB4) targeting the NKG2D receptor on natural killer (NK) cells and CD138 activates NK cells and has potent antitumor activity against human multiple myeloma in vitro and in vivo. Blood 107(5):1955–1962

Waetzig GH, Rose-John S (2012) Hitting a complex target: an update on interleukin-6 trans-signalling. Expert Opin Ther Targets 16(2):225–236

Wang SY, Veeramani S, Racila E, Cagley J, Fritzinger DC, Vogel CW et al (2009) Depletion of the C3 component of complement enhances the ability of rituximab-coated target cells to activate human NK cells and improves the efficacy of monoclonal antibody therapy in an in vivo model. Blood 114(26):5322–5330

Wang F, Ren J, Qiu XC, Wang LF, Zhu Q, Zhang YQ et al (2010) Selective cytotoxicity to HER2-positive tumor cells by a recombinant e23sFv-TD-tBID protein containing a furin cleavage sequence. Clin Cancer Res 16(8):2284–2294

Weiden PL, Breitz HB, Press O, Appelbaum JW, Bryan JK, Gaffigan S et al (2000) Pretargeted radioimmunotherapy (PRIT) for treatment of non-Hodgkin's lymphoma (NHL): initial phase I/II study results. Cancer Biother Radiopharm 15(1):15–29

Weiner LM, Surana R, Wang S (2010) Monoclonal antibodies: versatile platforms for cancer immunotherapy. Nat Rev Immunol 10(5):317–327

Wels W, Harwerth IM, Mueller M, Groner B, Hynes NE (1992) Selective inhibition of tumor cell growth by a recombinant single-chain antibody-toxin specific for the erbB-2 receptor. Cancer Res 52(22):6310–6317

Wels W, Beerli R, Hellmann P, Schmidt M, Marte BM, Kornilova ES et al (1995) EGF receptor and p185erbB-2-specific single-chain antibody toxins differ in their cell-killing activity on tumor cells expressing both receptor proteins. Int J Cancer 60(1):137–144

Weng WK, Levy R (2003) Two immunoglobulin G fragment C receptor polymorphisms independently predict response to rituximab in patients with follicular lymphoma. J Clin Oncol 21(21):3940–3947

Witzig TE, Flinn IW, Gordon LI, Emmanouilides C, Czuczman MS, Saleh MN et al (2002) Treatment with ibritumomab tiuxetan radioimmunotherapy in patients with rituximab-refractory follicular non-Hodgkin's lymphoma. J Clin Oncol 20(15):3262–3269

Wright A, Morrison SL (1997) Effect of glycosylation on antibody function: implications for genetic engineering. Trends Biotechnol 15(1):26–32

Wu AM, Senter PD (2005) Arming antibodies: prospects and challenges for immunoconjugates. Nat Biotechnol 23(9):1137–1146

Wu C, Ying H, Grinnell C, Bryant S, Miller R, Clabbers A et al (2007) Simultaneous targeting of multiple disease mediators by a dual-variable-domain immunoglobulin. Nat Biotechnol 25(11):1290–1297

Xiang R, Lode HN, Dolman CS, Dreier T, Varki NM, Qian X et al (1997) Elimination of established murine colon carcinoma metastases by antibody-interleukin 2 fusion protein therapy. Cancer Res 57(21):4948–4955

Yamane-Ohnuki N, Kinoshita S, Inoue-Urakubo M, Kusunoki M, Iida S, Nakano R et al (2004) Establishment of FUT8 knockout Chinese hamster ovary cells: an ideal host cell line for producing completely defucosylated antibodies with enhanced antibody-dependent cellular cytotoxicity. Biotechnol Bioeng 87(5):614–622

Yeung YA, Leabman MK, Marvin JS, Qiu J, Adams CW, Lien S et al (2009) Engineering human IgG1 affinity to human neonatal Fc receptor: impact of affinity improvement on pharmacokinetics in primates. J Immunol 182(12):7663–7671

Yeung YA, Wu X, Reyes AE 2nd, Vernes JM, Lien S, Lowe J et al (2010) A therapeutic anti-VEGF antibody with increased potency independent of pharmacokinetic half-life. Cancer Res 70(8):3269–3277

Younes A, Bartlett NL, Leonard JP, Kennedy DA, Lynch CM, Sievers EL et al (2010) Brentuximab vedotin (SGN-35) for relapsed CD30-positive lymphomas. N Engl J Med 363(19):1812–1821

Zalevsky J, Leung IW, Karki S, Chu SY, Zhukovsky EA, Desjarlais JR et al (2009) The impact of Fc engineering on an anti-CD19 antibody: increased Fcgamma receptor affinity enhances B-cell clearing in nonhuman primates. Blood 113(16):3735–3743

Zalevsky J, Chamberlain AK, Horton HM, Karki S, Leung IW, Sproule TJ et al (2010) Enhanced antibody half-life improves in vivo activity. Nat Biotechnol 28(2):157–159

Zeidler R, Reisbach G, Wollenberg B, Lang S, Chaubal S, Schmitt B et al (1999) Simultaneous activation of T cells and accessory cells by a new class of intact bispecific antibody results in efficient tumor cell killing. J Immunol 163(3):1246–1252

Index

F. Nimmerjahn (ed.), *Molecular and Cellular Mechanisms of Antibody Activity*, DOI 10.1007/978-1-4614-7107-3, © Springer Science+Business Media New York 2013

Printed by Publishers' Graphics LLC
MLSI130603.11.24.49